HAESE MATHEMATICS

Specialists in mathematics education

Mathematics 4

PYP 4 Write-on Workbook

for use with

IB Primary Years Programme

Michael Haese Mark Humphries Philippa Sawyers

MATHEMATICS PYP 4

Michael Haese B.Sc.(Hons.), Ph.D.
Mark Humphries B.Sc.(Hons.)
Philippa Sawyers B.Ed.(Hons.)

Published by Haese Mathematics
152 Richmond Road, Marleston, SA 5033, AUSTRALIA
Telephone: +61 8 8210 4666
Email: info@haesemathematics.com
Web: www.haesemathematics.com

National Library of Australia Card Number & ISBN 978-1-922416-79-7

First Edition 2024

Cartoon artwork by James Hobbs, Yi-Tung Huang, and Le Quynh Chi Ta.

Artwork by Hannah Coleman and Brian Houston.

Cover art by Le Quynh Chi Ta.

Computer software by Patrick French, Ben Hensley, Huda Kharrufa, Rachel Lee, and Han Zong Ng.

Audio recorded by Eloise Quinn-Valentine.

Production work by Hannah Coleman, Michael Mampusti, and Ngoc Vo.

Typeset in Australia by Charlotte Frost and Deanne Gallasch. Typeset in Times Roman 13.

Printed in China by Prolong Press Limited.

Acknowledgements: While every attempt has been made to trace and acknowledge copyright, the authors and publishers apologise for any accidental infringement where copyright has proved untraceable. They would be pleased to come to a suitable agreement with the rightful owner.

FOREWORD

Mathematics PYP 4 has been designed and written for the International Baccalaureate Primary Years Programme (IB PYP). The workbook and online interactive material provide an engaging and structured package, allowing students to explore and develop their confidence in Mathematics.

Our aim with this spiral bound write-on workbook is to provide the students and teachers with an alternative to photocopied worksheets, which in turn will give the students an organised record of their work throughout the year.

The book contains a variety of exercises ranging from basic to advanced, to cater for a range of student abilities and interests. The material is presented in a clear, easy-to-follow style to aid comprehension and retention, especially for English Language Learners. Each chapter ends with a set of Revision questions.

Important information and key notes are highlighted, and worked examples provide clear instruction and relevant explanations. Discussions, Activities, and Puzzles are used throughout the chapters to develop understanding. All of these features can be viewed through Snowflake, our online digital textbook platform.

We have endeavoured to provide a stimulating workbook and accompanying interactive material. Our aim is to develop and encourage student understanding and to grow an appreciation and love for Mathematics.

We welcome your feedback:

Email: info@haesemathematics.com
Web: www.haesemathematics.com

ONLINE FEATURES

Each workbook comes with a 12 month subscription to the online edition and its range of interactive features. This can be accessed through the **SNOWFLAKE** online learning platform via a web browser or our offline viewer.

To activate your electronic workbook, please contact Haese Mathematics by emailling info@haesemathematics.com with your proof of purchase such as a copy of your receipt.

For general queries regarding **SNOWFLAKE** and online subscriptions:

- Visit our help page: https://snowflake.haesemathematics.com.au/help
- Contact Haese Mathematics: info@haesemathematics.com

WATCH LISTEN LEARN

Watch Listen Learn is an exciting feature of this book.

This icon in a worked example or exercise denotes an active online link.

Simply click the icon (or anywhere in the example or exercise box) to access the Watch Listen Learn with a teacher's voice explaining each concept.

Play any line as often as you like. See how the basic processes come alive using movement and colour on the screen.

CLICKABLE ICONS

You will see these icons throughout the workbook. Click the icon to access online content.

Printable Worksheets containing extra questions or support material.

Listening Activities.

Video Demonstrations providing instruction.

Activities and Games.

TABLE OF CONTENTS

CHAPTER 1: NUMBER

Discussion

In our number system, we write numbers using symbols called **digits**.

The digits we use are 0, 1, 2, 3, 4, 5, 6, 7, 8, and 9.

The **place** of a digit in a number determines its value.

	thousands	hundreds	tens	units
one unit				1
one ten			1	0
one hundred		1	0	0
one thousand	1	0	0	0

1 Why are the zeros important when we write 1000?

2 How many:

- *units* make up one *ten*? ....
- How many *tens* make up one *hundred*? ....
- How many *hundreds* make up one *thousand*? ....

We can write a number in several forms.

numeral form 2 4 7 6

2 thousands 4 hundreds 7 tens 6 units

expanded form $2000 + 400 + 70 + 6$

words two thousand, four hundred, and seventy six

Exercise 1

numeral form 3 5 8

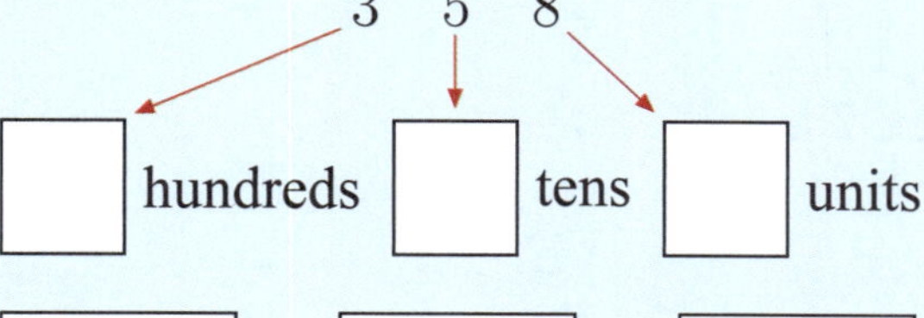

expanded form

words ....

Exercise 2

Partition each number using place values:

a 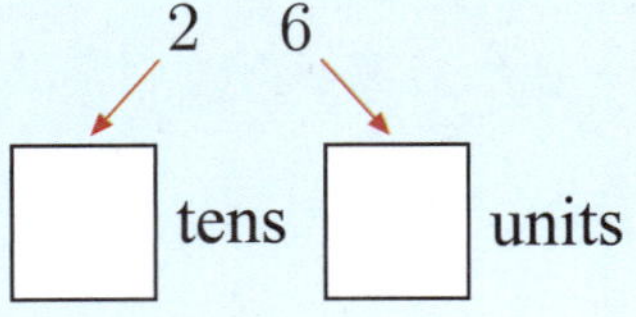

b 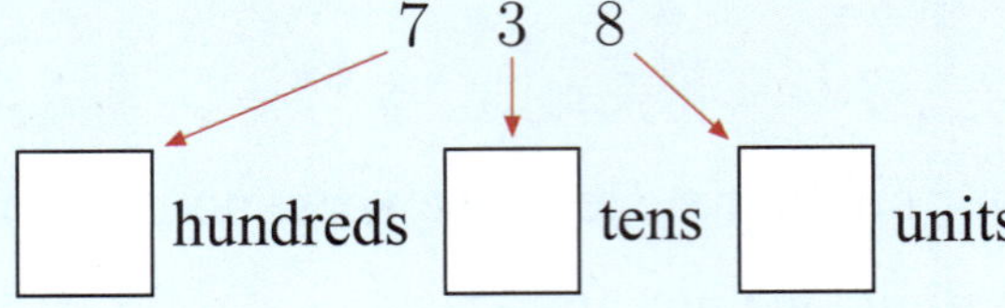

c 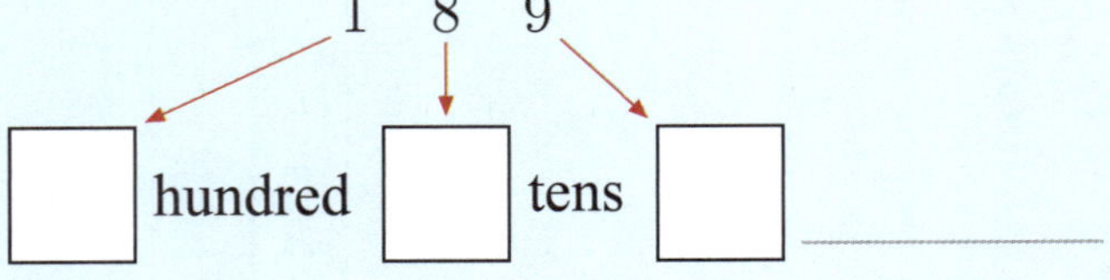

d 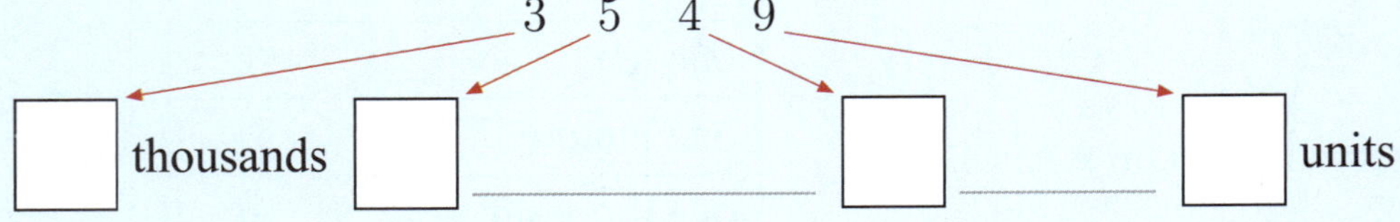

e 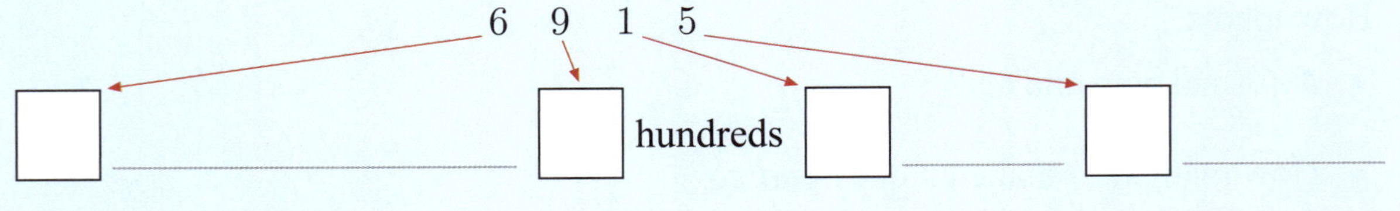

Exercise 3

Write each number as a numeral:

a $50 + 3 =$ ________

b $80 + 4 =$ ________

c $100 + 20 + 1 =$ ________

d $200 + 50 + 7 =$ ________

e $300 + 60 + 7 =$ ________

f $7000 + 300 + 40 + 1 =$ ________

g $2000 + 500 + 70 + 5 =$ ________

h $9000 + 800 + 20 + 6 =$ ________

Exercise 4

Write down the value of the digit 7 in:

a 76 ________

b 237 ________

c 725 ________

d 879 ________

e 1657 ________

f 7219 ________

Exercise 5

Write down the value of the digit 3 in:

a 183 ....
b 326 ....
c 3497 ....
d 8632 ....
e 5384 ....
f 6193 ....

Exercise 6

Write each number in expanded form:

a 39 = ...
b 78 = ...
c 127 = ...
d 253 = ...
e 1352 = ...
f 8624 = ...

Exercise 7

Write each number in words:

a 84 = ...
b 56 = ...
c 192 = ...
d 417 = ...
e 2468 = ...
f 5167 = ...

The digit **zero** or 0 is used to show an empty place value.

3409 = 3000 + 400 + 9
or three thousand, four hundred, and nine

↑ no tens

Exercise 8

Write each number as a numeral:

a $100 + 6 =$ ____________

b $200 + 40 =$ ____________

c $1000 + 200 + 4 =$ ____________

d $2000 + 40 + 5 =$ ____________

e $2000 + 600 + 30 =$ ____________

f $3000 + 4 =$ ____________

Exercise 9

Write each number in expanded form:

a $105 =$ ________________________

b $320 =$ ________________________

c $1027 =$ ________________________

d $2230 =$ ________________________

e $2409 =$ ________________________

f $4006 =$ ________________________

Exercise 10

Write each number in words:

a $606 =$ ____

b $230 =$ ____

c $7630 =$ ____

d $9088 =$ ____

e $1003 =$ ____

Exercise 11

Write each number as a numeral:

a four hundred and five = ____________

b seven hundred and sixty eight = ____________

c three thousand and ninety two = ____________

d nine thousand, two hundred, and forty seven = ____________

e six thousand, eight hundred, and twenty = ____________

Exercise 12

Complete:

Numeral	*Expanded form*	*Words*
239	....	two hundred and thirty nine
....	400 + 70 + 5	....
....	....	one thousand, two hundred, and ninety seven
....	6000 + 200 + 3	....
2374	....	....

Puzzle

Colour the matching tiles the same.

39	3000 + 900 + 30	three thousand and nine
309	3000 + 300 + 9	three thousand, nine hundred, and thirty
399	300 + 9	three hundred and ninety nine
390	3000 + 9	thirty nine
3930	300 + 90 + 9	three hundred and ninety
3309	30 + 9	three hundred and nine
3009	300 + 90	three thousand, three hundred, and nine

Game

You will need: a small group, two sets of number cards 0 - 9, pencil and paper, or mini white boards and markers

What to do:

1 Place all number cards face down on a table, then mix them up.

2 When everyone is ready, one person selects 4 cards and turns them over.

3 Each person must use the digits to make a 4 digit number which is *different* from everyone else in the group.

Write your number as a numeral, in expanded form, and in words.

For example: 2 0  6 5 These are the number cards.

6052
6000 + 50 + 2
six thousand and fifty two

This is written on your white board.

4 Check everyone else's work in your group. This is a team game!

5 Continue until each person in your group has selected a set of cards.

The next two place values are:

Ten thousands	10 000
Hundred thousands	100 000

3 6 2 5 0 8

3 hundred thousands 6 ten thousands

300 000 + 60 000 + 2000 + 500 + 8

three hundred and sixty two thousand, five hundred, and eight

Exercise 13

Write each number as a numeral:

a 20 000 + 9000 + 300 + 20 + 1 = ...

b 60 000 + 500 + 70 + 8 = ...

c 100 000 + 5000 + 400 + 80 = ...

d 700 000 + 500 + 2 = ...

e 300 000 + 80 000 + 600 + 5 = ...

Exercise 14

Write down the value of the digit 4 in:

a 2476 ....

b 4518 ....

c 54 321 ....

d 42 100 ....

e 540 267 ....

f 478 120 ....

Activity

Click to practise listening to and reading numbers.

Exercise 15

Write each number in expanded form:

a 91 263 = ...

b 20 851 = ...

c 408 000 = ...

d 134 406 = ...

e 750 845 = ...

Exercise 16

Write each number in words:

a 21 612 = ...

b 90 502 = ...

c 505 050 = ...

Exercise 17

Write each number in words:

a 133 748 = ...

....

....

b 840 711 = ...

....

....

This is an **abacus**. Each column represents a different place value.

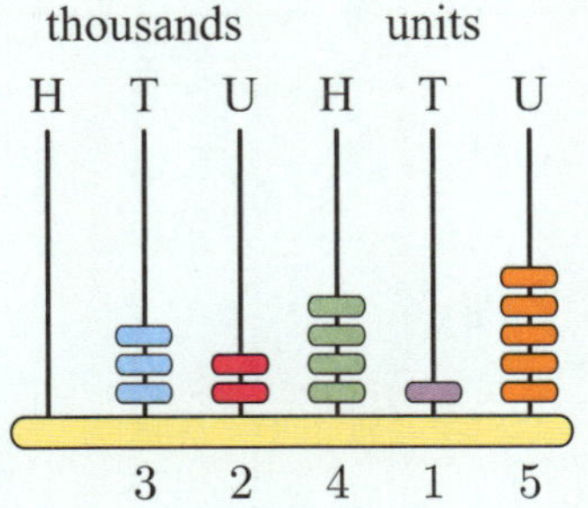

thirty two thousand,
four hundred,
and fifteen

Exercise 18

What number does each abacus show? Write your answer as a numeral.

a

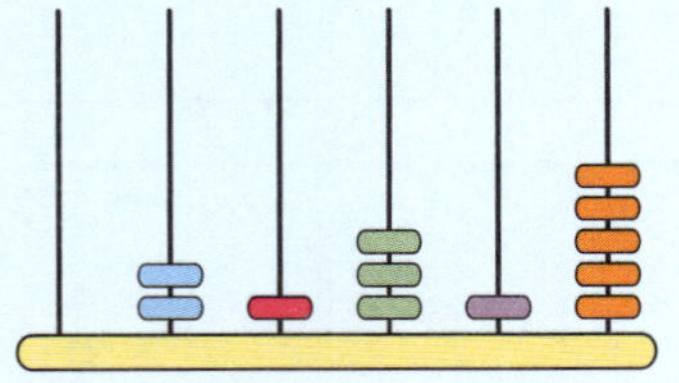

....

b

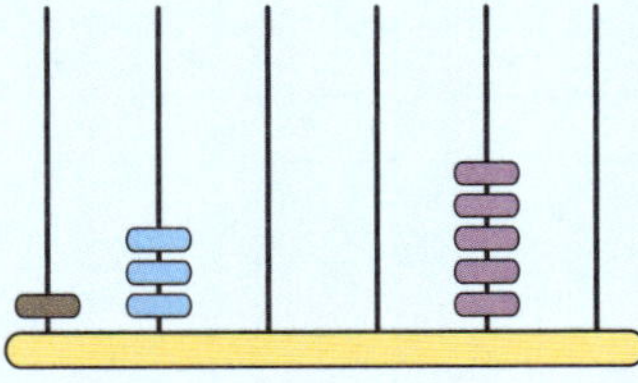

....

c

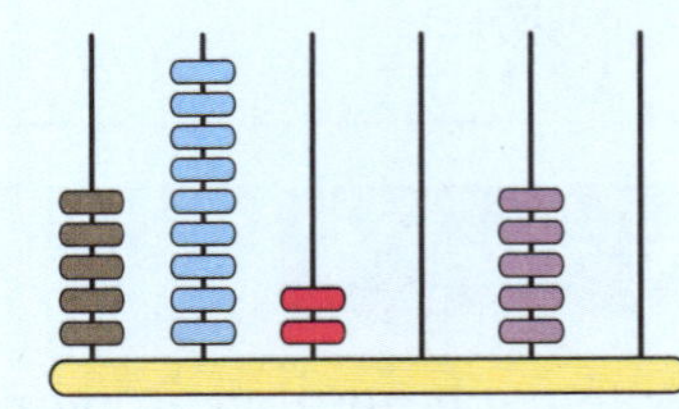

....

Exercise 19

What number does each abacus show? Write your answer in words.

a

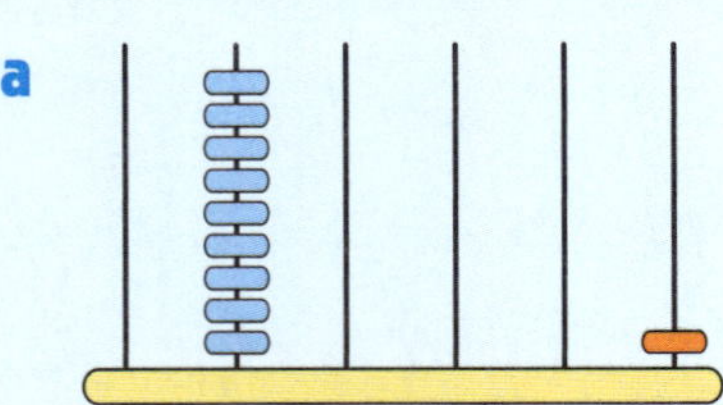

....

....

Exercise 20

Show the number on the abacus:

a 24 001

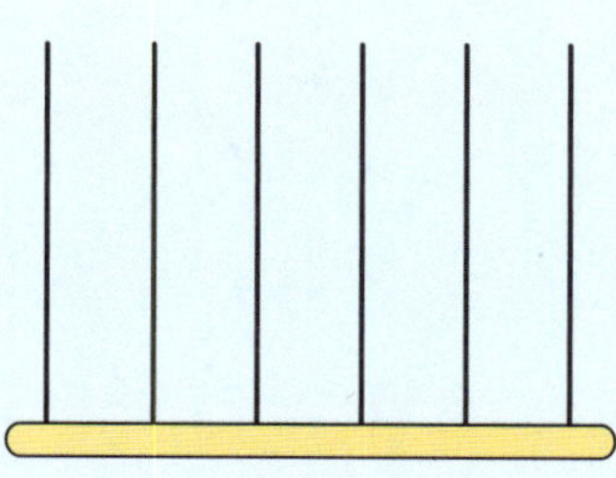

b ten thousand, one hundred, and twenty two

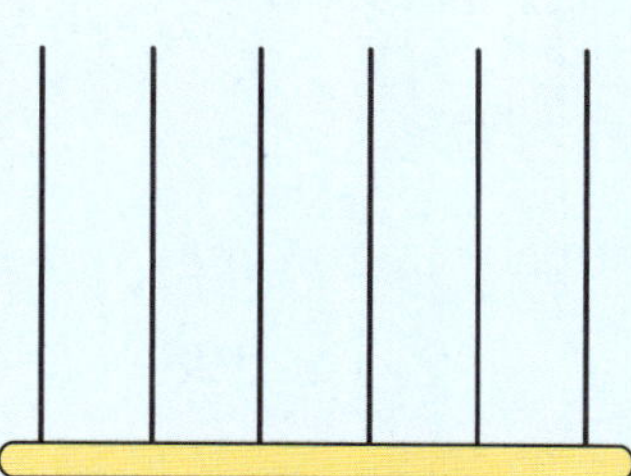

c $30\,000 + 5000 + 2$

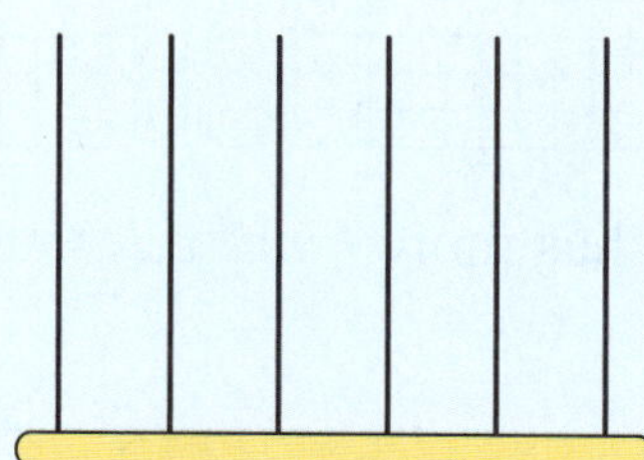

d 62 458

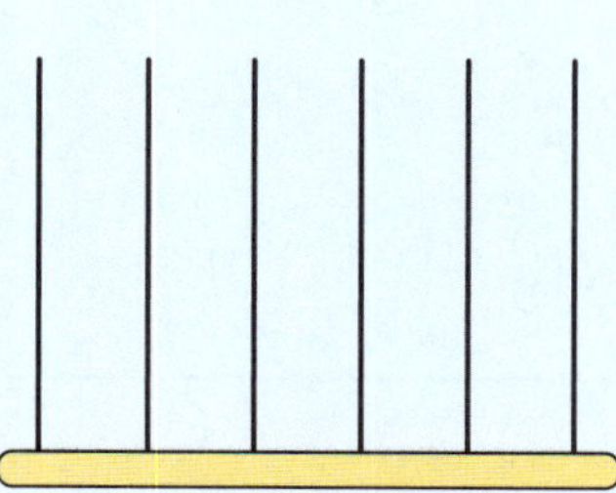

e fifty two thousand, four hundred, and one

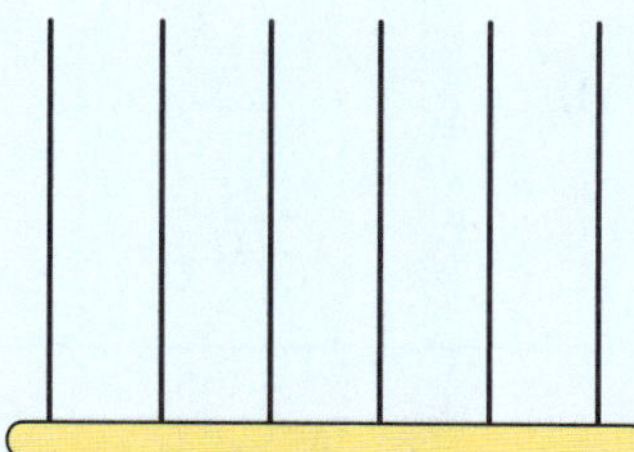

f $800\,000 + 7000 + 50 + 4$

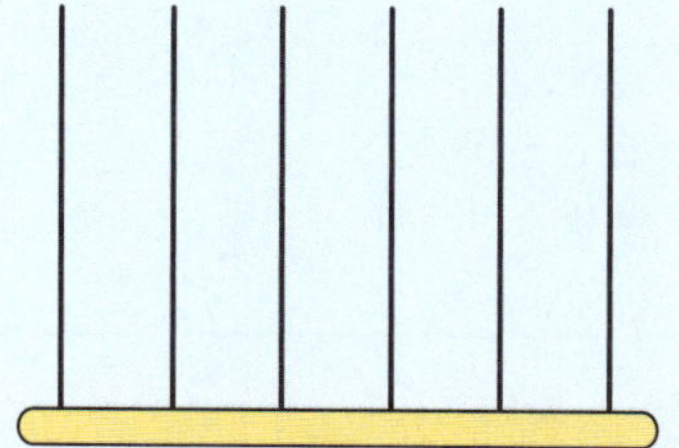

$<$ means "less than".

$>$ means "greater than".

Exercise 21

Complete using $<$ or $>$.

a 8 4

b 3 7

c 6 1

d 0 2

e 5 8

f 7 6

Exercise 22

Complete using $<$ or $>$.

a 30 ____ 10 **b** 20 ____ 50 **c** 90 ____ 60

d 70 ____ 40 **e** 50 ____ 30 **f** 40 ____ 80

To *compare* numbers, look at the largest place value first.

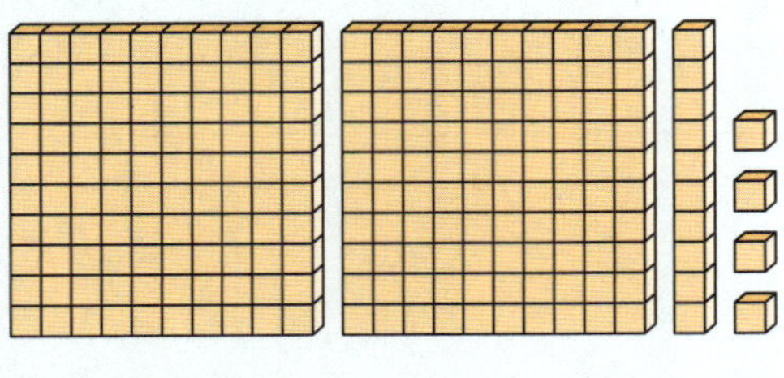

214 has 2 hundreds
1 ten
and 4 units.

198 has 1 hundred
9 tens
and 8 units.

214 has more *hundreds* than 198, so 214 is *greater than* 198.

$$214 > 198$$

Exercise 23

Complete using $<$ or $>$.

a 28 ____ 14 **b** 19 ____ 32 **c** 37 ____ 25

d 69 ____ 48 **e** 91 ____ 75 **f** 56 ____ 71

Exercise 24

Complete using $<$ or $>$.

a 18 ____ 9 **b** 86 ____ 121 **c** 125 ____ 96

Exercise 25

Complete using $<$ or $>$.

a 215 ____ 341 **b** 562 ____ 478 **c** 735 ____ 902

d 645 ____ 482 **e** 3160 ____ 4257 **f** 6218 ____ 5916

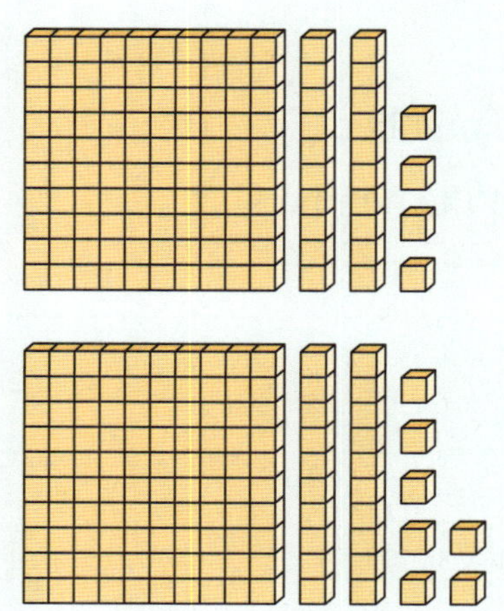

124 has 1 hundred
2 tens
and 4 units.

127 has 1 hundred
2 tens
and 7 units.

124 and 127 have the same number of hundreds
and the same number of tens.

124 has fewer units than 127, so 124 is less than 127.

$$124 < 127$$

Exercise 26

Complete using $<$ or $>$.

a 41 ____ 43 **b** 37 ____ 35 **c** 25 ____ 28

d 55 ____ 53 **e** 84 ____ 87 **f** 27 ____ 21

g 63 ____ 65 **h** 94 ____ 99 **i** 77 ____ 74

Exercise 27

Complete using $<$ or $>$.

a 256 ____ 238 **b** 147 ____ 196 **c** 521 ____ 550

d 726 ____ 722 **e** 381 ____ 387 **f** 934 ____ 935

g 2614 ____ 2485 **h** 3907 ____ 3953 **i** 8172 ____ 8177

24 31 19

19 < 24 and 19 < 31. 19 is the *lowest* or *least* of the numbers.

31 > 24 and 31 > 19. 31 is the *highest* or *greatest* of the numbers.

Exercise 28

Choose the lowest and highest number from each group.

a 7, 1, 5 lowest = ________ highest = ________

b 38, 29, 46 lowest = ________ highest = ________

c 118, 79, 131, 86 lowest = ________ highest = ________

d 234, 198, 257, 254 lowest = ________ highest = ________

Exercise 29

Choose the least and greatest number from each group.

a 48, 51, 36 least = ________ greatest = ________

b 9, 11, 4, 14 least = ________ greatest = ________

c 84, 91, 95, 83 least = ________ greatest = ________

d 108, 97, 105, 112 least = ________ greatest = ________

Ascending order means from lowest to highest.

Descending order means from highest to lowest.

The numbers 21, 15, 12, and 27 all have two digits.

We look at the tens first, then at the units.

15 and 12 have the smallest number of tens. 12 has fewer units, so 12 is the lowest number, and 15 is next.

21 and 27 have the same number of tens. 21 has fewer units than 27, so 21 is the next lowest number. 27 is the highest number.

In ascending order, we have: 12, 15, 21, 27.

Exercise 30

Arrange in ascending order:

a 19, 31, 20, 14

b 38, 29, 36, 25

c 81, 60, 67, 73, 64

d 136, 230, 118, 209

e 4186, 978, 3257, 4092

f seventeen, twenty two, eleven, nineteen

Puzzle

3	6	1	8	4

Arrange these cards to make a 5 digit number which is:

a as low as possible

b as high as possible

Exercise 31

Arrange in descending order:

a 41, 28, 36, 39

b 148, 130, 161, 155

c 2185, 3640, 2897, 2714

d 1088, 968, 1182, 1084

e thirty seven, forty eight, thirty three, fifty

f ninety three, one hundred and ten, ninety one, one hundred and two

Listening Activity

Click and listen to the instructions:

1 ______________________ **2** 

3 ______________________ **4** ______________________

5 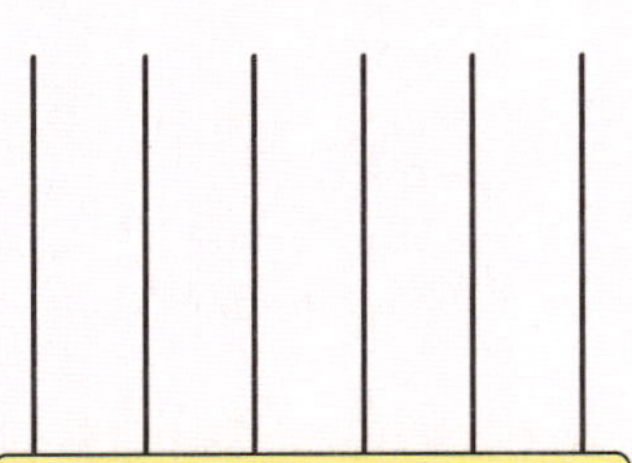

6 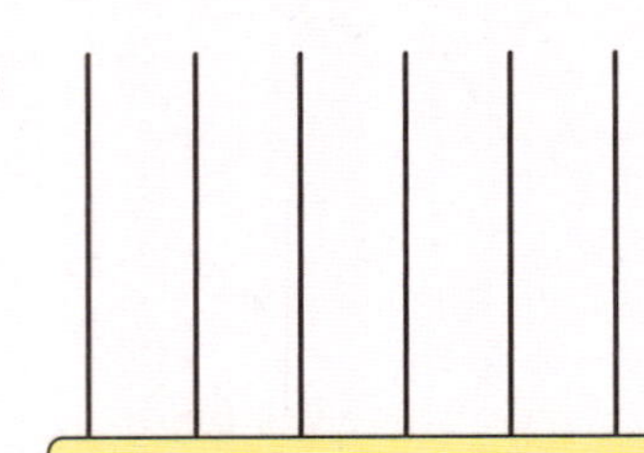

7 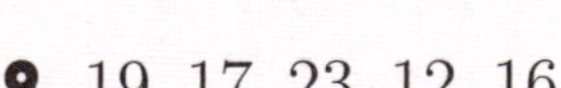

8 ______________________

9 19, 17, 23, 12, 16

10 37, 57, 28, 45, 36

______________________ ______________________

Revision

1 Write each number as a numeral:

a $300 + 50 + 8 =$ ____________ **b** $5000 + 700 + 2 =$ ____________

c $40\,000 + 3000 + 500 + 80 + 3 =$ ____________

d $500\,000 + 60\,000 + 4000 + 10 + 5 =$ ____________

2 Write down the value of the digit 8 in:

a 581 ____________ **b** 28 094 ____________

c 3876 ____________ **d** 842 619 ____________

3 Write each number in expanded form:

a $703 =$ ____________ **b** $2010 =$ ____________

c $54\,691 =$ ______________________________

d $307\,208 =$ ______________________________

4 Write each number in words:

a $682 =$ ______________________________

b $3109 =$ ______________________________

c 18 047 = ...

....

d 624 308 = ...

....

5 Write each number as a numeral:

a three thousand, six hundred, and seven = ...

b forty seven thousand, five hundred, and sixteen = ...

c nine hundred and two thousand, and fifty nine = ...

6 Complete:

Numeral	*Expanded form*	*Words*
....	600 000 + 4000 + 80 + 8	....
....	....	thirteen thousand, five hundred, and ninety one
721 054	....	....

7 What number does each abacus show? Write your answer as a numeral.

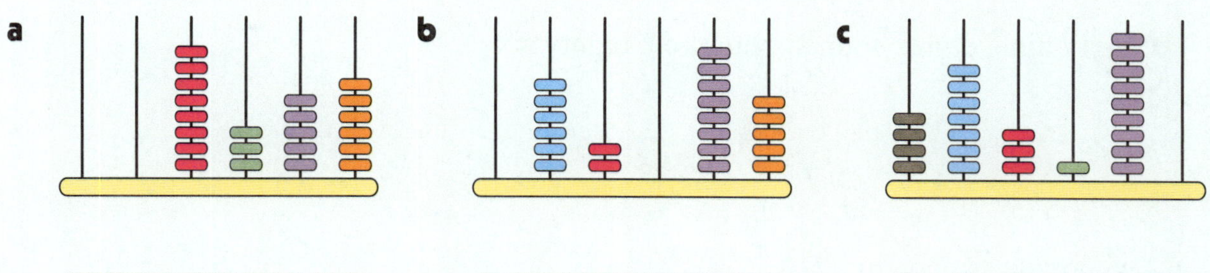

a .... **b** .... **c** ....

8 Complete using $<$ or $>$.

a 5 ____ 7 **b** 8 ____ 6 **c** 60 ____ 90

d 80 ____ 30 **e** 38 ____ 29 **f** 66 ____ 82

9 Show the number on the abacus:

a 3169

b sixty seven thousand and twenty eight

c $400\,000 + 20\,000 + 500 + 10 + 9$

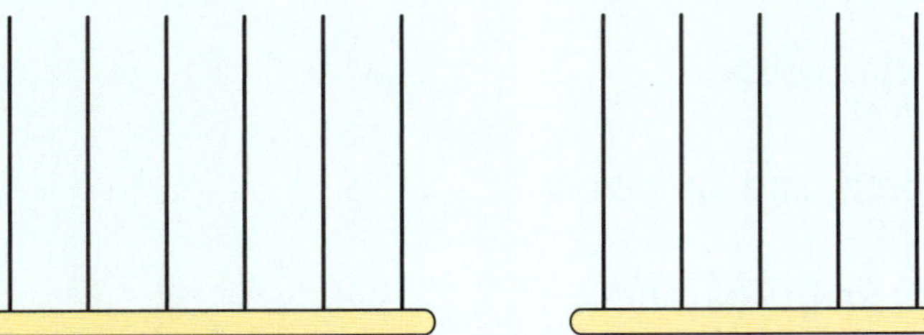

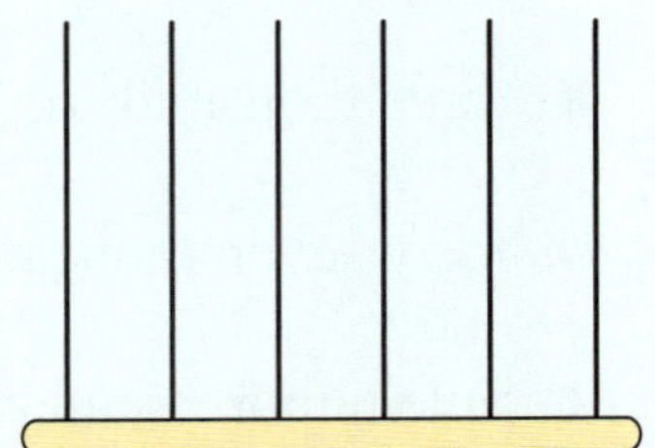

10 Complete using $<$ or $>$.

a 74 ____ 122 **b** 884 ____ 1006 **c** 212 ____ 196

d 327 ____ 285 **e** 4517 ____ 6401 **f** 9136 ____ 7659

11 Complete using $<$ or $>$.

a 85 ____ 88 **b** 116 ____ 111 **c** 325 ____ 314

d 863 ____ 857 **e** 4290 ____ 4257 **f** 5482 ____ 5395

12 Arrange in ascending order:

a 59, 52, 61, 47, 49 ______________________

b 608, 594, 625, 623 ______________________

c twenty seven, two thousand and three, three hundred and eighteen, seventy two

____

____

13 Arrange in descending order:

a 356, 294, 385, 359 ______________________

b sixty nine, eighty four, six hundred, eighty six

____

____

c 5829, 985, 5892, 5928 ______________________

CHAPTER 2: ADDITION

Mental additions are done without a pencil and paper, or apparatus.

Exercise 1

Write down the result of each addition:

a $6 + 2 =$ ______ **b** $3 + 5 =$ ______ **c** $8 + 4 =$ ______

d 3 add $8 =$ ______ **e** 7 add $6 =$ ______ **f** 5 add $6 =$ ______

g 9 plus $7 =$ ______ **h** 4 plus $9 =$ ______ **i** 8 plus $8 =$ ______

j $6 + 0 =$ ______ **k** $5 + 10 =$ ______ **l** $0 + 17 =$ ______

Exercise 2

a Add together 7 and 4. ______

b Find the total of 9 and 5. ______

c Find the sum of 3 and 3. ______

d The total of 10 and 7 is ______.

e The number 6 more than 8 is ______.

Exercise 3

Write pairs of numbers that add to 10.

$10 = \square + \square$ $10 = \square + \square$ $10 = \square + \square$

$10 = \square + \square$ $10 = \square + \square$ $10 = \square + \square$

$10 = \square + \square$ $10 = \square + \square$ $10 = \square + \square$

$10 = \square + \square$ $10 = \square + \square$

Puzzle

If I roll two ordinary dice, the possible totals are:

2, 3, 4, ______, ______, ______, ______,

______, ______, ______, and ______.

Complete the table:

Total	*Number of ways*	*Sum*
2	1	$1+1$
3	2	$1+2$ or $2+1$
4		

Exercise 4

Complete each addition:

a $20 + 4 =$ ______

b $70 + 3 =$ ______

c $60 + 7 =$ ______

d $100 + 9 =$ ______

e $200 + 30 =$ ______

f $600 + 50 + 4 =$ ______

g $700 + 10 + 6 =$ ______

h $300 + 40 + 5 =$ ______

Exercise 5

Complete these additions.

a $5 + 3 =$ ______
$50 + 30 =$ ______

b $2 + 7 =$ ______
$20 + 70 =$ ______

c $4 + 5 =$ ______
$40 + 50 =$ ______

d $8 + 2 =$ ______
$80 + 20 =$ ______

e $7 + 4 =$ ______
$70 + 40 =$ ______

f $6 + 9 =$ ______
$60 + 90 =$ ______

g $2 + 3 =$ ______
$20 + 30 =$ ______
$200 + 300 =$ ______

h $4 + 6 =$ ______
$40 + 60 =$ ______
$400 + 600 =$ ______

i $8 + 5 =$ ______
$80 + 50 =$ ______
$800 + 500 =$ ______

Exercise 6

Fill in the missing number.

a $3 +$ ______ $= 9$

b ______ $+ 6 = 8$

c $7 +$ ______ $= 10$

d ______ $+ 4 = 12$

e $5 +$ ______ $= 14$

f ______ $+ 8 = 16$

g $10 +$ ______ $= 17$

h ______ $+ 9 = 15$

i $9 +$ ______ $= 19$

j ______ $+ 8 = 19$

k $50 +$ ______ $= 70$

l ______ $+ 60 = 130$

m $4 +$ ______ $= 24$

n ______ $+ 200 = 800$

o $90 +$ ______ $= 170$

Exercise 7

Write down the sum:

a $1+4+2=$ ______ **b** $3+3+1=$ ______ **c** $4+2+3=$ ______

d $2+5+2=$ ______ **e** $4+2+5=$ ______ **f** $6+4+2=$ ______

g $3+5+4=$ ______ **h** $2+8+6=$ ______ **i** $7+3+4=$ ______

Exercise 8

Write down sets of *three different* numbers which total 10.

______ , ______ , and ______ ______ , ______ , and ______

______ , ______ , and ______ ______ , ______ , and ______

______ , ______ , and ______ ______ , ______ , and ______

______ , ______ , and ______ ______ , ______ , and ______

Exercise 9

Complete these additions.

a

$3+4=$ ______

$13+4=$ ______

$23+4=$ ______

$53+4=$ ______

b

$7+3=$ ______

$17+3=$ ______

$37+3=$ ______

$67+3=$ ______

Discussion

How can you *mentally* add:

a $25+3$ **b** $22+8$?

Exercise 10

Write down the sum:

a $14+2=$ ______ **b** $11+5=$ ______ **c** $3+16=$ ______

d $4+15=$ ______ **e** $12+6=$ ______ **f** $7+11=$ ______

g $22+3=$ ______ **h** $25+2=$ ______ **i** $4+31=$ ______

j $34+5=$ ______ **k** $6+42=$ ______ **l** $51+8=$ ______

Exercise 11

Write down the sum:

a $16+4=$ ______ **b** $12+8=$ ______ **c** $7+13=$ ______

d $15+5=$ ______ **e** $21+9=$ ______ **f** $6+24=$ ______

g $27+3=$ ______ **h** $34+6=$ ______ **i** $1+49=$ ______

j $44+6=$ ______ **k** $8+52=$ ______ **l** $63+7=$ ______

Exercise 12

Complete each addition:

a $4+8+3=$ ______ **b** $7+9+2=$ ______ **c** $5+6+8=$ ______

d $9+3+4=$ ______ **e** $18+5+3=$ ______ **f** $13+9+4=$ ______

Puzzle

Use the numbers 1 to 9 to make the sum of *every* row, column, and diagonal the same.

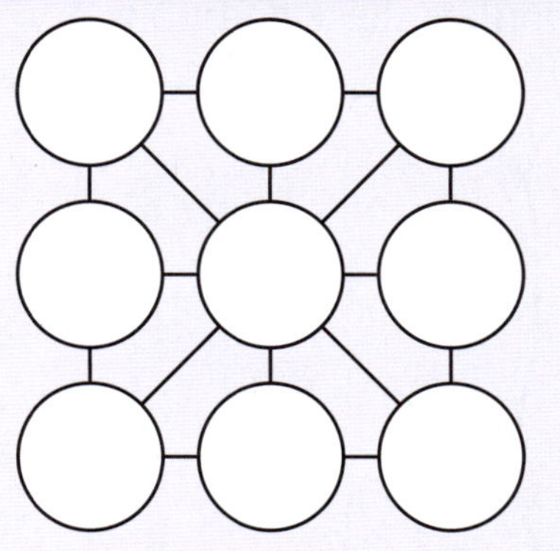

Exercise 13

Find the sum:

a $20 + 40 + 30 =$ ....

b $70 + 20 + 50 =$ ....

c $30 + 70 + 40 =$ ....

d $80 + 50 + 30 =$ ....

e $110 + 50 + 30 =$ ....

f $140 + 60 + 50 =$ ....

Exercise 14

Find the sum:

a $20 + 30 + 5 =$ ....

b $40 + 30 + 8 =$ ....

c $10 + 3 + 50 =$ ....

d $60 + 2 + 30 =$ ....

e $6 + 40 + 30 =$ ....

f $70 + 4 + 30 =$ ....

g $80 + 50 + 7 =$ ....

h $60 + 8 + 90 =$ ....

Exercise 15

Complete these additions.

a

b

Exercise 16

Complete these additions.

a

b

$4 + 7 =$ ....

$14 + 7 =$ ....

$34 + 7 =$ ....

$54 + 7 =$ ....

Discussion

How can you *mentally* add $38 + 7$?

Exercise 17

Complete these additions.

a $17 + 6 =$ ______ **b** $15 + 8 =$ ______ **c** $5 + 19 =$ ______

d $8 + 13 =$ ______ **e** $24 + 7 =$ ______ **f** $28 + 9 =$ ______

g $16 + 6 =$ ______ **h** $3 + 38 =$ ______ **i** $34 + 8 =$ ______

j $7 + 45 =$ ______ **k** $48 + 4 =$ ______ **l** $8 + 59 =$ ______

m $57 + 7 =$ ______ **n** $9 + 69 =$ ______ **o** $65 + 8 =$ ______

Exercise 18

Complete these additions.

a $8 + 7 + 9 =$ ______ **b** $9 + 5 + 7 =$ ______ **c** $8 + 8 + 6 =$ ______

d $13 + 8 + 7 =$ ______ **e** $15 + 8 + 4 =$ ______ **f** $6 + 17 + 8 =$ ______

g $9 + 14 + 5 =$ ______ **h** $26 + 9 + 3 =$ ______ **i** $8 + 25 + 9 =$ ______

Exercise 19

Complete these additions.

a $1 + 7 =$ ______ **b** $6 + 4 =$ ______ **c** $4 + 9 =$ ______

$11 + 7 =$ ______ $6 + 14 =$ ______ $14 + 9 =$ ______

$11 + 17 =$ ______ $16 + 14 =$ ______ $14 + 19 =$ ______

Group Activity

Your teacher will separate the class into groups.

You will be given a set of mental additions.

Discuss amongst your group ways to perform each addition mentally.

Remember to *listen* to everyone's ideas. Someone else may have an idea which can help you!

Exercise 20

Write down the result of each addition:

a $15 + 12 =$ ______ **b** $18 + 11 =$ ______ **c** $13 + 17 =$ ______

d $16 + 13 =$ ______ **e** $19 + 11 =$ ______ **f** $16 + 15 =$ ______

g $18 + 14 =$ ______ **h** $19 + 19 =$ ______ **i** $15 + 17 =$ ______

Exercise 21

Complete these additions.

a $4 + 6 + 4 =$ ______ **b** $2 + 8 + 7 =$ ______ **c** $3 + 9 + 7 =$ ______

d $4 + 7 + 3 =$ ______ **e** $9 + 8 + 2 =$ ______ **f** $12 + 4 + 8 =$ ______

g $13 + 7 + 5 =$ ______ **h** $16 + 14 + 2 =$ ______ **i** $13 + 9 + 17 =$ ______

j $5 + 6 + 15 =$ ______ **k** $18 + 7 + 12 =$ ______ **l** $13 + 11 + 9 =$ ______

Exercise 22

Find the sum:

a $4 + 7 + 3 + 5 =$ ______ **b** $8 + 1 + 2 + 6 =$ ______

c $9 + 6 + 10 + 4 =$ ______ **d** $10 + 9 + 6 + 1 =$ ______

Exercise 23

Describe what is happening in each sequence:

a 6 → 12 → 18 → 24 → 30

I start with ...., and add .... each time.

b 4 → 11 → 18 → 25 → 32

....

c 5 → 14 → 23 → 32 → 41

....

d 3 → 15 → 27 → 39 → 51

....

Exercise 24

Add 4 each time:

a 3 → ☐ → ☐ → ☐ → ☐ → ☐ → ☐

b 6 → ☐ → ☐ → ☐ → ☐ → ☐ → ☐

Add 5 each time:

c 11 → ☐ → ☐ → ☐ → ☐ → ☐ → ☐

Add 8 each time:

d 3 → ☐ → ☐ → ☐ → ☐ → ☐ → ☐

Add 10 each time:

e 2 → ☐ → ☐ → ☐ → ☐ → ☐ → ☐

Add 9 each time:

f 7 → ☐ → ☐ → ☐ → ☐ → ☐ → ☐

Exercise 25

Sally sells ice cream at the beach.

This table shows what she sold yesterday:

Flavour	*Small*	*Medium*	*Large*
chocolate	12	9	10
vanilla	8	16	9

Write an addition to solve each problem:

a Sally sold ...... + ... = ... large ice creams.

b Sally sold ...... + ... = ... small ice creams.

c Sally sold ...... + ... = ... medium ice creams.

d Sally sold ...... + ... + ... = ... chocolate ice creams.

e Sally sold ...... + ... + ... = ... vanilla ice creams.

Exercise 26

Complete using $=$ or $\neq$:

a $7 + 9$ 15

b $20 + 2$ 202

c $50 + 70$ 120

d $300 + 700$ 100

e $5 + 9 + 3$ 17

f $64 + 5$ 69

g $75 + 5$ 80

h $7 + 6 + 7$ 19

i $21 + 4 + 7$ 33

j $40 + 5 + 20$ 110

k $30 + 60 + 40$ 130

l $19 + 8$ 27

m $6 + 9 + 7$ 22

n $6 + 12 + 3$ 20

o $24 + 7 + 6$ 35

p $18 + 15$ 23

q $13 + 6 + 15$ 33

r $5 + 14 + 11$ 30

Find the missing number: $3+8=\_\_\_+7$

I know that $3+8=11$.

What number do I need to add to 7, to give 11?

I know that $4+7=11$.

So, the missing number is 4. $3+8=4+7$

Exercise 27

Fill in the missing number.

a $6+3=\_\_\_+5$

b $10+7=8+\_\_\_$

c $9+6=\_\_\_+7$

d $12+7=9+\_\_\_$

e $13+6=\_\_\_+4$

f $8+\_\_\_=7+12$

g $\_\_\_+9=13+8$

h $9+16=14+\_\_\_$

i $11+\_\_\_=7+16$

j $7+17=\_\_\_+14$

Listening Activity

Listen to the questions, and write your answers in the spaces provided.

1 ___ **2** ___ **3** ___

4 ___ **5** ___ **6** ___

7 ___ **8** ___ **9** ___

10 ___ **11** ___ **12** ___

Exercise 28

Complete using $<$, $=$, or $>$:

a $16+8$ ___ 22

b $400+300$ ___ 700

c $5+67$ ___ 70

d $14+8+3$ ___ 25

e $50+90+70$ ___ 200

f $6+12+8$ ___ 28

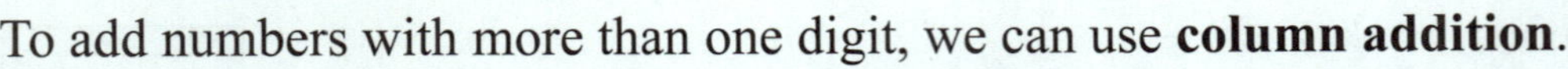

To add numbers with more than one digit, we can use **column addition**.

We write the numbers in columns so the place values line up. We then add each column, starting with the units and working from right to left.

	T	U
	3	4
+	5	2
	8	6

1 Set out your addition.

2 Add the units:

$4 + 2 = 6$

3 Add the tens:

$3 + 5 = 8$

So, $34 + 52 = 86$.

Exercise 29

Complete these additions.

a

	T	U
	3	1
+	2	5

b

	T	U
	2	6
+	6	3

c

	T	U
	4	4
+	2	3

d

	H	T	U
	1	6	2
+		3	5

e

	H	T	U
	2	4	2
+	1	3	7

f

	H	T	U
	3	1	8
+	4	6	1

Exercise 30

Set out these additions and find the answer.

a $64 + 15$

b $47 + 52$

c $73 + 25$

Exercise 31

Set out these additions and find the answer.

a $106 + 52$

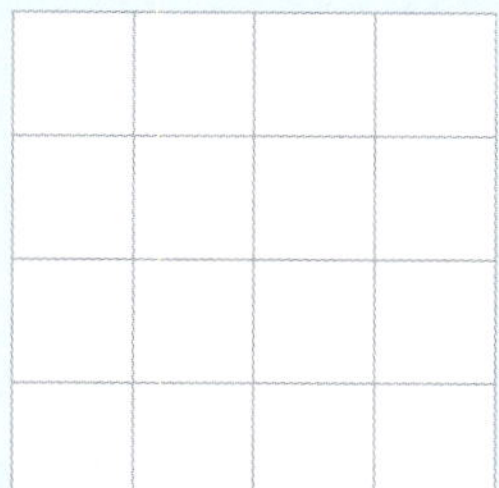

b $235 + 140$

c $546 + 313$

If there are 10 or more units, we exchange 10 units for 1 ten and **carry** the ten into the tens column.

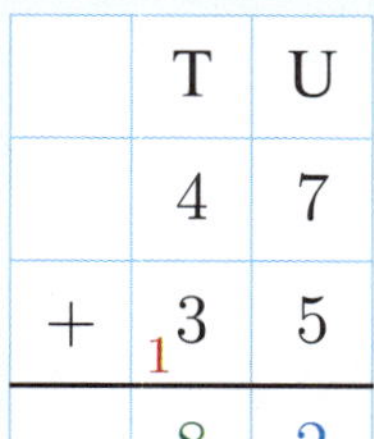

	T	U
	4	7
+	[1] 3	5
	8	2

1 Set out your addition.

2 Add the units:

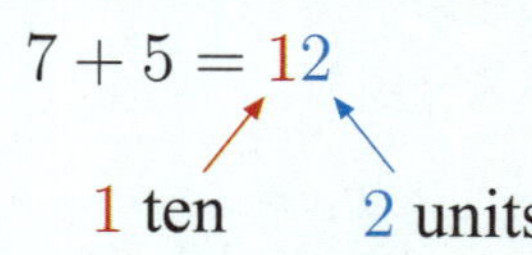

"Carry" the 1 ten into the tens column.
Write the 2 units below the line.

3 Add the tens:

$4 + 3 + 1 = 8$

So, $47 + 35 = 82$.

Exercise 32

Complete these additions.

a

	T	U
	2	4
+	1	9

b

	T	U
	3	5
+	1	8

c

	T	U
	4	6
+	2	8

d

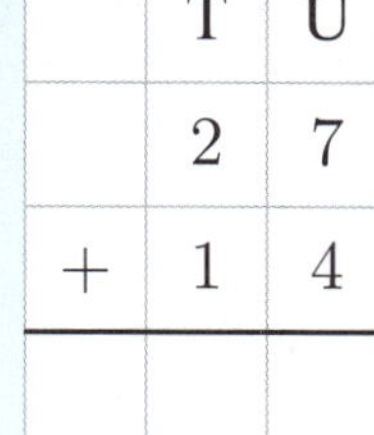

	T	U
	2	7
+	1	4

e

	T	U
	3	5
+	5	7

f

	T	U
	5	8
+	2	3

g

	T	U
	4	6
+	3	7

h

	T	U
	5	9
+	2	2

Exercise 33

Complete these additions.

a

	H	T	U
	1	4	5
+		3	8

b

	H	T	U
	2	3	7
+	1	2	9

c

	H	T	U
	1	6	5
+	1	2	7

Exercise 34

Set out these additions and find the answer.

a $27 + 49$

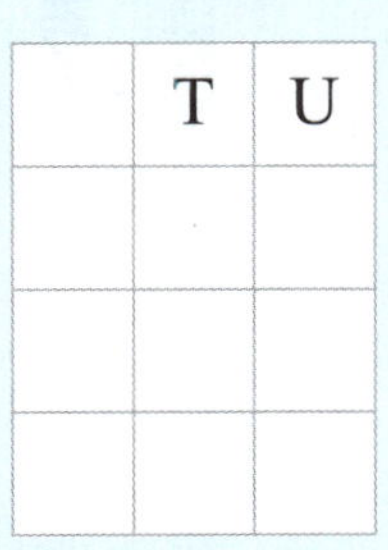

b $44 + 36$

	T	U

c $57 + 15$

d $108 + 24$

e $232 + 59$

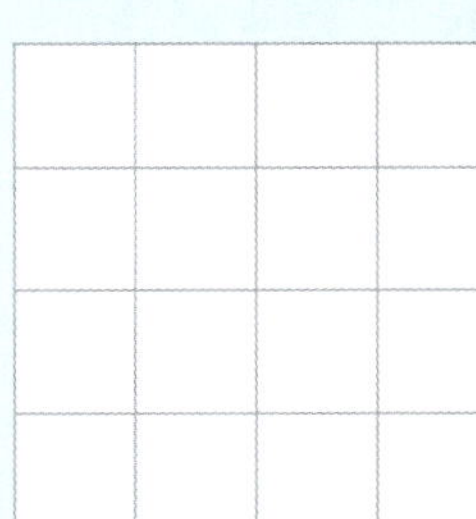

f $125 + 208$

If there are 10 or more tens, we exchange 10 tens for 1 hundred and **carry** the hundred into the hundreds column.

	H	T	U
	1	3	6
+	1	9	2
	2	2	8

1 Set out your addition.

2 Add the units:

$6 + 2 = 8$

3 Add the tens:

$3 + 9 = 12$

"Carry" the 1 into the hundreds column.
Write the 2 below the line.

4 Add the hundreds:

$1 + 1 = 2$

So, $136 + 92 = 228$.

Exercise 35

Complete these additions.

a

	H	T	U
		3	6
+		8	1

b

	H	T	U
		5	3
+		6	5

c

	H	T	U
		7	1
+		6	8

d

	H	T	U
		4	6
+		7	5

e

	H	T	U
		8	2
+		3	9

f

	H	T	U
	1	5	6
+		9	3

g

	H	T	U
	2	8	5
+	1	6	4

h

	H	T	U
	1	3	8
+	2	8	6

i

	H	T	U
	1	6	8
+	1	3	7

Exercise 36

Set out these additions and find the answer.

a $67 + 49$

	H	T	U

b $76 + 83$

	H	T	U

c $49 + 76$

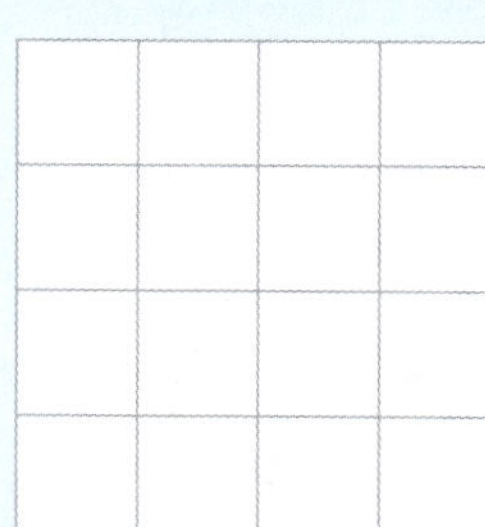

d $84 + 97$

e $83 + 79$

f $59 + 68$

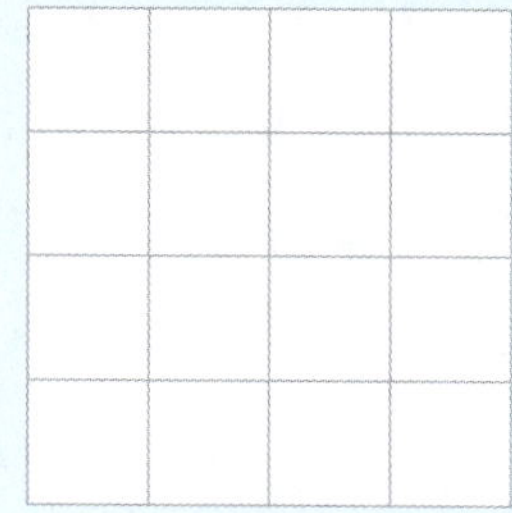

Exercise 37

Set out these additions and find the answer.

a $129 + 37$

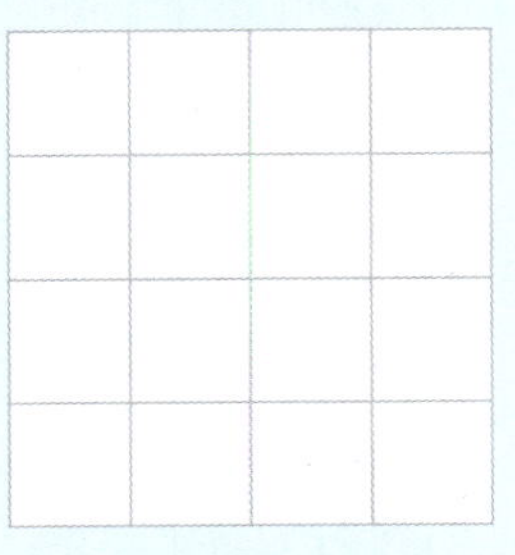

b $448 + 65$

c $249 + 85$

d $265 + 113$

e $278 + 208$

f $311 + 179$

Exercise 38

Complete these additions.

a

	T	U
	2	1
	3	4
+	1	0

b

	T	U
	6	5
	2	3
+		8

c

	H	T	U
	1	2	8
	2	9	1
+		7	3

Emil scored 19 points in one game, and 22 points in the next.

How many points did Emil score in total?

The word "total" tells us to add.

In total, Emil scored 41 points.

	T	U
	1	9
+	$_1$2	2
	4	1

When you are given a worded problem, write your answer in a sentence.

Exercise 39

Circle the words that tell you to **add**, then work out the answer.

a Amy made 16 Chinese dumplings, and Jing Tong made 22.

How many Chinese dumplings were made altogether?

Altogether, the girls made ______ Chinese dumplings.

b Ryley walked 45 steps to his classroom, and then 33 steps to the playground.

Find the total number of steps Ryley walked.

____

c Lucy has 25 lemons on her tree, and Carys has 36 lemons on her tree.

How many lemons are there altogether?

____

d Lewis bounced the basketball 56 times, but Michael could only manage 36 bounces.

In total, how many times was the ball bounced?

____

e In an archery competition, Meg scored 128 points in her first round, and 173 points in her second round.

Find the total number of points Meg scored in her rounds.

____

f Pedro has 27 marbles, Gary has 22 marbles, and Kiet has 16 marbles.

How many marbles are there altogether?

____

Exercise 40

Some of these additions have incorrect answers. Find the errors and correct them.

a

H	T	U
	6	9
+	$_1$3	7
1	0	6

b

H	T	U
	7	4
+	$_1$4	7
2	1	1

c

H	T	U
	9	6
+	$_1$3	7
1	2	3

d

H	T	U
	7	5
+	$_1$4	9
1	2	4

Revision

1 Complete each addition:

a $2+5=$ ______ **b** $4+3=$ ______ **c** $7+8=$ ______

d $11+0=$ ______ **e** $6+12=$ ______ **f** $9+6=$ ______

g $14+5=$ ______ **h** $7+19=$ ______ **i** $26+4=$ ______

j $6+85=$ ______ **k** $15+13=$ ______ **l** $63+9=$ ______

2 Find the total:

a $3+5+1=$ ______ **b** $7+4+2=$ ______ **c** $4+8+6=$ ______

d $7+9+3=$ ______ **e** $5+8+12=$ ______ **f** $16+7+14=$ ______

3 Complete these additions.

a $2+4=$ ______ **b** $3+7=$ ______ **c** $7+5=$ ______

$20+40=$ ______ $30+70=$ ______ $70+50=$ ______

$200+400=$ ______ $300+700=$ ______ $700+500=$ ______

4 Fill in the missing number.

a $6+$ ______ $=8$ **b** ______ $+5=14$ **c** $8+$ ______ $=11$

d ______ $+7=18$ **e** $13+$ ______ $=21$ **f** ______ $+18=33$

5 Find the sum:

a $3+6+7+2=$ ______ **b** $9+6+1+4=$ ______

6 Find the sum:

a $30 + 60 + 4 =$ ________ **b** $60 + 70 + 20 =$ ________

c $80 + 3 + 70 =$ ________ **d** $130 + 90 + 60 =$ ________

7 Describe what is happening in the sequence.

7 → 15 → 23 → 31 → 39 → 47

I start with ______, and add ______ each time.

8 **a** Add 4 each time:

5 → ☐ → ☐ → ☐ → ☐ → ☐ → ☐

b Add 7 each time:

2 → ☐ → ☐ → ☐ → ☐ → ☐ → ☐

9 Fill in the missing number.

a $5 + 7 =$ ______ $+ 3$ **b** $8 +$ ______ $= 11 + 4$

c ______ $+ 13 = 6 + 16$ **d** $14 + 9 = 8 +$ ______

10 Complete these additions.

a

	H	T	U
		8	3
+		1	6

b

	H	T	U
	2	2	7
+		6	6

c

	H	T	U
		8	1
+		5	8

d

	H	T	U
	2	0	7
+	1	7	9

e

	H	T	U
	3	5	4
+	2	8	7

f

	H	T	U
	1	8	9
+	3	1	6

11 Complete using $<$, $=$, or $>$:

a $17 + 4$ ____ 20 **b** $3 + 2 + 5$ ____ 10 **c** $60 + 50$ ____ 120

d $7 + 25$ ____ 32 **e** $12 + 18$ ____ 28 **f** $7 + 19 + 3$ ____ 30

12 Set out these additions and find the answer.

a $65 + 33$

b $57 + 24$

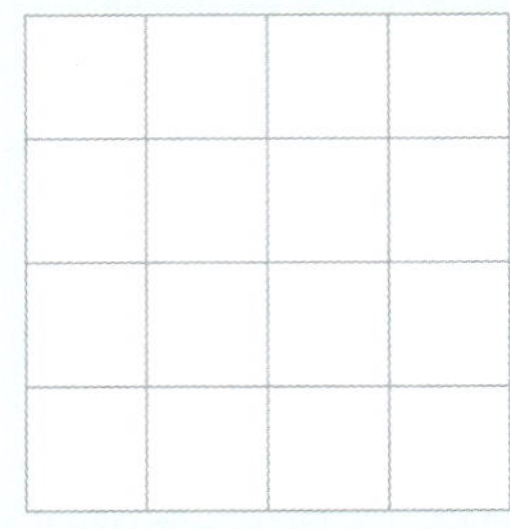

c $85 + 32$

d $106 + 118$

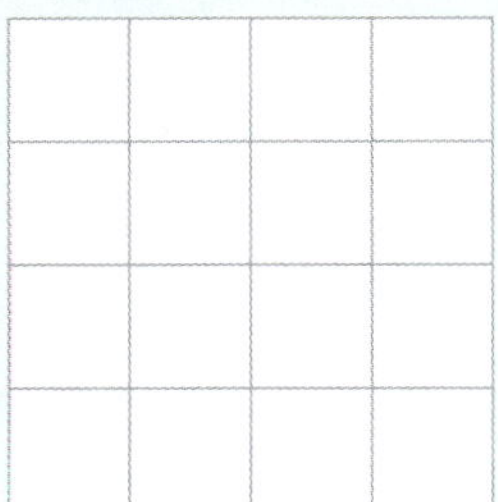

e $253 + 79$

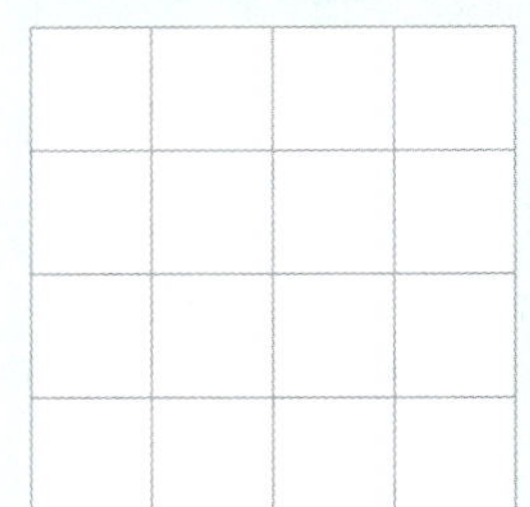

f $614 + 196$

13 Complete these additions.

a

	H	T	U
		4	8
		1	3
+		6	5

b

	H	T	U
	2	5	6
		8	3
+	1	0	8

14 Circle the words that tell you to **add**, then work out the answer.

a Bernard has 39 books. Georgia has 46 books.

How many books are there altogether?

....

b Jenny scored 116 points and 135 points in two games of ten-pin bowling.

How many points did Jenny score in total?

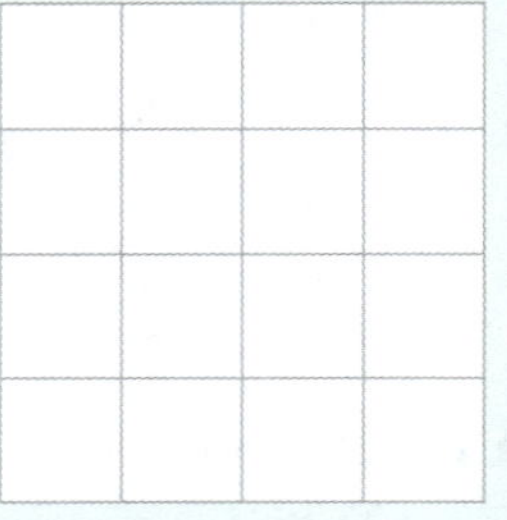

....

CHAPTER 3: SUBTRACTION

$10 - 3 = 7$

"ten minus three equals seven"

Mental subtractions are done without a pencil and paper, or apparatus.

Exercise 1

Write down the result of each subtraction:

a $5 - 2 =$ ______ **b** $7 - 3 =$ ______ **c** $10 - 1 =$ ______

d $6 - 5 =$ ______ **e** $8 - 2 =$ ______ **f** $9 - 4 =$ ______

g $7 - 7 =$ ______ **h** $10 - 6 =$ ______ **i** $9 - 8 =$ ______

j $14 - 14 =$ ______ **k** $15 - 0 =$ ______ **l** $10 - 8 =$ ______

Addition and subtraction are related.

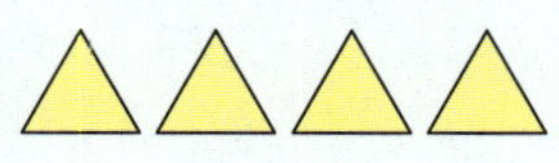

4

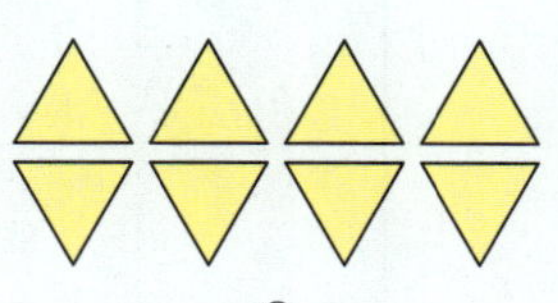

8

$4 + 8 = 12$

$8 + 4 = 12$

$12 - 4 = 8$

$12 - 8 = 4$

Exercise 2

Complete:

a

$4 + 7 =$ ______

$7 + 4 =$ ______

______ $- 4 =$ ______

______ $- 7 =$ ______

b

$8 + 5 =$ ______

$5 + 8 =$ ______

______ $- 8 =$ ______

______ $- 5 =$ ______

Exercise 3

Complete:

a
$8 + 9 = \ldots$
so $\ldots - 8 = \ldots$
and $\ldots - 9 = \ldots$

b
$7 + 5 = \ldots$
so $\ldots - 7 = \ldots$
and $\ldots - 5 = \ldots$

c
$12 + 7 = \ldots$
so $\ldots - 12 = \ldots$
and $\ldots - 7 = \ldots$

d
$10 + 18 = \ldots$
so $\ldots - 10 = \ldots$
and $\ldots - 18 = \ldots$

Exercise 4

Complete:

a
$9 + \ldots = 14$
so $14 - 9 = \ldots$
and $14 - \ldots = 9$

b
$\ldots + 3 = 8$
so $8 - \ldots = \ldots$
and $8 - \ldots = \ldots$

c
$7 + \ldots = 9$
so $9 - \ldots = \ldots$
and $9 - \ldots = \ldots$

d
$\ldots + 5 = 12$
so $12 - \ldots = \ldots$
and $12 - \ldots = \ldots$

Exercise 5

Solve each subtraction mentally:

a $11 - 3 = \ldots$ **b** $13 - 4 = \ldots$ **c** $16 - 7 = \ldots$

d $12 - 6 = \ldots$ **e** $15 - 8 = \ldots$ **f** $14 - 9 = \ldots$

g $13 - 7 = \ldots$ **h** $17 - 8 = \ldots$ **i** $12 - 9 = \ldots$

Exercise 6

Complete:

a 13 minus 8 = ______

b 17 subtract 9 = ______

c 14 take away 7 = ______

d Find the difference: 11 − 8 = ______

e Take 5 away from 12: 12 − 5 = ______

f Subtract 6 from 13: ______ − ______ = ______

g 8 less than 14: ______ − ______ = ______

The **difference** between two numbers is the distance between them on the number line.

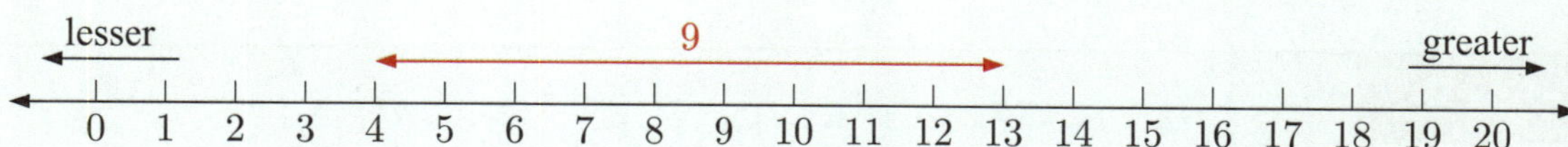

To find the difference between two numbers, we subtract the *lesser* number from the *greater* number.

The difference between 4 and 13 is 13 − 4 = 9.

Exercise 7

For each pair of numbers, write a subtraction to find their difference:

a 8 and 5

______ − ______ = ______

b 3 and 10

______ − ______ = ______

c 15 and 6

______ − ______ = ______

d 8 and 16

______ − ______ = ______

e 9 and 12

______ − ______ = ______

f 18 and 9

______ − ______ = ______

g 20 and 17

______ − ______ = ______

h 11 and 16

______ − ______ = ______

Exercise 8

Complete these subtractions.

a $9 - 7 =$ ______
$90 - 70 =$ ______
$900 - 700 =$ ______

b $8 - 4 =$ ______
$80 - 40 =$ ______
$800 - 400 =$ ______

c $7 - 2 =$ ______
$70 - 20 =$ ______
$700 - 200 =$ ______

d $11 - 5 =$ ______
$110 - 50 =$ ______
$1100 - 500 =$ ______

e $14 - 8 =$ ______
$140 - 80 =$ ______
$1400 - 800 =$ ______

f $16 - 9 =$ ______
$160 - 90 =$ ______
$1600 - 900 =$ ______

Exercise 9

Complete each subtraction:

a $12 - 10 =$ ______ **b** $17 - 10 =$ ______ **c** $21 - 10 =$ ______

d $35 - 10 =$ ______ **e** $59 - 10 =$ ______ **f** $86 - 10 =$ ______

g $38 - 20 =$ ______ **h** $43 - 20 =$ ______ **i** $66 - 20 =$ ______

j $52 - 30 =$ ______ **k** $94 - 40 =$ ______ **l** $105 - 10 =$ ______

m $83 - 50 =$ ______ **n** $111 - 60 =$ ______ **o** $96 - 70 =$ ______

Exercise 10

Complete these subtractions.

a $8 - 2 =$ ______
$18 - 2 =$ ______
$28 - 2 =$ ______
$38 - 2 =$ ______

b $10 - 7 =$ ______
$20 - 7 =$ ______
$40 - 7 =$ ______
$70 - 7 =$ ______

c $14 - 9 =$ ______
$24 - 9 =$ ______
$44 - 9 =$ ______
$84 - 9 =$ ______

Discussion

How can we use the patterns in the previous Exercise to perform subtractions mentally?

If we do not have enough units to subtract, we *exchange* 1 .... for 10 ....

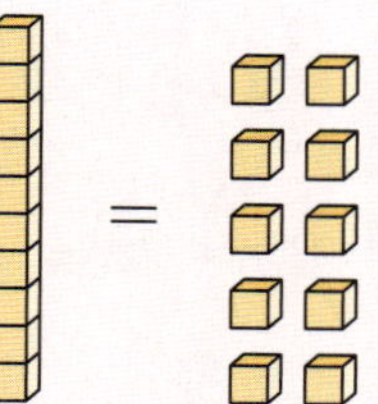

Exercise 11

Solve each subtraction mentally:

a $19 - 7 =$ .... **b** $18 - 5 =$ .... **c** $23 - 3 =$ ....

d $20 - 6 =$ .... **e** $36 - 4 =$ .... **f** $30 - 9 =$ ....

g $45 - 5 =$ .... **h** $50 - 2 =$ .... **i** $52 - 4 =$ ....

j $89 - 6 =$ .... **k** $47 - 5 =$ .... **l** $60 - 3 =$ ....

m $43 - 8 =$ .... **n** $98 - 6 =$ .... **o** $70 - 8 =$ ....

p $90 - 5 =$ .... **q** $92 - 8 =$ .... **r** $67 - 7 =$ ....

s $74 - 7 =$ .... **t** $100 - 7 =$ .... **u** $103 - 9 =$ ....

Puzzle

The *difference* between 12 and a number is 9.

The number could be .... or ....

Exercise 12

Complete these subtractions.

a $6 - 3 =$ .... **b** $9 - 4 =$ .... **c** $8 - 2 =$ ....

$16 - 13 =$ .... $19 - 14 =$ .... $18 - 12 =$ ....

$26 - 23 =$ .... $29 - 24 =$ .... $28 - 22 =$ ....

$36 - 33 =$ .... $39 - 34 =$ .... $38 - 32 =$ ....

Exercise 13

Find the answer mentally.

a $14-11=$ ______ **b** $17-12=$ ______ **c** $19-16=$ ______

d $23-22=$ ______ **e** $28-21=$ ______ **f** $27-24=$ ______

g $35-33=$ ______ **h** $36-32=$ ______ **i** $38-31=$ ______

Exercise 14

Complete using $=$ or $\neq$.

a $12-6$ ☐ 6 **b** $14-9$ ☐ 6 **c** $17-8$ ☐ 11

d $7-4$ ☐ $15-12$ **e** $20-3$ ☐ $9+7$ **f** $14-5$ ☐ $2+7$

g $6+5$ ☐ $20-8$ **h** $15+6$ ☐ $27-6$ **i** $27-9$ ☐ $4+14$

Exercise 15

Find the missing numbers.

a $16-4=12$

so $16-$ ______ $=4$

b $26-7=$ ______

so $26-$ ______ $=7$

c $31-8=$ ______

so $31-$ ______ $=8$

d $54-9=$ ______

so $54-$ ______ $=9$

Listening Activity

Listen to the questions, and write the answers in the spaces provided.

1 ________ **2** ________ **3** ________

4 ________ **5** ________ **6** ________

7 ________ **8** ________

9 ________ **10** ________

Exercise 16

Find the missing number.

a $10 - \_\_\_\_ = 6$ **b** $\_\_\_\_ - 8 = 7$ **c** $12 - \_\_\_\_ = 9$

d $\_\_\_\_ - 11 = 0$ **e** $18 - \_\_\_\_ = 9$ **f** $\_\_\_\_ - 5 = 10$

g $16 - \_\_\_\_ = 4$ **h** $\_\_\_\_ - 3 = 17$ **i** $19 - \_\_\_\_ = 15$

Exercise 17

Fill in the missing number.

a $\_\_\_\_ - 30 = 40$ **b** $100 - \_\_\_\_ = 90$ **c** $\_\_\_\_ - 60 = 20$

d $800 - \_\_\_\_ = 300$ **e** $\_\_\_\_ - 400 = 200$ **f** $900 - \_\_\_\_ = 100$

Exercise 18

Fill in the missing number.

a $7 - 3 = \_\_\_\_ - 5$ **b** $\_\_\_\_ - 4 = 2 + 5$

c $12 - \_\_\_\_ = 9 - 6$ **d** $8 + 6 = 19 - \_\_\_\_$

e $13 - 8 = \_\_\_\_ - 6$ **f** $15 - \_\_\_\_ = 3 + 7$

g $\_\_\_\_ - 11 = 12 - 9$ **h** $18 + 7 = 35 - \_\_\_\_$

i $19 - 12 = \_\_\_\_ - 6$ **j** $23 - \_\_\_\_ = 4 + 12$

Exercise 19

Subtract 3 each time:

a 19 → ☐ → ☐ → ☐ → ☐ → ☐

b 24 → ☐ → ☐ → ☐ → ☐ → ☐

Subtract 5 each time:

c 32 → ☐ → ☐ → ☐ → ☐ → ☐

d 43 → ☐ → ☐ → ☐ → ☐ → ☐

Subtract 7 each time:

e 42 → ☐ → ☐ → ☐ → ☐ → ☐

Subtract 10 each time:

f 80 → ☐ → ☐ → ☐ → ☐ → ☐

Subtract 9 each time:

g 105 → ☐ → ☐ → ☐ → ☐ → ☐

Exercise 20

Describe what is happening in each sequence:

a 27 → 22 → 17 → 12 → 7

I start with ...., and subtract .... each time.

b 35 → 29 → 23 → 17 → 11

....

c 47 → 39 → 31 → 23 → 15

....

d 62 → 51 → 40 → 29 → 18

....

Exercise 21

Circle the words which tell you to add or subtract, then answer the question.

a Laura has 7 crayons. Her brother Noah has 12 crayons. How many more crayons does Noah have than Laura?

Noah has ______ more crayons than Laura.

b Faruk has 13 lemons and 9 limes. How many fruits does he have altogether?

Altogether, Faruk has ______ fruits.

c Carla baked 18 cookies. She gave 6 cookies to her mum. How many cookies does Carla have left?

Carla has ______ cookies left.

d A music lesson lasts for 50 minutes. Joseph spends the first 10 minutes warming up. How much time is remaining in the lesson?

There are ______ minutes remaining in the lesson.

e Mateo jumped 40 hurdles in training yesterday, and 30 hurdles today. How many hurdles has Mateo jumped in total?

Mateo has jumped ______ hurdles in total.

f Tammy did 38 sit-ups yesterday. Today she did 9 fewer. How many sit-ups did Tammy do today?

Tammy did ______ sit-ups today.

g Samia has 10 plastic chairs. Her friend Hamia brings another 14 chairs for a party. How many chairs are there altogether?

Altogether, there are ______ chairs for the party.

h Philippe picked 86 peaches on Saturday and 9 fewer peaches on Sunday. How many peaches did Philippe pick on Sunday?

Philippe picked ______ peaches on Sunday.

To subtract numbers with more than one digit, we can use **column subtraction**.

We write the numbers in columns so the place values line up.

We then subtract each column, starting with the units.

36 – 15

	T	U
	3	6
–	1	5
	2	1

So, 36 – 15 = 21.

1 Set out your subtraction.

2 Subtract the units:

$6 - 5 = 1$

We write 1 below the line in the units column.

3 Subtract the tens:

$3 - 1 = 2$

We write 2 below the line in the tens column.

Exercise 22

Complete these subtractions.

a 36 – 13

	T	U
	3	6
–	1	3

b 47 – 27

	T	U
	4	7
–	2	7

c 58 – 24

	T	U
	5	8
–	2	4

d 76 – 25

	T	U
	7	6
–	2	5

e 99 – 68

	T	U
	9	9
–	6	8

f 87 – 76

	T	U
	8	7
–	7	6

g 143 – 22

	H	T	U
	1	4	3
–		2	2

h 158 – 34

	H	T	U
	1	5	8
–		3	4

i 175 – 44

	H	T	U
	1	7	5
–		4	4

Exercise 23

Complete these subtractions.

a 256 − 125

	H	T	U
	2	5	6
−	1	2	5

b 367 − 134

	H	T	U
	3	6	7
−	1	3	4

c 983 − 251

	H	T	U
	9	8	3
−	2	5	1

Exercise 24

Set out these subtractions and find the answer.

a 76 − 45

	T	U

b 97 − 55

c 85 − 42

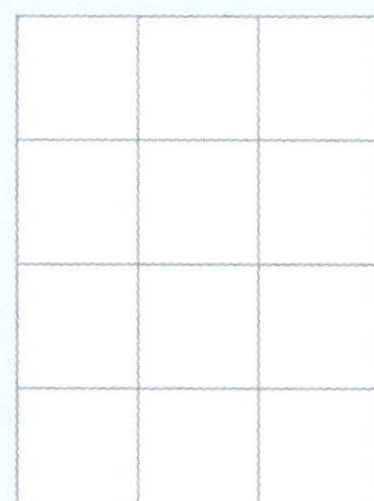

d 78 − 35

e 87 − 43

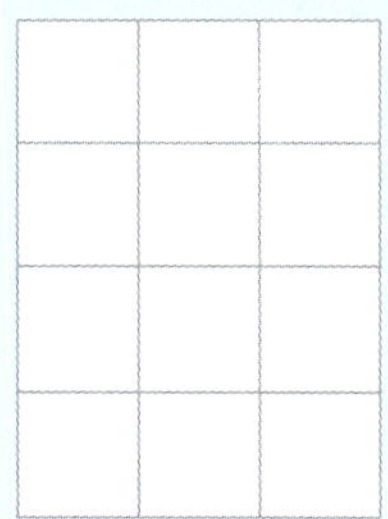

f 99 − 46

g 137 − 35

	H	T	U

h 276 − 161

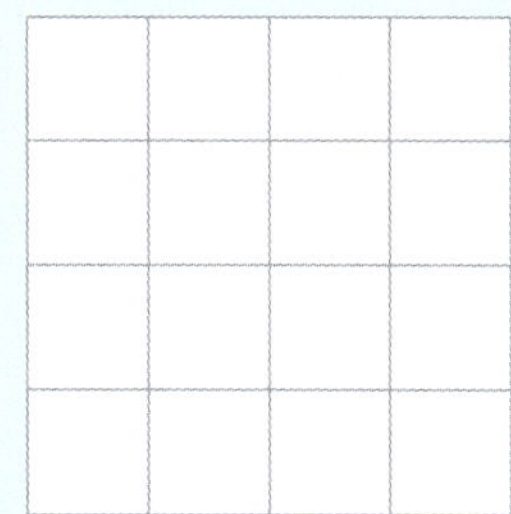

i 575 − 263

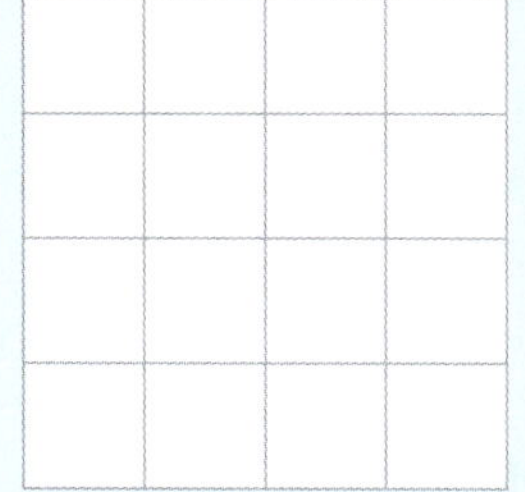

	T	U
	7	2
−	2	6

We *always* begin a subtraction in the units column.

We *always* subtract the bottom number from the top number.

But what do we do with $2 - 6$?

To complete this subtraction we **exchange** 1 ten for 10 units

	T	U
	6	12
	~~7~~	~~2~~
−	2	6
	4	6

72 is now written as 6 tens and 12 units.

$12 - 6 = 6$ units.

$6 - 2 = 4$ tens.

So, $72 - 26 = 46$.

Exchange means "change over".

Exercise 25

Complete these subtractions. You will need to use exchanging.

a $73 - 19$

	T	U
	7	3
−	1	9

b $81 - 17$

	T	U
	8	1
−	1	7

c $65 - 27$

	T	U
	6	5
−	2	7

d $180 - 29$

	H	T	U
	1	8	0
−		2	9

e $290 - 145$

	H	T	U
	2	9	0
−	1	4	5

f $360 - 149$

	H	T	U
	3	6	0
−	1	4	9

Exercise 26

Set out these subtractions and find the answer.

a 61 – 25

	T	U

b 64 – 37

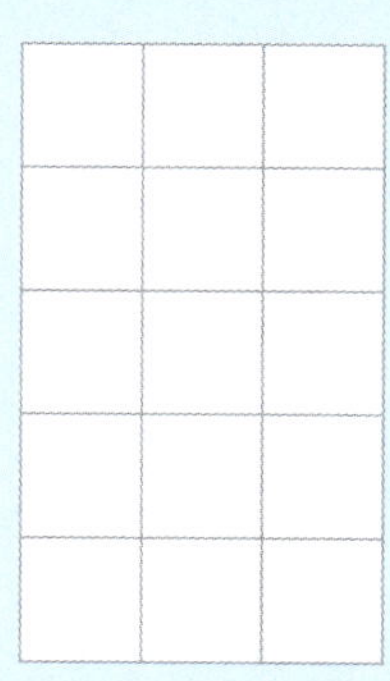

c 71 – 49

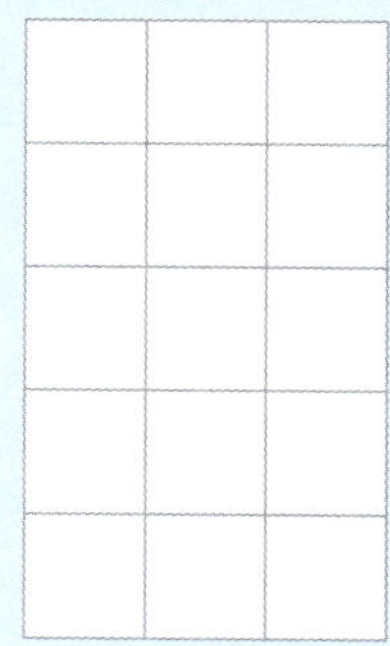

d 251 – 126

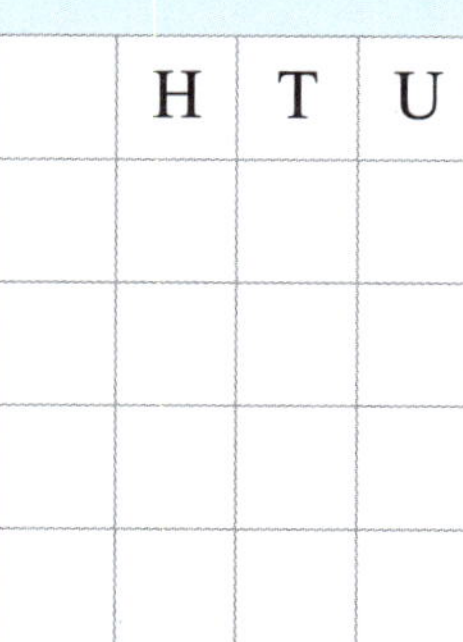

e 481 – 152

f 940 – 327

Discussion

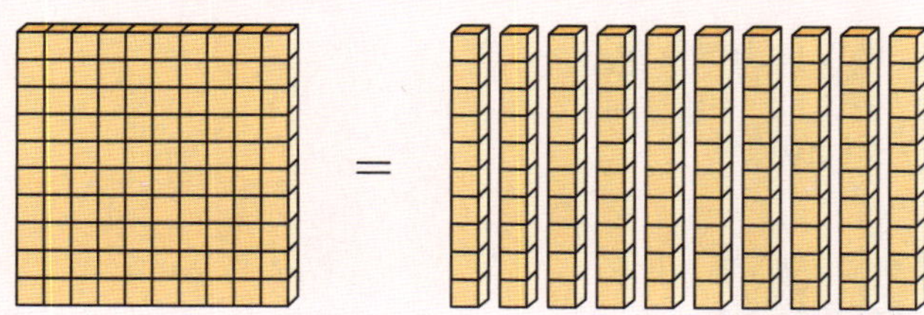

How can we find 218 – 153?

8 – 3 = 5 units.

But what do we do with 1 – 5 tens?

	H	T	U
	2	1	8
–	1	5	3
			5

To complete this subtraction we ______________

1 hundred for 10 ________.

We can now complete the subtraction.

	H	T	U
	1	11	
	~~2~~	~~1~~	8
–	1	5	3
		6	5

218 – 153 = ________

Exercise 27

Complete these subtractions.

a 165 – 83

	H	T	U
	1	6	5
–		8	3

b 227 – 190

	H	T	U
	2	2	7
–	1	9	0

c 319 – 174

	H	T	U
	3	1	9
–	1	7	4

Exercise 28

Circle the words that tell you to subtract, then solve each problem:

a Thomas has 39 lollies in his bag.
He eats 15 lollies.
How many lollies does Thomas have left in his bag?

Thomas has ☐ lollies left in his bag.

b Nick buys a bag of 60 cockles to take fishing.
He gives 39 cockles to his father.
How many cockles does Nick have left in his bag?

Nick has  cockles left in his bag.

c Sarah needs 128 counters for her mathematics lesson.
She has collected 43 counters.
How many more counters does Sarah need for her lesson?

Sarah needs  more counters for her mathematics lesson.

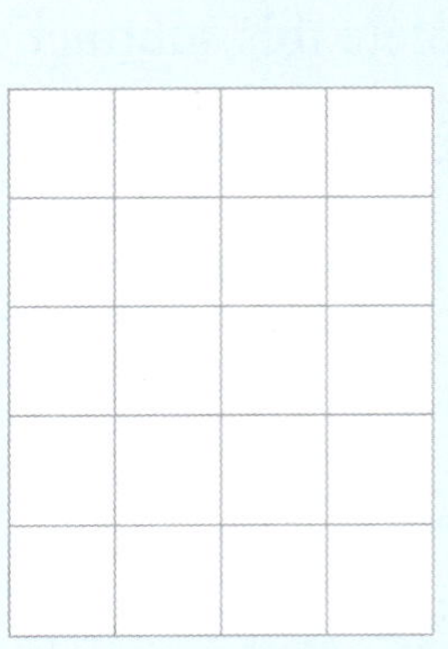

Exercise 29

Solve each problem using an addition and a subtraction:

a Lydia rode her bicycle for 45 minutes then ran for 28 minutes.

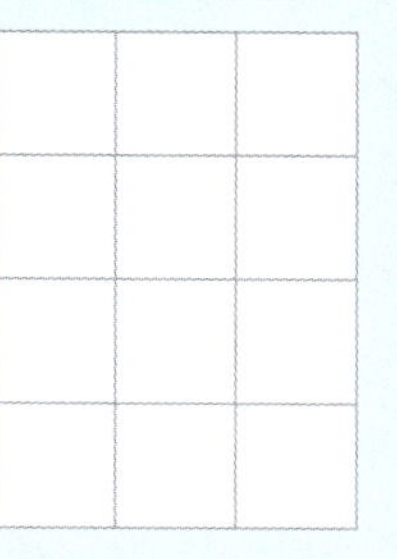

i Lydia exercised for ☐ minutes in total.

ii Lydia rode for ☐ minutes more than she ran.

b Mustafa has 209 oranges on his fruit stall. He sells 13 oranges to Mrs Habib, 12 to Mrs Khan, and 8 to Mrs Singh.

i Mustafa has sold ☐ oranges altogether.

ii Mustafa has ☐ oranges left on his stall.

Exercise 30

Complete these subtractions by exchanging.

a

	H	T	U
	4	2	1
−	2	6	5

b

	H	T	U
	9	7	3
−	4	7	8

c

	H	T	U
	6	0	5
−	3	9	6

Revision

1 **a** 9 minus 6 = ______ **b** 16 subtract 8 = ______

c 14 take away 9 = ______ **d** Find the difference: 11 − 7 = ______

e Take 3 away from 12: ______ − ______ = ______

f Subtract 5 from 13: ______ − ______ = ______

2 Complete:

a $16 + 7 =$ ______
so ______ $- 16 =$ ______
and ______ $- 7 =$ ______

b ______ $+ 6 = 15$
so $15 - 6 =$ ______
and $15 -$ ______ $= 6$

c $17 =$ ______ $+ 9$
so $17 - 9 =$ ______
and $17 -$ ______ $= 9$

d $22 - 8 =$ ______
so $22 -$ ______ $= 8$

3 Complete each subtraction:

a $18 - 9 =$ ______ **b** $15 - 7 =$ ______ **c** $16 - 10 =$ ______

d $20 - 6 =$ ______ **e** $25 - 8 =$ ______ **f** $69 - 4 =$ ______

g $47 - 8 =$ ______ **h** $50 - 3 =$ ______ **i** $62 - 9 =$ ______

4 For each pair of numbers, write a subtraction to find their difference:

a 53 and 20
______ $-$ ______ $=$ ______

b 6 and 82
______ $-$ ______ $=$ ______

5 Complete each subtraction:

a $70 - 50 =$ ______ **b** $900 - 300 =$ ______ **c** $120 - 40 =$ ______

d $1000 - 200 =$ ______ **e** $160 - 80 =$ ______ **f** $1500 - 800 =$ ______

6 Complete each subtraction:

a $74 - 10 =$ ______ **b** $29 - 20 =$ ______ **c** $53 - 10 =$ ______

d $89 - 30 =$ ______ **e** $105 - 40 =$ ______ **f** $97 - 80 =$ ______

7 Find the answer mentally.

a $18 - 13 =$ ______ **b** $26 - 21 =$ ______ **c** $29 - 25 =$ ______

8 Find the missing number.

a ______ $- 7 = 12$ **b** $22 -$ ______ $= 16$ **c** ______ $- 11 = 5$

d $23 -$ ______ $= 18$ **e** ______ $- 20 = 60$ **f** $600 -$ ______ $= 300$

9 Find the missing number.

a $4 + 7 = 18 -$ ______

b $13 -$ ______ $= 19 - 12$

c ______ $- 8 = 8 + 5$

d $16 - 5 =$ ______ $- 9$

10 **a** Subtract 4 each time:

27 → ☐ → ☐ → ☐ → ☐ → ☐

b Subtract 6 each time:

38 → ☐ → ☐ → ☐ → ☐ → ☐

11 Describe what is happening in the sequence.

55 → 47 → 39 → 31 → 23 → 15

I start with ______, and subtract ______ each time.

12 Linda has 19 pins. She uses 15 pins to put together a dress. How many pins does Linda have left?

Linda has ______ pins left.

13 Complete these subtractions.

a

	H	T	U
		5	6
−		2	5

b

	H	T	U
	1	6	5
−	1	4	4

c

	H	T	U
	2	8	7
−	1	6	0

d

	H	T	U
		7	2
−		3	8

e

	H	T	U
	1	6	0
−		4	7

f

	H	T	U
	3	1	8
−	2	2	0

g

	H	T	U
	2	1	5
−	1	6	4

h

	H	T	U
	3	1	4
−	1	6	8

i

	H	T	U
	5	1	2
−	2	1	9

14 Set out these subtractions and find the answer.

a 68 – 43

b 347 – 121

c 83 – 18

d 264 – 172

e 164 – 88

f 222 – 139

15 Solve each problem:

a Sammy picks 85 apricots from his tree.

He gives 47 apricots to his parents.

Sammy has ☐ apricots left.

b Charlotte wants to watch a movie which is 134 minutes long, and a documentary which is 57 minutes long.

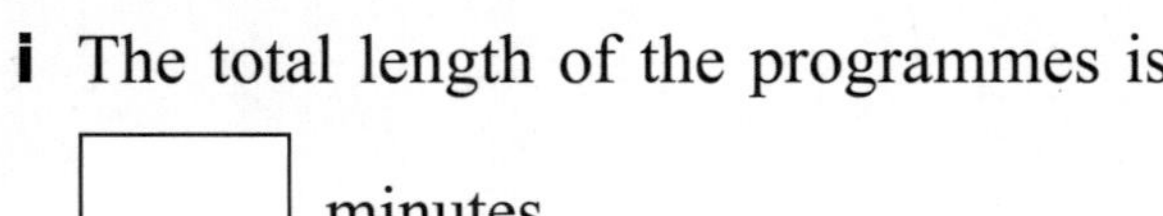

i The total length of the programmes is ☐ minutes.

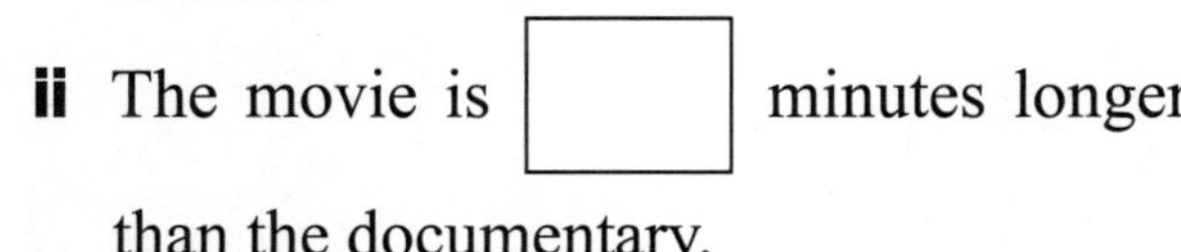

ii The movie is ☐ minutes longer than the documentary.

CHAPTER 4: MULTIPLICATION

Multiplication is a quick way to add the same number many times.

$$\underbrace{6+6+6+6+6}_{\text{5 lots of 6}} = 5 \times 6$$

This multiplication can be described in words as:

- five times six
- five lots of six
- 5 multiplied by 6
- the product of 5 and 6.

Exercise 1

Write as a multiplication and find the answer:

a $3+3+3+3 =$ ______ $\times\, 3 =$ ______

b $6+6+6 =$ ______ $\times$ ______ $=$ ______

c $7+7+7+7+7 =$ ____________ $=$ ______

d $8+8+8+8+8+8 =$ ____________ $=$ ______

Exercise 2

Write as an addition and find the answer:

a $3 \times 5 =$ ______ $+$ ______ $+$ ______ $=$ ______

b $4 \times 8 =$ ______ $+$ ______ $+$ ______ $+$ ______ $=$ ______

c $3 \times 11 =$ ____________________ $=$ ______

d $5 \times 4 =$ ________________________________ $=$ ______

Exercise 3

Write as a multiplication:

a six multiplied by seven ______________

b the product of nine and twelve ______________

This **multiplication table** shows the **times tables** you should already know by heart.

×	1	2	3	4	5	6	7	8	9	10
1	1	2	3	4	5	6	7	8	9	10
2	2	4	6	8	10	12	14	16	18	20
3	3	6	9	12	15	18	21	24	27	30
4	4	8	12	16	20	24	28	32	36	40
5	5	10	15	20	25	30	35	40	45	50
6	6	12	18	24	30	36	42	48	54	60
7	7	14	21	28	35	42	49	56	63	70
8	8	16	24	32	40	48	56	64	72	80
9	9	18	27	36	45	54	63	72	81	90
10	10	20	30	40	50	60	70	80	90	100

The shaded lines show that $5 \times 8 = 40$.

Click the icon alongside for help with your times tables.

Exercise 4

Find each product:

a $3 \times 2 =$ ______ **b** $6 \times 6 =$ ______ **c** $7 \times 4 =$ ______

d $8 \times 6 =$ ______ **e** $5 \times 9 =$ ______ **f** $10 \times 3 =$ ______

g $9 \times 8 =$ ______ **h** $6 \times 5 =$ ______ **i** $3 \times 9 =$ ______

j $4 \times 5 =$ ______ **k** $7 \times 8 =$ ______ **l** $5 \times 10 =$ ______

m $8 \times 8 =$ ______ **n** $6 \times 9 =$ ______ **o** $9 \times 9 =$ ______

Exercise 5

Complete:

a four times five equals ______ **b** 6 lots of 2 is ______

c the product of 4 and 8 is ______ **d** seven multiplied by five is ______

e 7 lots of 9 is ______ **f** 6 times 7 equals ______

g eight multiplied by three equals ______ **h** the product of 10 and 7 is ______

Exercise 6

Complete these multiplication flowers.

a

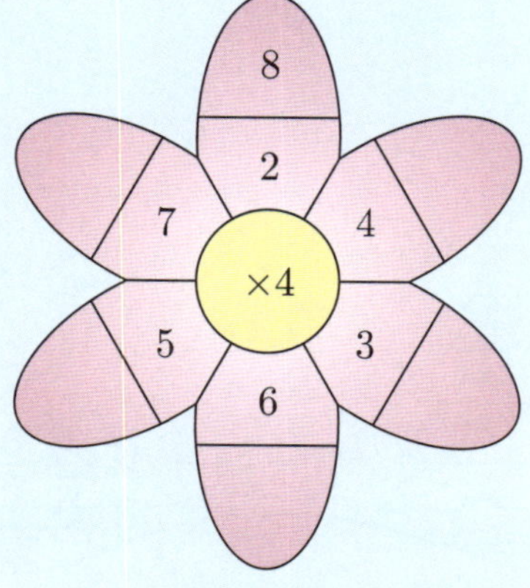

b

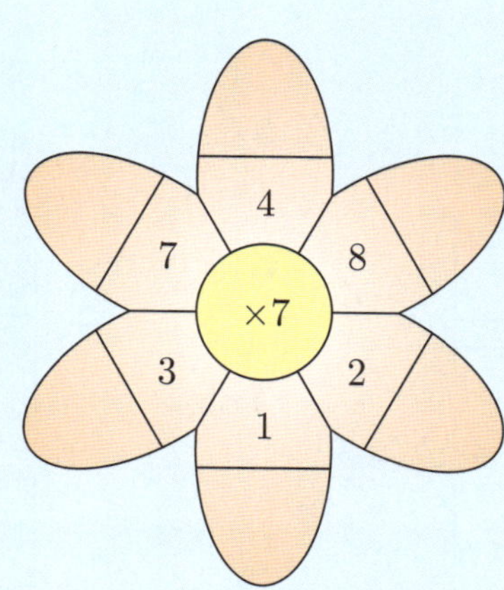

Exercise 7

Find the missing numbers.

a $3 \times \square = 12$

b $4 \times \square = 20$

c $\square \times 6 = 30$

d The product of 5 and $\square$ is 15.

e $\square$ multiplied by 4 is 28.

f Seven times $\square$ equals thirty five.

g 6 lots of $\square$ is 42.

h 8 multiplied by $\square$ is 72.

i $4 \times \square = 36$

j The product of $\square$ and 8 is 32.

Exercise 8

Fill in the multiplication wheels.

a

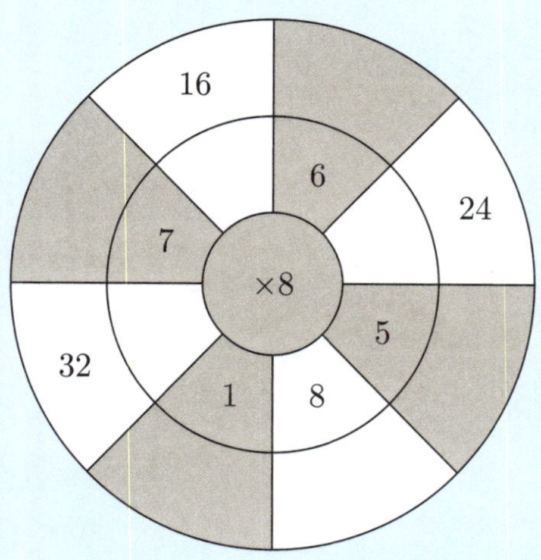

b

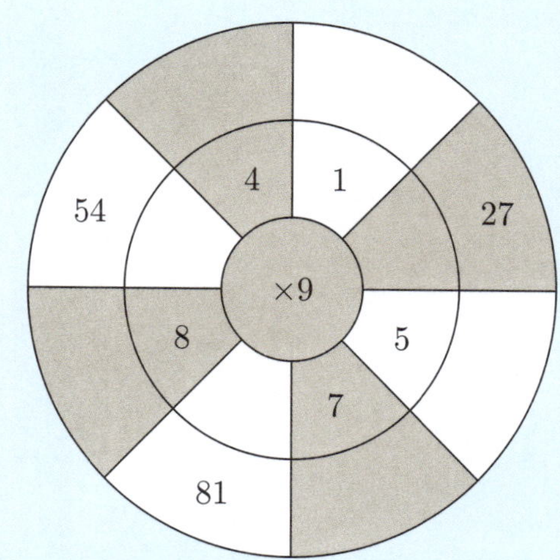

Puzzle

Complete the multiplications. Match your answers to the letters in the table to find where Alina lives.

a $4 \times 9 =$ ☐ **b** $4 \times 10 =$ ☐

c $9 \times 2 =$ ☐ **d** $7 \times 5 =$ ☐

e $8 \times 4 =$ ☐ **f** $8 \times 5 =$ ☐

g $7 \times 9 =$ ☐ **h** $6 \times 8 =$ ☐

i $7 \times 7 =$ ☐

E	O	K	F	U	M	C	B	T	N	S	A	D	H
54	20	36	15	49	32	21	81	18	63	45	40	48	35

___ ___ ___ ___ ___ ___ ___ ___ ___
a **b** **c** **d** **e** **f** **g** **h** **i**

Can you find where Alina lives on a map?

What country does she live in? __________

Exercise 9

Six multiplications below are incorrect.

Find and correct the mistakes.

Write a tick ✓ if the multiplication is correct.

a $5 \times 8 = 40$ __________

b $6 \times 7 = 43$ __________

c $3 \times 9 = 27$ __________

d $4 \times 6 = 24$ __________

e $8 \times 4 = 34$ __________

f $6 \times 5 = 30$ __________

g $7 \times 7 = 48$ __________

h $6 \times 9 = 56$ __________

i $7 \times 4 = 34$ __________

j $8 \times 8 = 64$ __________

k $7 \times 5 = 30$ __________

l $3 \times 7 = 21$ __________

When multiplying, changing the order will not change the answer.

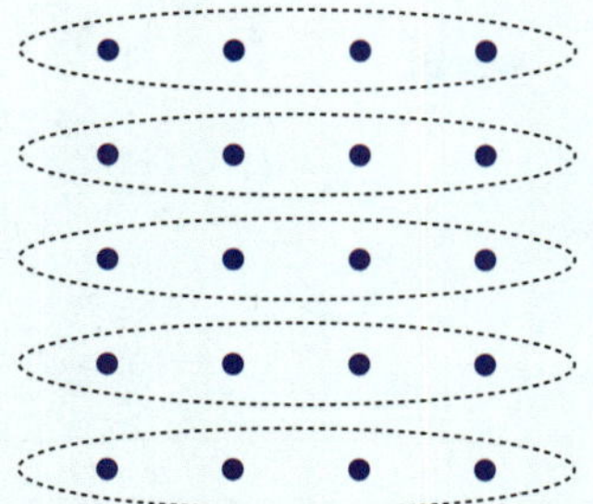

5 lots of $4 = 20$

$5 \times 4 = 20$

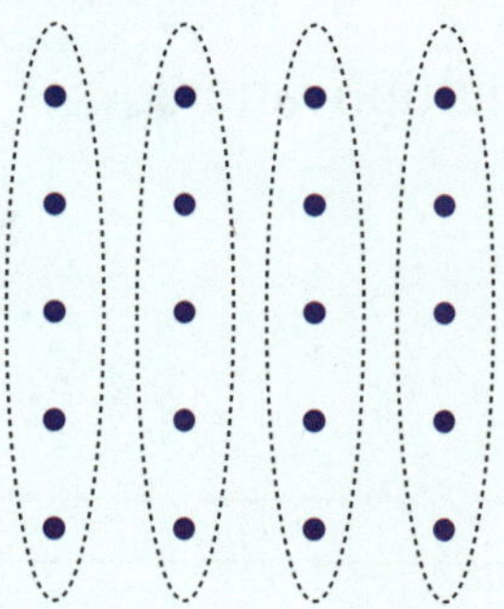

4 lots of $5 = 20$

$4 \times 5 = 20$

Exercise 10

Complete the multiplications.

a $3 \times 5 = 15$

so $5 \times 3 = \square$

b $6 \times 3 = \square$

so $3 \times 6 = \square$

c $6 \times 4 = \square$

so $4 \times \square = \square$

d $5 \times 2 = \square$

so $\square \times \square = \square$

e $7 \times 3 = \square$

so $\square \times \square = \square$

f $9 \times 4 = \square$

so $\square \times \square = \square$

g $5 \times 7 = \square$

so $\square \times \square = \square$

h $6 \times 8 = \square$

so $\square \times \square = \square$

Discussion

If I multiply any number by 1, the result is ....

If I multiply any number by 0, the result is ....

Exercise 11

Complete each multiplication:

a $10 \times 0 =$ ______ **b** $18 \times 1 =$ ______ **c** $0 \times 12 =$ ______

d $1 \times 20 =$ ______ **e** ______ $\times 1 = 11$ **f** $6 \times$ ______ $= 0$

Exercise 12

Write down multiplications which have the answer:

a 20 ______ ______ ______ ______

b 16 ______ ______ ______

c 36 ______ ______ ______

d 24 ______ ______ ______ ______

e 18 ______ ______ ______ ______

f 30 ______ ______ ______ ______

$$3 = 1 + 1 + 1 = 3 \times 1$$
$$30 = 10 + 10 + 10 = 3 \times 10$$
$$300 = 100 + 100 + 100 = 3 \times 100$$
$$3000 = 1000 + 1000 + 1000 = 3 \times 1000$$

Exercise 13

Complete:

a $70 = 7 \times$ ______

$700 = 7 \times$ ______

$7000 = 7 \times$ ______

b $90 = 9 \times$ ______

$900 = 9 \times$ ______

$9000 = 9 \times$ ______

c $4 \times 10 =$ ______

$4 \times 100 =$ ______

$4 \times 1000 =$ ______

d $6 \times 10 =$ ______

$6 \times 100 =$ ______

$6 \times 1000 =$ ______

Exercise 14

Add 11 each time:

11 → ☐ → ☐ → ☐ → ☐ ↓ ☐ ← ☐ ← ☐ ← ☐ ← ☐

Complete the 11 times table.

$1 \times 11 =$ ______

$2 \times 11 =$ ______

$3 \times 11 =$ ______

$4 \times 11 =$ ______

$5 \times 11 =$ ______

$6 \times 11 =$ ______

$7 \times 11 =$ ______

$8 \times 11 =$ ______

$9 \times 11 =$ ______

$10 \times 11 =$ ______

Exercise 15

Add 12 each time:

12 → ☐ → ☐ → ☐ → ☐ ↓ ☐ ← ☐ ← ☐ ← ☐ ← ☐

Complete the 12 times table.

$1 \times 12 =$ ______

$2 \times 12 =$ ______

$3 \times 12 =$ ______

$4 \times 12 =$ ______

$5 \times 12 =$ ______

$6 \times 12 =$ ______

$7 \times 12 =$ ______

$8 \times 12 =$ ______

$9 \times 12 =$ ______

$10 \times 12 =$ ______

Discussion

What patterns can you see in the 11 and 12 times tables?

Lucy has 4 baskets and places 8 eggs in each.

How many eggs does Lucy have altogether?

Lucy has 4 lots of 8 eggs.

$4 \times 8 = 32$

So, Lucy has 32 eggs altogether.

Exercise 16

Answer each multiplication question in a sentence:

a Nicholas practises guitar for 4 hours each week.
After 5 weeks, how many hours has Nicholas practised?

☐ × ☐ = ☐

....

b Tom has 3 containers and keeps 7 toy cars in each.
How many toy cars does Tom have altogether?

☐ × ☐ = ☐

....

c Jasmin's photograph book has 9 pages.
If Jasmin places 6 photographs on each page, how many photographs will fit into the book?

☐ × ☐ = ☐

....

d Dana wrote 8 short stories.
Each story was 7 pages long.
How many pages did Dana write?

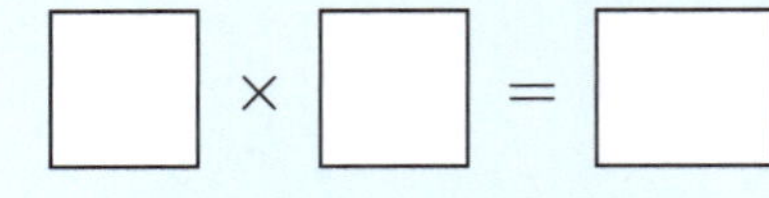

....

e Leah has 5 punnets of sweet corn seedlings. Each punnet holds 12 seedlings.
How many sweet corn seedlings does Leah have altogether?

....

The results of the times tables gives us **multiples** of a number.

The multiples of 4 are 4, 8, 12, 16, 20, ...

$1 \times 4 = 4$
$2 \times 4 = 8$
$3 \times 4 = 12$
$4 \times 4 = 16$
$5 \times 4 = 20$
$\vdots$

Exercise 17

The multiples of 3 are:

3, 6, ____, ____, ____, ____, ____, ____, ____, ____, ...

The multiples of 7 are:

7, ____, ____, ____, ____, ____, ____, ____, ____, ____, ...

The multiples of 8 are:

____, ____, ____, ____, ____, ____, ____, ____, ____, ____, ...

The multiples of 11 are:

____, ____, ____, ____, ____, ____, ____, ____, ____,

____, ...

The multiples of 12 are:

____, ____, ____, ____, ____, ____, ____, ____, ____,

____, ...

Exercise 18

Draw a circle ◯ around the multiples of 6.

Draw a square □ around the multiples of 5.

54 16 48 45
24 30 26 42
51 27 15 39
35 18

The number that is a multiple of both 5 *and* 6 is ____.

Puzzle

1 I am a multiple of 3.
I am also a multiple of 7.
I am less than 30.
What number am I?

____

2 I am a multiple of 4.
I am also a multiple of 9.
I am greater than 20, but less than 40.
What number am I?

____

3 I am a multiple of 11.
I am more than 40, but less than 60.
I am *not* a multiple of 5.
What number am I?

____

4 I am a multiple of 6.
I am also a multiple of 8.
I am *less* than 48.
What number am I?

____

When we **double** a number, we **multiply it by 2**.

The new number is **twice** the old number.

$$\text{Double } 10 = 2 \times 10$$
$$= 20$$

Exercise 19

Double each number:

a double 2 = ________ **b** twice 6 = ________ **c** double 3 = ________

d twice 7 = ________ **e** double 10 = ________ **f** twice 8 = ________

g double 9 = ________ **h** twice 4 = ________ **i** double 5 = ________

When we **treble** or **triple** a number, we **multiply it by 3**.

The new number is **triple** the old number.

Exercise 20

Treble each number:

a triple 2 = ________ **b** triple 5 = ________ **c** triple 3 = ________

d triple 6 = ________ **e** triple 4 = ________ **f** triple 7 = ________

g triple 9 = ________ **h** triple 8 = ________ **i** triple 10 = ________

Exercise 21

Complete:

a
double 2 = 4
double 20 = 40
double 200 = ...
double 2000 = ...

b
double 1 = ...
double 10 = ...
double 100 = ...
double 1000 = ...

c
double 3 = ...
double 30 = ...
double 300 = ...
double 3000 = ...

d
double 8 = 16
double 80 = 160
double 800 = ...

e
double 5 = ...
double 50 = ...
double 500 = ...

f
double 9 = ...
double 90 = ...
double 900 = ...

Activity

Complete each chain:

a 2 $\xrightarrow{+3}$ 5 $\xrightarrow{\times 3}$ ☐ $\xrightarrow{-8}$ ☐ $\xrightarrow{\text{double}}$ ☐ $\xrightarrow{+9}$ ☐

b 5 $\xrightarrow{-2}$ ☐ $\xrightarrow{\text{triple}}$ ☐ $\xrightarrow{+2}$ ☐ $\xrightarrow{\times 7}$ ☐

c 6 $\xrightarrow{-3}$ ☐ $\xrightarrow{\text{double}}$ ☐ $\xrightarrow{+4}$ ☐ $\xrightarrow{\times 6}$ ☐ $\xrightarrow{\text{double}}$ ☐

d 1 $\xrightarrow{+3}$ ☐ $\xrightarrow{\times 2}$ ☐ $\xrightarrow{\text{double}}$ ☐ $\xrightarrow{+4}$ ☐ $\xrightarrow{\text{double}}$ ☐

Listening Activity

Listen to the questions, and write your answers in the spaces provided.

1 .... **2** .... **3** ....

4 .... **5** ....

6 .... **7** ....

8 ....

The **factors** of a number can be multiplied together in pairs to give the number.

We know that $1 \times 6 = 6$ and $2 \times 3 = 6$.

There are no other factor pairs of 6.

So, the factors of 6 are 1, 2, 3, and 6.

Exercise 22

The factor pairs of 12 are:

________, ________, and ________.

The factors of 12 are:

____, ____, ____, ____, ____, and ____.

Every whole number has the factor pair $1 \times$ the number.

Exercise 23

The factor pairs of 16 are:

________, ________, and ________.

The factors of 16 are:

____, ____, ____, ____, and ____.

Exercise 24

By writing down factor pairs, list the factors of:

a 4

The factors of 4 are:

____, ____, and ____.

b 18

The factors of 18 are:

____, ____, ____, ____, ____, and ____.

We know that 10 units make 1 ten,
10 tens make 1 hundred,
and 10 hundreds make 1 thousand.

When we multiply by 10, each digit moves one place to the *left*.

We write a zero to fill in the empty units place.

Th	H	T	U
	3	2	9
3	2	9	0

$329 \times 10 = 3290$

Exercise 25

Complete each multiplication:

a $6 \times 10 =$ ______ **b** $10 \times 10 =$ ______ **c** $16 \times 10 =$ ______

d $50 \times 10 =$ ______ **e** $49 \times 10 =$ ______ **f** $25 \times 10 =$ ______

g $79 \times 10 =$ ______ **h** $318 \times 10 =$ ______ **i** $507 \times 10 =$ ______

j $640 \times 10 =$ ______ **k** $826 \times 10 =$ ______ **l** $4712 \times 10 =$ ______

Exercise 26

Find the missing number.

a $\square \times 10 = 120$ **b** $\square \times 10 = 170$ **c** $\square \times 10 = 30$

d $\square \times 10 = 260$ **e** $\square \times 10 = 430$ **f** $10 \times \square = 200$

g $10 \times \square = 1400$ **h** $10 \times \square = 6010$ **i** $10 \times \square = 5900$

Exercise 27

Complete these multiplications.

a $27 \times 10 =$ ______
$270 \times 10 =$ ______
$2700 \times 10 =$ ______

b ______ $\times 10 = 420$
______ $\times 10 = 4200$
______ $\times 10 = 42\,000$

Exercise 28

Complete:

a $10 \xrightarrow{+1} \square \xrightarrow{+1} \square$

$11 \times 1 =$ ......

$12 \times 1 =$ ......

b $20 \xrightarrow{+2} \square \xrightarrow{+2} \square$

$11 \times 2 =$ ......

$12 \times 2 =$ ......

c $30 \xrightarrow{+3} \square \xrightarrow{+3} \square$

$11 \times 3 =$ ......

$12 \times 3 =$ ......

d $40 \xrightarrow{+4} \square \xrightarrow{+4} \square$

$11 \times 4 =$ ......

$12 \times 4 =$ ......

e $50 \xrightarrow{+5} \square \xrightarrow{+5} \square$

$11 \times 5 =$ ......

$12 \times 5 =$ ......

f $60 \xrightarrow{+6} \square \xrightarrow{+6} \square$

$11 \times 6 =$ ......

$12 \times 6 =$ ......

g $70 \xrightarrow{+7} \square \xrightarrow{+7} \square$

$11 \times 7 =$ ......

$12 \times 7 =$ ......

h $80 \xrightarrow{+8} \square \xrightarrow{+8} \square$

$11 \times 8 =$ ......

$12 \times 8 =$ ......

i $90 \xrightarrow{+9} \square \xrightarrow{+9} \square$

$11 \times 9 =$ ......

$12 \times 9 =$ ......

j $100 \xrightarrow{+10} \square \xrightarrow{+10} \square$

$11 \times 10 =$ ......

$12 \times 10 =$ ......

k $110 \xrightarrow{+11} \square \xrightarrow{+11} \square$

$11 \times 11 =$ ......

$12 \times 11 =$ ......

l $120 \xrightarrow{+12} \square \xrightarrow{+12} \square$

$11 \times 12 =$ ......

$12 \times 12 =$ ......

Exercise 29

Complete each factor pair:

a $3 \times \square = 24$ **b** $\square \times 4 = 32$ **c** $7 \times \square = 28$

d $\square \times 7 = 63$ **e** $4 \times \square = 40$ **f** $\square \times 9 = 81$

g $5 \times \square = 30$ **h** $\square \times 8 = 72$ **i** $12 \times \square = 60$

j $\square \times 1 = 21$ **k** $11 \times \square = 99$ **l** $\square \times 19 = 19$

Exercise 30

Complete the multiplication table.

×	1	2	3	4	5	6	7	8	9	10	11	12
1												
2												
3												
4												
5												
6												
7												
8												
9												
10												
11												
12												

To multiply numbers greater than those in our times tables, we use **column multiplication**.

16×5

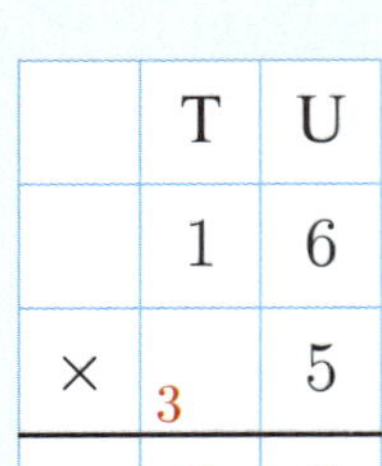

	T	U
	1	6
$\times$	3	5
	8	0

1 Set out the multiplication.
Write the numbers in columns so the place values line up.

2 $6 \times 5 = 30$
We "carry" the 3 tens into the tens column.
There are 0 units remaining, so we write 0 below the line.

3 $10 \times 5 = 50$, plus the 3 tens or 30 carried over is 80.
We write 8 tens below the line.

So, $16 \times 5 = 80$.

Exercise 31

Complete each multiplication:

a 14×2

	T	U
	1	4
$\times$		2

b 23×3

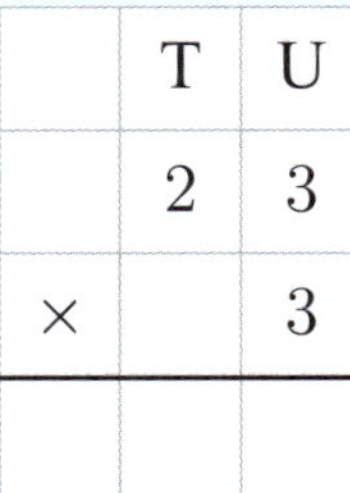

	T	U
	2	3
$\times$		3

c 12×4

	T	U
	1	2
$\times$		4

d 17×3

	T	U
	1	7
$\times$		3

e 19×5

	T	U
	1	9
$\times$		5

f 17×5

	T	U
	1	7
$\times$		5

g 18×4

	T	U
	1	8
$\times$		4

h 16×6

	T	U
	1	6
$\times$		6

i 29×3

	T	U
	2	9
$\times$		3

38×4

	H	T	U
		3	8
$\times$		3	4
	1	5	2

1 Set out the multiplication.

2 $8 \times 4 = 32$

We "carry" the 3 tens into the tens column.

We write the remaining 2 units below the line.

3 $30 \times 4 = 120$, plus the 3 tens or 30 carried over is 150.

We write the 5 tens below the line, and "carry" the 1 hundred.

So, $38 \times 4 = 152$.

Exercise 32

Complete each multiplication:

a 42×3

	H	T	U
		4	2
$\times$			3

b 51×4

	H	T	U
		5	1
$\times$			4

c 65×3

	H	T	U
		6	5
$\times$			3

d 68×4

	H	T	U
		6	8
$\times$			4

e 74×3

	H	T	U
		7	4
$\times$			3

f 24×6

	H	T	U
		2	4
$\times$			6

g 36×7

	H	T	U
		3	6
$\times$			7

h 47×8

	H	T	U
		4	7
$\times$			8

i 23×9

	H	T	U
		2	3
$\times$			9

Exercise 33

Set out each multiplication, and find the answer:

a 37×6

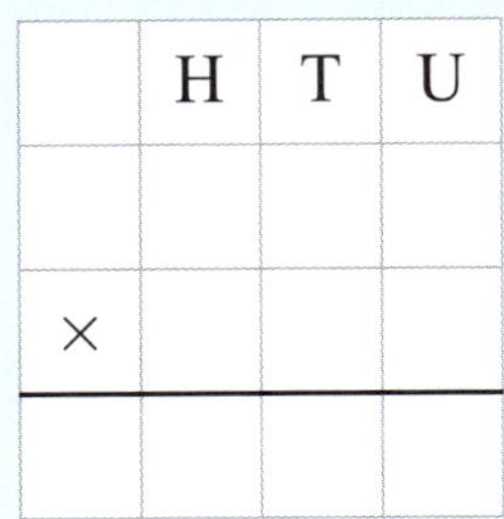

b 28×7

c 45×5

d 42×9

e 54×8

f 63×7

Exercise 34

Write a multiplication to solve each problem:

a Terry buys 5 packets which each contain 24 jelly beans.

How many jelly beans does Terry have?

Terry has ☐ jelly beans.

b Molly's dog eats 25 dog treats every day for 6 days.

How many treats does Molly's dog eat?

Molly's dog eats ☐ dog treats.

c Tirzah is making cupcakes for a birthday party.
Her baking tray can hold 16 cupcakes.
She bakes 9 trays of cupcakes.

How many cupcakes has Tirzah made?

Tirzah has made ☐ cupcakes.

Exercise 35

Solve each problem. You will need addition as well as multiplication.

a Maureen drinks 8 cups of tea each day.

Caroline drinks 5 cups of tea each day.

In total, the ladies drink ☐ cups of tea each day.

In one week, the ladies will drink ☐ cups of tea.

b Dmitri plants tulip bulbs in his garden each year.

If Dmitri digs 6 rows and plants 47 bulbs in each row, he will plant ☐ tulip bulbs.

If Dmitri then plants 1 *more* row, he will have planted ☐ tulip bulbs in total.

c Tash has cartons of eggs for sale.

She has 7 cartons with 12 eggs in each, and 15 cartons with 6 eggs in each.

How many eggs does Tash have for sale?

Tash has ☐ eggs for sale.

Challenge — 400 dumplings

In this Challenge, you may need to use addition, subtraction, and multiplication.

Yiu Min and Claudia need to make 400 dumplings for the Chinese New Year.

Yiu Min makes 38 dumplings in 1 hour, while Claudia makes 47 dumplings.

Yiu Min works for 4 hours. Claudia works for 5 hours.

Write a sentence to answer each question:

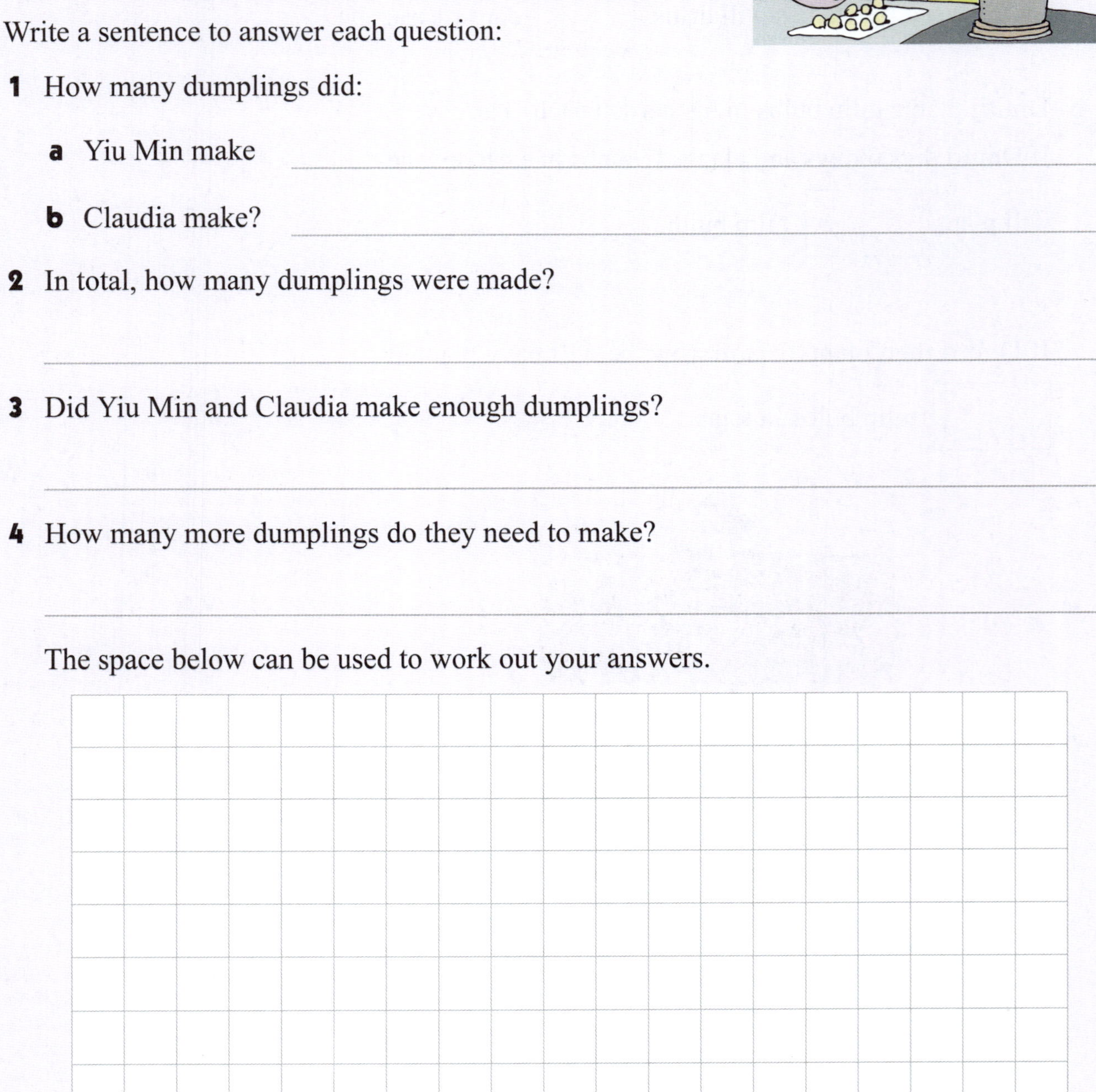

1 How many dumplings did:

- **a** Yiu Min make ____________________
- **b** Claudia make? ____________________

2 In total, how many dumplings were made?

____

3 Did Yiu Min and Claudia make enough dumplings?

____

4 How many more dumplings do they need to make?

____

The space below can be used to work out your answers.

Revision

1 Write as a multiplication and find the answer:

a $6+6+6+6+6=$ ______ $\times$ ______ $=$ ______

b the product of eight and five $=$ ______ $\times$ ______ $=$ ______

2 Complete:

a 3 lots of 11 is ______

b the product of 9 and 7 is ______

c 6 multiplied by 12 is ______

d 7 times 8 equals ______

3 Find each product:

a $5\times 2=$ ______

b $3\times 6=$ ______

c $7\times 11=$ ______

d $9\times 3=$ ______

e $4\times 7=$ ______

f $12\times 8=$ ______

4 Complete:

a $6\times$ ______ $=36$

b ______ $\times 9=45$

c $10\times$ ______ $=80$

5 Complete:

a

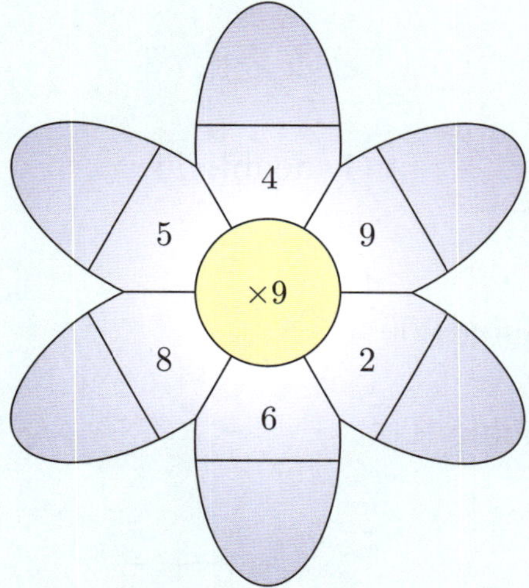

b

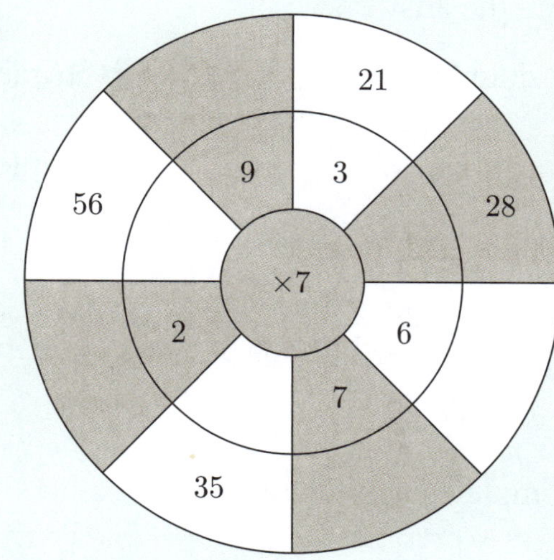

6 Write down multiplications which have the answer 12.

______ ______ ______

______ ______ ______

7 Jerome has 4 shelves. Each shelf holds 5 books.
How many books are on the shelves?

☐ $\times$ ☐ $=$ ☐

______

8 Complete:

a $16 \times 1 =$ ______ **b** $0 \times 25 =$ ______ **c** ______ $\times 1 = 21$

9 Complete:

a $80 = 8 \times$ ______

$800 = 8 \times$ ______

$8000 = 8 \times$ ______

b $2 \times 10 =$ ______

$2 \times 100 =$ ______

$2 \times 1000 =$ ______

10 The multiples of 6 are:

6, ____, ____, ____, ____, ____, ____, ____, ____,

____, ____, ____, ...

11 **a** I am a multiple of 7.
I am also a multiple of 9.
I am less than 80.
What number am I?

______

b I am a multiple of 4,
but *not* a multiple of 12.
I am between 30 and 40.
What number am I?

______

12 Find the answers:

a double $5 =$ ______ **b** triple $3 =$ ______ **c** double $7 =$ ______

d triple $6 =$ ______ **e** triple $8 =$ ______ **f** double $11 =$ ______

13 Double each number:

a double $30 =$ ______ **b** double $900 =$ ______

c double $20 =$ ______ **d** double $1000 =$ ______

14 Complete each factor pair:

a $6 \times \square = 48$ **b** $\square \times 7 = 49$ **c** $1 \times \square = 28$

d $\square \times 11 = 132$ **e** $33 \times \square = 33$ **f** $\square \times 8 = 64$

15 The factor pairs of 20 are:

______, ______, and ______.

The factors of 20 are:

____, ____, ____, ____, ____, and ____.

16 Complete each multiplication:

a $15 \times 10 =$ ______ **b** $82 \times 10 =$ ______ **c** $145 \times 10 =$ ______

d ______ $\times 10 = 340$ **e** ______ $\times 10 = 4650$ **f** $10 \times$ ______ $= 870$

17 Complete:

a $16 \times 10 =$ ______ **b** $160 \times 10 =$ ______ **c** $1600 \times 10 =$ ______

18 Complete each multiplication:

a 14×7

	T	U
	1	4
×		7

b 26×3

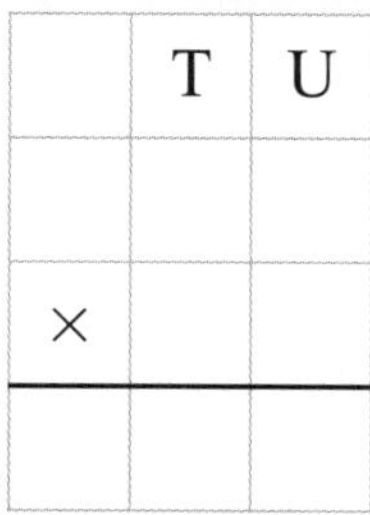

c 37×2

19 Peggy eats 3 meals each day.

How many meals will she eat on a 21 day holiday?

Peggy will eat ☐ meals on the holiday.

20 Complete each multiplication:

a 38×5

	H	T	U
		3	8
×			5

b 64×7

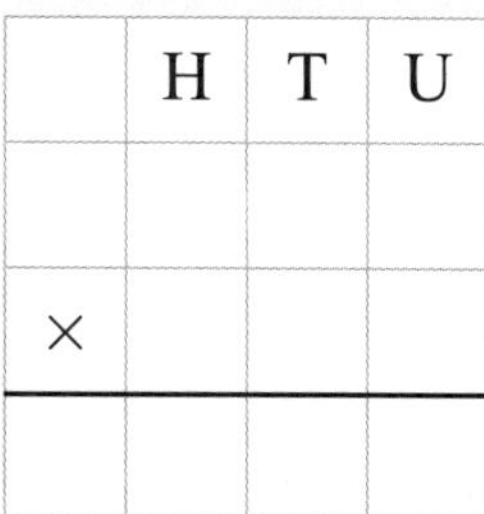

c 83×6

d 97×4

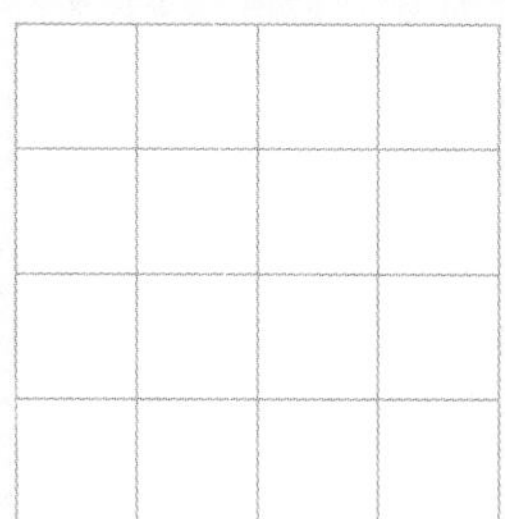

e 79×8

f 56×9

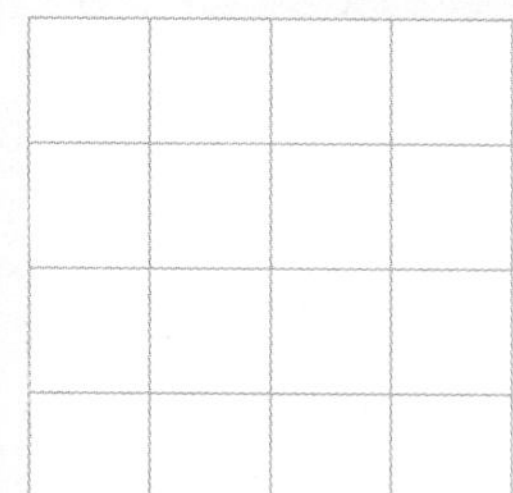

21 At a wedding reception there are 16 tables.

Each table has 8 guests.

How many guests are at the reception?

There are ☐ guests at the reception.

22 Leon plants 38 lines of strawberries with 10 plants in each line.

☐ $\times$ ☐ $=$ ☐

In total, Leon planted ☐ strawberry plants.

23 Megan has 4 lemon trees. She picks 67 lemons from each tree.

Megan picked ☐ lemons in total.

24 At his restaurant, Ming has 7 stacks with 26 plates in each. There are 48 plates already on the tables.

There are ☐ plates in the stacks.

There are ☐ plates in total.

CHAPTER 5: DATA HANDLING

We often collect information about the world around us.

The information we collect is called **data**.

We can collect data:

- by *watching* what is happening
- by *talking* to people and asking questions
- by *measuring* quantities such as length and time
- using *questionnaires* or *surveys*.

The study of data is called **statistics**.

Discussion

What data does your teacher collect about you?

Discuss *why* your teacher collects this information.

Mrs Selby asked her students what their favourite indoor sport is.

She recorded the data using a **tally**.

The **frequency** of each sport is the number of students who chose it.

Sport	*Tally*	*Frequency*
Badminton	卌 |	6
Basketball	卌 卌	10
Netball	卌 |||	8
Squash	|	1
Table tennis	||||	4
Volleyball	||	2
	Total	31

The **mode** is the most common response.

The mode of this data is "basketball".

Exercise 1

This table records how a group of students travel to school.

	Tally	*Frequency*
Car	卌 卌 |	
Bus	||||	
Walk	|||	
Train	|	
Cycle	卌 |||	
	Total	

a Complete the frequency column.

b Suggest a title for the first column.

c What is the mode of the data?

d How many students take the bus to school?

e How many students walk or cycle to school?

Exercise 2

Caroline asked her kindergarten class to tell her their favourite fruit.

a Complete this table showing their answers.

Fruit	*Tally*	....
Apple	|||	3
Banana	卌 |	
Peach	||||	
Orange		7
Strawberry	卌 卌	
	Total	

b What is the mode of the data?

c How many students chose banana?

d How many *more* students chose orange than chose peach?

Discussion

When you use a tally to count objects, how can you make sure you do not count the same object twice?

Exercise 3

Stuart asked 30 people to name an ocean.

He recorded these responses:

At	I	P	P	Ar	At	S	I
P	At	I	Ar	P	Ar	I	P
At	Ar	P	P	At	I	P	S
Ar	P	At	I	S	At		

At = Atlantic
I = Indian
P = Pacific
Ar = Arctic
S = Southern

a Complete the table.

Ocean	*Tally*	*Frequency*
Atlantic		
	Total	

b What is the mode of the data?

c What was the least common response?

d How many people named an ocean beginning with "A"?

Group Activity — Class Party

What to do:

1. To plan a class party, your teacher needs information about:
   - food
   - drinks
   - entertainment.

   As a class, decide on appropriate questions for your teacher to ask you. For each question, write a list of choices for your response.
2. Your teacher will prepare tables so that every student in the class can record their answers. Each student should draw *one* tally mark for each question.
3. Record the questions asked and the tallies of results below.

Food question: ....

....

Food	*Tally*	*Frequency*
	Total	

Drinks question: ....

....

Drink	*Tally*	*Frequency*
	Total	

Entertainment question: ....

....

Entertainment	*Tally*	*Frequency*
	Total	

4 The most common answers were:

food .... drink ....

entertainment ....

A **pictograph** uses pictures to show data.
The **key** tells us what each picture represents.

Exercise 4

Linden drew this pictograph to show the number of butterflies he has counted in the school garden.

Butterflies in the school garden at lunch time

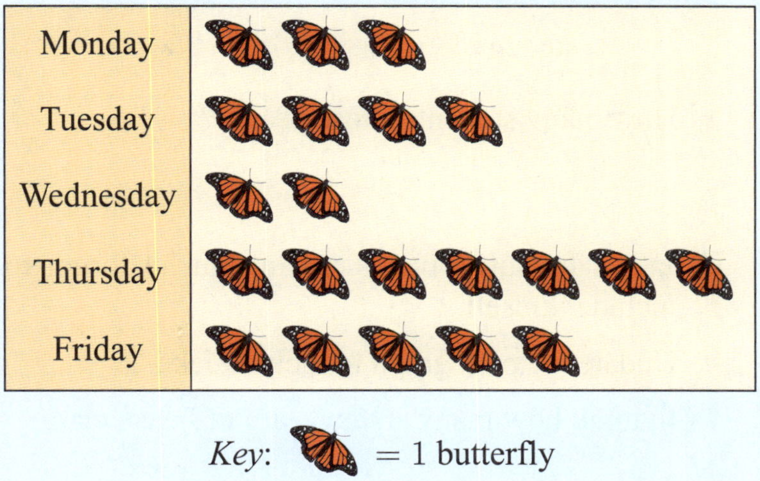

Monday

Tuesday

Wednesday

Thursday

Friday

Key: = 1 butterfly

a How many butterflies did Linden see on Tuesday?

....

b On which day did Linden see the least number of butterflies?

....

c How many butterflies did Linden see altogether?

....

Exercise 5

Chloe recorded the type of ski trail used by skiers coming down a slope.

Draw a pictograph to display the results.

Trail type	*Tally*
Beginner	||||
Intermediate	~~||||~~ |
Advanced	~~||||~~
Expert	||

Ski trails used

Beginner	
Intermediate	
Advanced	
Expert	

Key: [skier symbol] = 1 skier

Exercise 6

Jake asked the students in his class to name their favourite nut.

He drew this pictograph to display the results.

Favourite nut

Cashew	🯅 🯅 🯅
Peanut	🯅 ½
Walnut	🯅 🯅 🯅 🯅
Almond	🯅 🯅 ½
Pecan	🯅

Key: 🯅 = 2 students

a How many students chose cashew?

$2 + 2 + 2 =$ ....

b What does [half figure] represent?

.... student

c How many students chose peanut?

....

d Jake's favourite nut is pecan, but he forgot to include himself.

Update the pictograph to include Jake.

e In total, how many students are in Jake's class?

....

Exercise 7

Rosie counted the pies sold at her bakery today.

Pie	*Number*
Apple	6
Blueberry	7
Cherry	2
Peach	5

a Draw a pictograph to display the results.

Pies sold

Key: (pie) = 2 pies

b What is the mode of the data? ....

Activity

What to do:

1. Choose *one* of the tables of results from the Class Party Activity on pages **88 - 89**.
2. Draw a pictograph to display the results.

....

Key: (stick figure) = 2 people

On page **86**, we saw how Caroline collected data from her kindergarten class about their favourite fruit.

Here is a **bar graph** of Caroline's data.

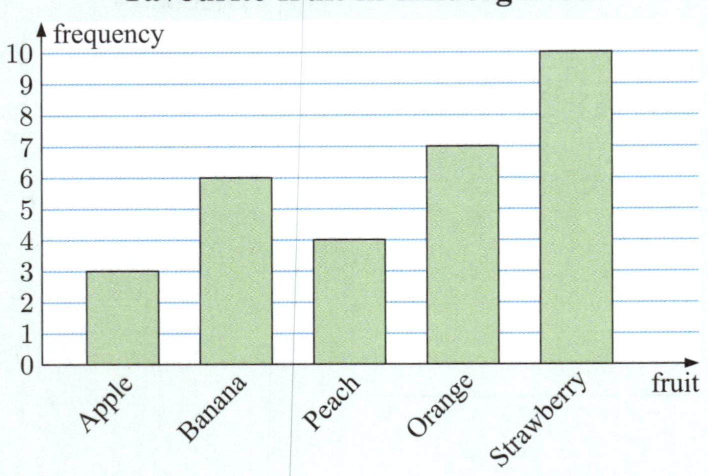

The *height* of each bar shows the *frequency* of each choice.

3 children chose "apple" as their favourite fruit, so the bar for apple is 3 units high.

The mode is "strawberry". It has the highest bar.

Exercise 8

This bar graph shows the colours of cars passing the school in a 10 minute period.

a Give the graph an appropriate title.

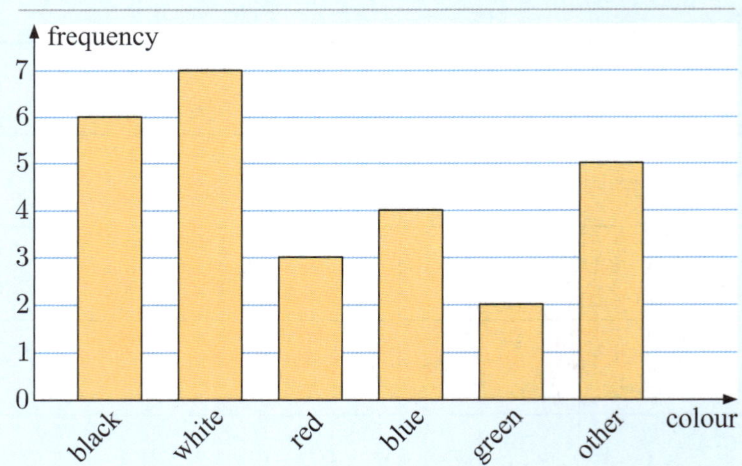

b What is the mode of the data?

c How many red cars passed the school?

d In total, how many cars passed the school in the 10 minutes?

Exercise 9

Lydia asked the students in her class how many pets they had.

She recorded these responses:

2	1	2	0	3	1	2	0	1
3	2	0	1	2	3	4	1	2
2	0	3	2	0	4	0	2	3

a Organise the data in a tally.

Number of pets	*Tally*	*Frequency*
0		
1		
2		
3		
4		

b Complete this bar graph to display the data.

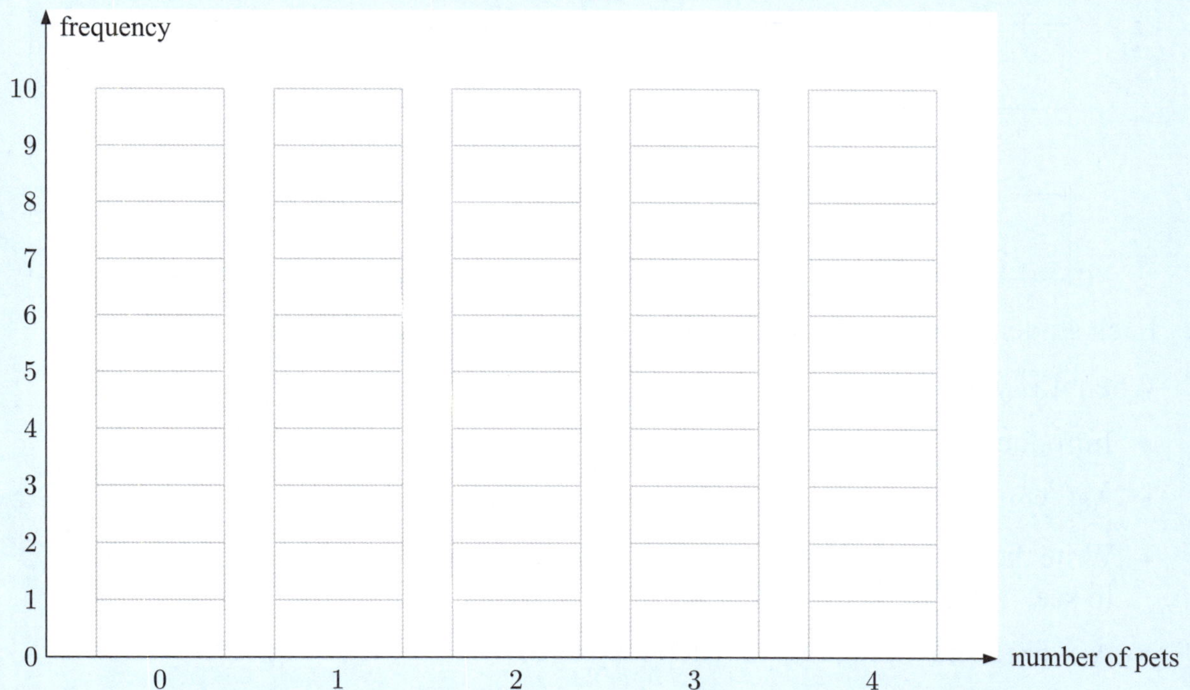

c What is the mode of the data? ________ pets

Activity — Collecting and displaying data

What to do:

1 Write a question you would like to ask the students in your class.

You might like to ask about:

- favourite pets
- favourite holiday activity
- how they travel to school.
- favourite subject
- least favourite food

The question I will ask is:

....

....

2 In the table below, write **6** choices you will give the students as possible answers to your question.

Answer	*Tally*	*Frequency*

3 Each student should ask their question to the class in turn.

When it is your turn:

- Introduce yourself.
- Ask your question.
- Write the choices of response on the whiteboard for everyone to see.
- Ask for a show of hands for each response.
- Record the hands using the tally.

4 Complete the frequency column of your table.

5 How many students answered your question? ....

6 Draw a bar graph to display your results.

frequency

15
14
13
12
11
10
9
8
7
6
5
4
3
2
1
0

7 Describe what you discovered from your question.

Activity

Data can also be displayed by entering information into a **spreadsheet**.

Click on the icons for a spreadsheet template and for a demonstration.

Use a spreadsheet to make a bar graph of your results in the previous Activity.

Print your bar graph for use in a class display.

Billy organised the toys on his shelf using a **Carroll diagram**:

	vehicle	not a vehicle
blue		
not blue		

Exercise 10

Anthea drew this Carroll diagram to display the weather in her city last week.

	windy	not windy
rainy	Monday Wednesday	Thursday
not rainy	Sunday Tuesday	Friday Saturday

a Was it rainy on Thursday? ....

b Was it windy on Friday? ....

c On how many days was it windy but not rainy? ....

d Describe the weather on Saturday. ....

e List the days on which it was rainy.

....

Exercise 11

Place the numbers 1 to 10 into this Carroll diagram.

	factor of 10	not a factor of 10
factor of 6		
not a factor of 6		

Exercise 12

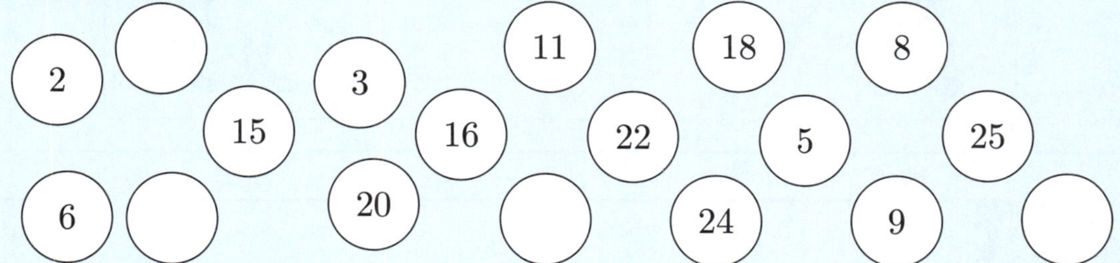

a Place the numbers in the circles into this Carroll diagram.

	multiple of 3	not a multiple of 3
less than 10		
10 or more		

b Write down one more number of your choice in each box of the Carroll diagram.

c Copy your chosen numbers into the empty circles.

Game

Can you move all the tortoises into the correct box of the Carroll diagram?

Click the icon to play the game.

Look again at the toys on Billy's shelf:

Billy can also organise the toys using a **Venn diagram**:

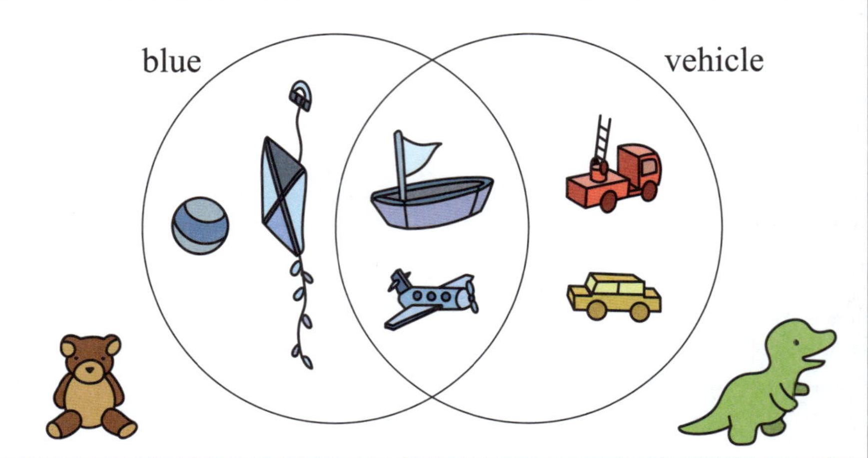

Exercise 13

Jane and Sara visited a wildlife park. They drew a Venn diagram of the animals they saw.

a Did Jane see the lion? ______

b Who saw the crocodile? ______

c List the animals seen by Jane but not Sara.

______

d How many animals were not seen by either child? ______

e Who saw more animals? ______

Exercise 14

Christopher and Elizabeth have placed the first five letters of the alphabet in this Venn diagram.

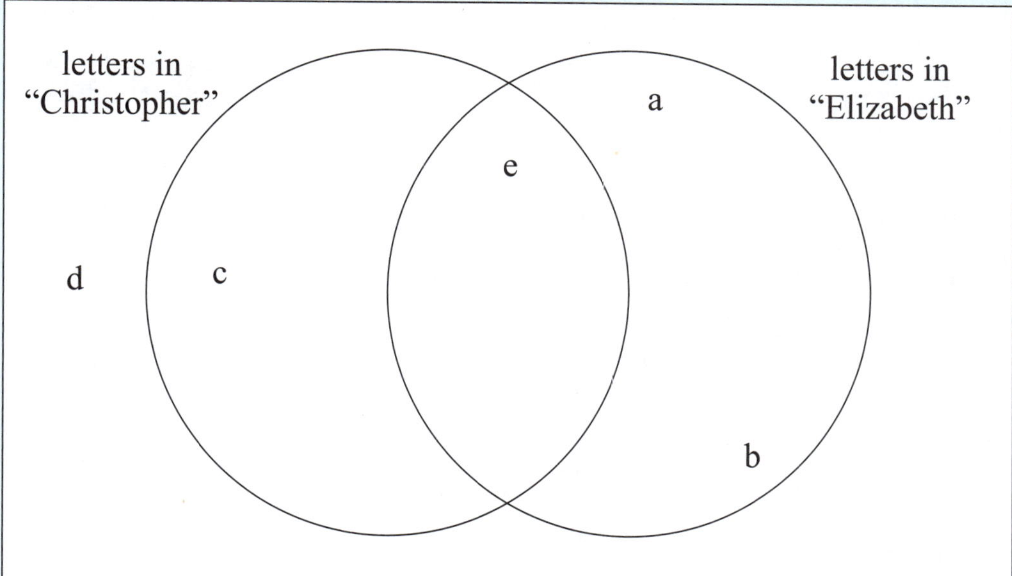

Place the remaining letters of the alphabet in the Venn diagram.

Puzzle

Mr Williams placed 12 objects in a Venn diagram, but he did not include the labels.

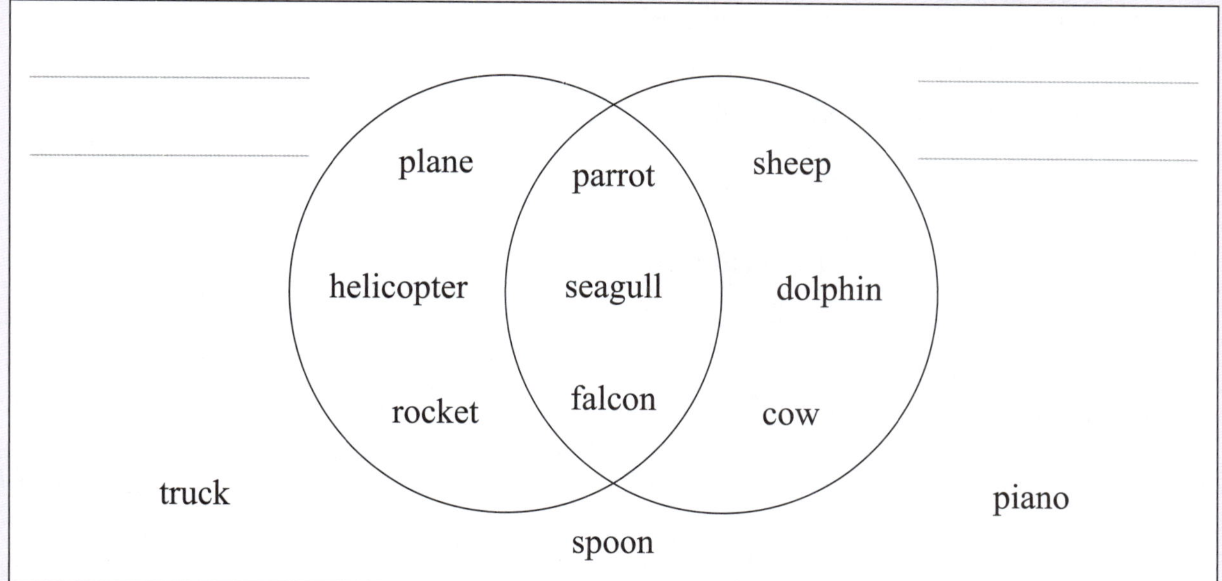

Can you work out what the labels should be?

Revision

1 A group of students was asked the following question:

"How many siblings do you have?"

Number of siblings	*Tally*	*Frequency*
0	||	2
1		9
2	~~||||~~ ~~||||~~	
3	||||	
4	||	
5 or more	|	
	Total	

a Complete the table.

b How many students have 3 siblings?

____

c What is the mode of the data?

__________ siblings

d In total, how many students answered this question?

____

2 Tom collected data about the types of vehicle that passed his school during recess. He displayed his data on a bar graph.

a How many trucks passed the school?

____

b How many buses passed the school?

____

c What is the mode of the data?

____

Vehicles passing the school

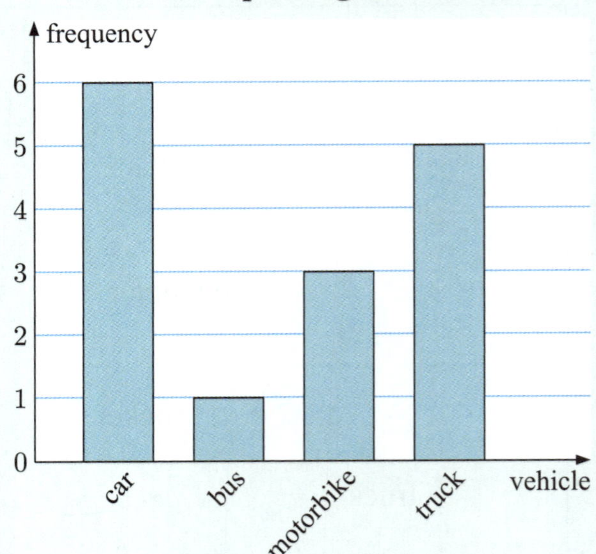

d How many *more* cars than motorbikes passed the school?

____

e How many vehicles passed the school altogether?

____

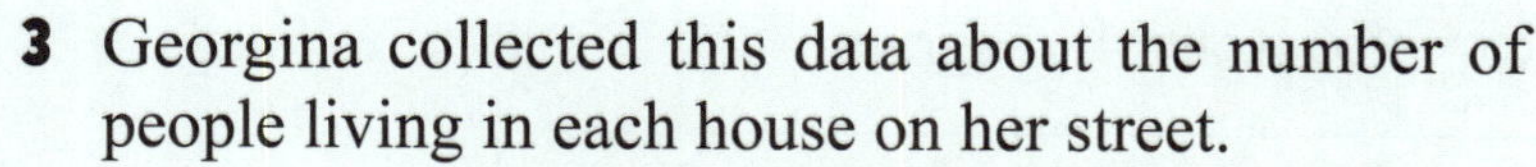

3 Georgina collected this data about the number of people living in each house on her street.

3	1	2	4	3	2	4	3
1	2	3	5	3	4	1	4
3	1	4					

a Organise the data in a tally.

Number of people	*Tally*	*Frequency*
1		
2		
3		
4		
5		

b How many houses have 1 person living in them? ....

c Complete this pictograph to display the data.

Number of people living in houses

1	
2	
3	
4	
5	

Key: = 2 houses

d What is the mode of the data?

.... people

4 Laura organised her school subjects into a Carroll diagram:

	is good at	is not good at
likes	Drama Sport Mathematics	English
does not like	Science Art	Music

a Is Laura good at Science? ....

b Does Laura like Music? ....

c Which subject does Laura like but is not good at?

....

d How many subjects is Laura good at? ....

5 
Ghana

Colombia

Italy

Cuba

Lebanon

Vietnam

Sweden

Pakistan

Place each country's name in this Venn diagram.

flag has a star

flag contains yellow

CHAPTER 6: DIVISION

Divya has 18 lollies.

She shares them **equally** with Yumiko and Cam.

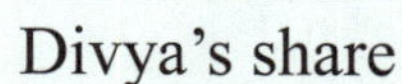

Divya's share

Yumiko's share

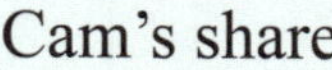

Cam's share

Each person now has 6 lollies.

Divya has **divided** the 18 lollies into 3 equal groups of 6.

$$18 \div 3 = 6$$

"eighteen divided by three equals six"

Exercise 1

What division is shown by the diagram?

a

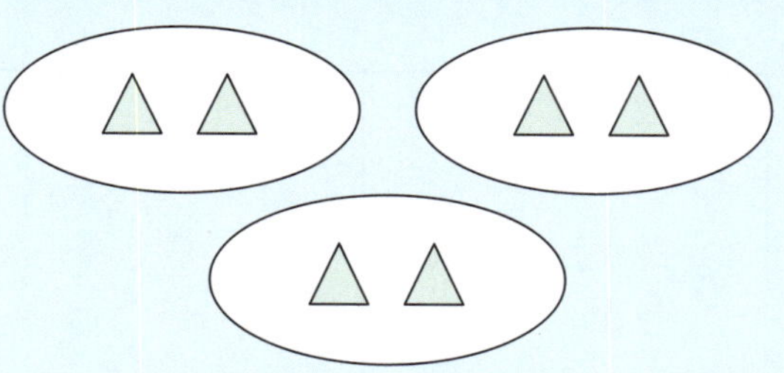

$\_\_\_\_\_ \div 3 = \_\_\_\_\_$

b

$\_\_\_\_\_ \div \_\_\_\_\_ = \_\_\_\_\_$

c

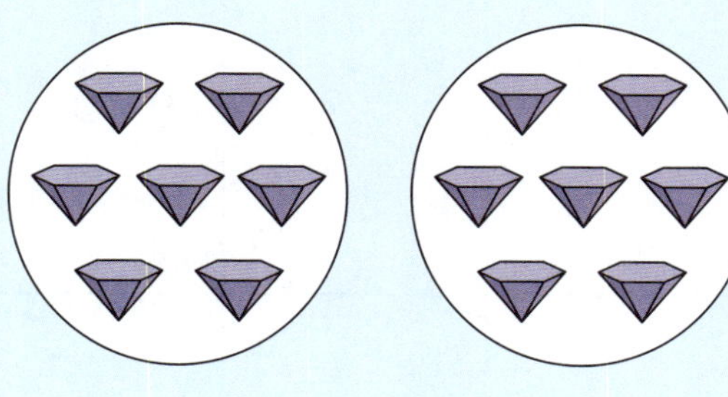

$\_\_\_\_\_ \div \_\_\_\_\_ = \_\_\_\_\_$

d

____

Exercise 2

Use a diagram to show the division and find the answer.

a $9 \div 3 =$ ______

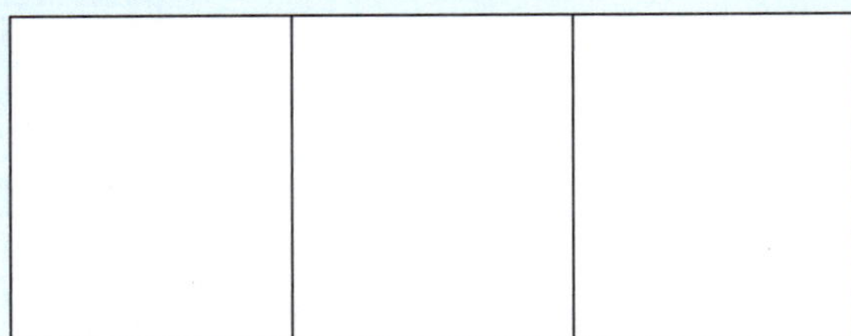

b $12 \div 2 =$ ______

c $12 \div 3 =$ ______

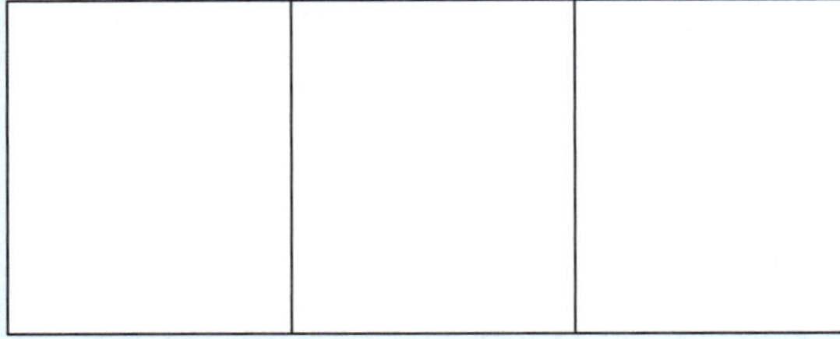

d $8 \div 4 =$ ______

e $15 \div 3 =$ ______

f $16 \div 2 =$ ______

g $16 \div 4 =$ ______

h $15 \div 5 =$ ______

Division is the **opposite** of multiplication.

$3 \times 6 = 18$
$18 \div 3 = 6$

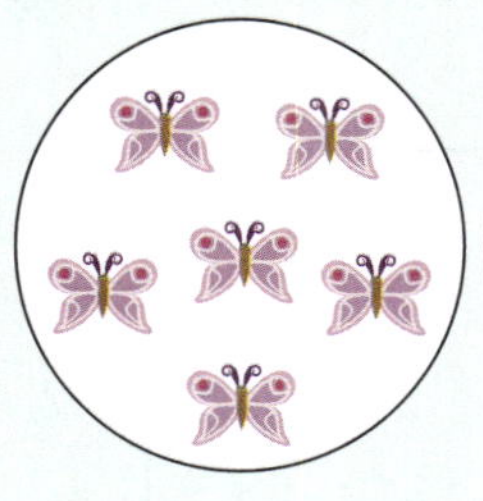
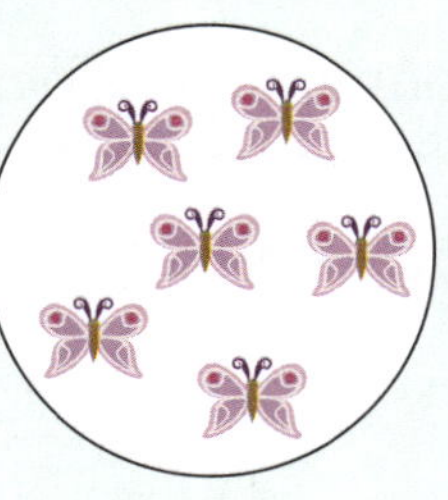
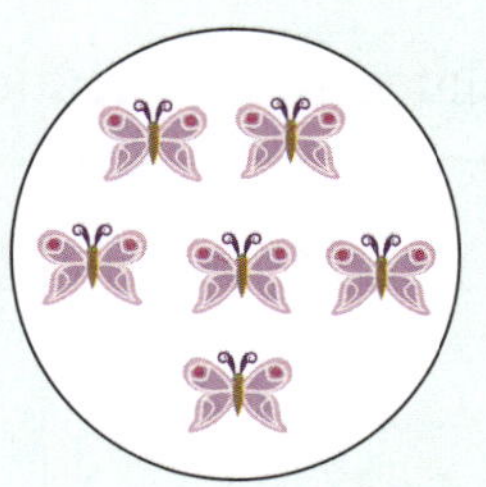

To find $18 \div 3$, look down the column for the 3 times table until you find 18.

$3 \times 6 = 18$
so $18 \div 3 = 6$

The more you practise your times tables, the easier division will be!

×	1	2	3	4	5	6	7	8	9	10
1	1	2	3	4	5	6	7	8	9	10
2	2	4	6	8	10	12	14	16	18	20
3	3	6	9	12	15	18	21	24	27	30
4	4	8	12	16	20	24	28	32	36	40
5	5	10	15	20	25	30	35	40	45	50
6	6	12	18	24	30	36	42	48	54	60
7	7	14	21	28	35	42	49	56	63	70
8	8	16	24	32	40	48	56	64	72	80
9	9	18	27	36	45	54	63	72	81	90
10	10	20	30	40	50	60	70	80	90	100

Exercise 3

Use the multiplication table to solve each division:

a $12 \div 4 =$ ______ **b** $20 \div 5 =$ ______ **c** $14 \div 2 =$ ______

d $27 \div 3 =$ ______ **e** $45 \div 5 =$ ______ **f** $30 \div 6 =$ ______

g $16 \div 4 =$ ______ **h** $25 \div 5 =$ ______ **i** $28 \div 7 =$ ______

j $32 \div 8 =$ ______ **k** $49 \div 7 =$ ______ **l** $72 \div 9 =$ ______

m $90 \div 10 =$ ______ **n** $63 \div 7 =$ ______ **o** $54 \div 9 =$ ______

We can use times tables to write related multiplications and divisions.

$7 \times 2 = 14$
$2 \times 7 = 14$
$14 \div 7 = 2$
$14 \div 2 = 7$

Exercise 4

Complete these related multiplications and divisions.

a
$2 \times 9 =$ ____
$9 \times 2 =$ ____
____ $\div 2 = 9$
____ $\div 9 = 2$

b
$7 \times 8 =$ ____
$8 \times$ ____ $=$ ____
____ $\div 7 =$ ____
____ $\div 8 =$ ____

c
$8 \times 4 =$ ____
$4 \times$ ____ $=$ ____
____ $\div 8 =$ ____
____ $\div 4 =$ ____

d
$3 \times$ ____ $= 15$
____ $\times 3 = 15$
$15 \div 3 =$ ____
$15 \div$ ____ $= 3$

e
$4 \times$ ____ $= 28$
____ $\times 4 =$ ____
____ $\div 4 =$ ____
$28 \div$ ____ $=$ ____

f
$6 \times$ ____ $= 54$
____ $\times 6 =$ ____
____ $\div 6 =$ ____
$54 \div$ ____ $=$ ____

Exercise 5

Complete the multiplication. Write down the related multiplication and divisions.

a
$4 \times 9 =$ ____
____ $\times$ ____ $=$ ____
____ $\div$ ____ $=$ ____
____ $\div$ ____ $=$ ____

b
$9 \times$ ____ $= 63$
____ $\times$ ____ $=$ ____
____ $\div$ ____ $=$ ____
____ $\div$ ____ $=$ ____

c
____ $\times 6 = 60$
____ $\times$ ____ $=$ ____
____ $\div$ ____ $=$ ____
____ $\div$ ____ $=$ ____

Exercise 6

Use your times tables to solve these mentally.

a $20 \div 2 =$ ____
b $18 \div 3 =$ ____
c $21 \div 3 =$ ____
d $28 \div 4 =$ ____
e $20 \div 5 =$ ____
f $21 \div 7 =$ ____
g $18 \div 6 =$ ____
h $30 \div 6 =$ ____
i $14 \div 7 =$ ____
j $16 \div 8 =$ ____
k $36 \div 6 =$ ____
l $27 \div 9 =$ ____
m $48 \div 6 =$ ____
n $49 \div 7 =$ ____
o $56 \div 8 =$ ____
p $70 \div 10 =$ ____
q $81 \div 9 =$ ____
r $35 \div 5 =$ ____

Exercise 7

a Complete these related multiplications and divisions.

$1 \times 8 =$ ______

$8 \times 1 =$ ______

$8 \div$ ______ $=$ ______

$8 \div$ ______ $=$ ______

b What is the result if we divide a number by itself? ______________

c What is the result if we divide a number by 1? ______________

Discussion

How can we use our times tables to complete these divisions?

- $42 \div$ ☐ $= 7$
- ☐ $\div 4 = 8$

Exercise 8

Use your times tables to complete each division:

a $16 \div$ ☐ $= 8$ **b** $21 \div$ ☐ $= 7$ **c** $28 \div$ ☐ $= 4$

d $35 \div$ ☐ $= 5$ **e** $54 \div$ ☐ $= 9$ **f** $48 \div$ ☐ $= 6$

g $27 \div$ ☐ $= 9$ **h** $81 \div$ ☐ $= 9$ **i** $32 \div$ ☐ $= 8$

Exercise 9

Use your times tables to complete each division:

a ☐ $\div 5 = 2$ **b** ☐ $\div 4 = 2$ **c** ☐ $\div 6 = 3$

d ☐ $\div 4 = 9$ **e** ☐ $\div 7 = 6$ **f** ☐ $\div 9 = 5$

g ☐ $\div 8 = 9$ **h** ☐ $\div 4 = 11$ **i** ☐ $\div 11 = 12$

Puzzle

Use these numbers only to write as many divisions as you can.

2	3	4	6	8	12	18	24

Exercise 10

Answer each question using a division or a multiplication:

a Mrs Jones has a box of 27 tennis balls. She shares them equally into 3 bags. How many balls are in each bag?

$\square \div \square = \square$

There are ______ balls in each bag.

b Dennis is a farrier. He makes horse shoes. He has 36 horse shoes in his truck when he visits Lucy's farm. How many horses can have new shoes?

______ horses can have new shoes.

c Maia has 15 pairs of shoes. She divides them equally onto 3 shelves in her closet. How many pairs of shoes are on each shelf?

$\square \div \square = \square$

There are ______ pairs of shoes on each shelf.

How many shoes are on each shelf?

$\square \times \square = \square$

There are ______ shoes on each shelf.

All multiples of 10 end with a zero.

When we *multiply* a whole number by 10, each digit moves one place to the left, and we write a zero on its end.

$15 \times 10 = 150$

When we *divide* any multiple of 10 by 10, we *remove* one zero from its end. Each digit moves one place to the right.

$150 \div 10 = 15$

Exercise 11

Complete each division:

a $20 \div 10 =$ ______ **b** $40 \div 10 =$ ______ **c** $60 \div 10 =$ ______

d $70 \div 10 =$ ______ **e** $110 \div 10 =$ ______ **f** $130 \div 10 =$ ______

g $240 \div 10 =$ ______ **h** $300 \div 10 =$ ______ **i** $470 \div 10 =$ ______

j $580 \div 10 =$ ______ **k** $900 \div 10 =$ ______ **l** $7100 \div 10 =$ ______

Exercise 12

Complete each division:

a $\square \div 10 = 3$ **b** $\square \div 10 = 8$ **c** $\square \div 10 = 9$

d $\square \div 10 = 12$ **e** $\square \div 10 = 16$ **f** $\square \div 10 = 20$

g $\square \div 10 = 35$ **h** $\square \div 10 = 41$ **i** $\square \div 10 = 200$

j $\square \div 10 = 65$ **k** $\square \div 10 = 77$ **l** $\square \div 10 = 810$

Exercise 13

Complete these divisions.

a $10 \div 10 =$ ______
$100 \div 10 =$ ______
$1000 \div 10 =$ ______

b ______ $\div 10 = 5$
______ $\div 10 = 50$
______ $\div 10 = 500$

c ______ $\div 10 = 86$
______ $\div 10 = 860$
______ $\div 10 = 8600$

Jana has 26 strawberries. She wants to share them equally between herself and her three friends.

Jana divides the strawberries into 4 bowls.

She places 6 strawberries into each bowl, but has 2 left over.

The two left over strawberries are called the **remainder**.

We write $26 \div 4 = 6$ remainder 2

so $26 \div 4 = 6 \text{ r } 2$

Exercise 14

What division is shown by the diagram?

a

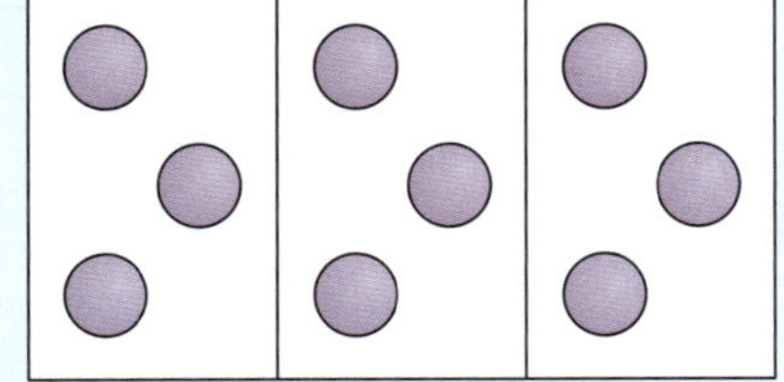

______ ÷ 3 = ______ r ______

b

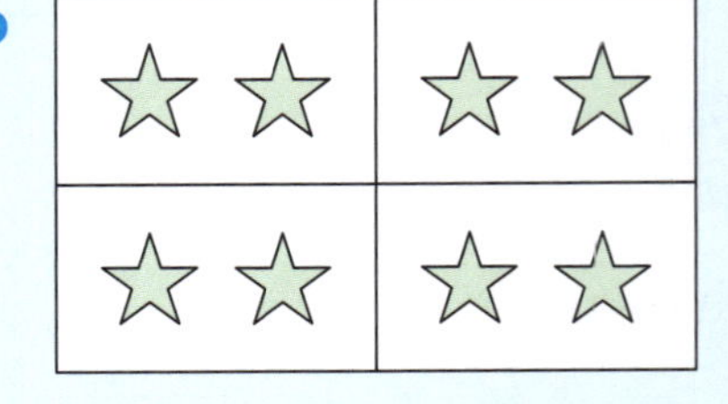

______ ÷ ______ = ______ r ______

c

______ ÷ ______ = ______ r ______

d

____

e

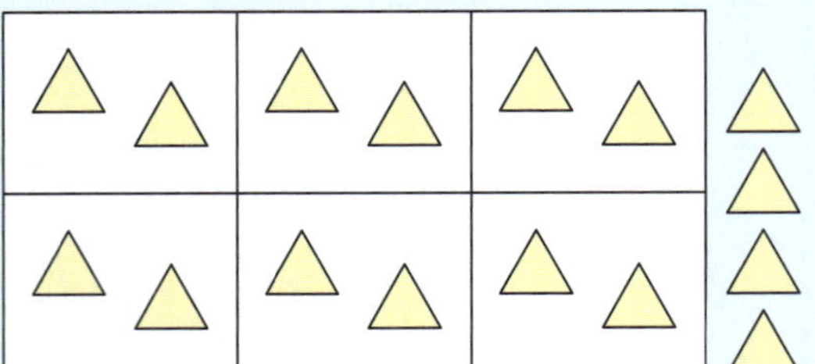

____

f

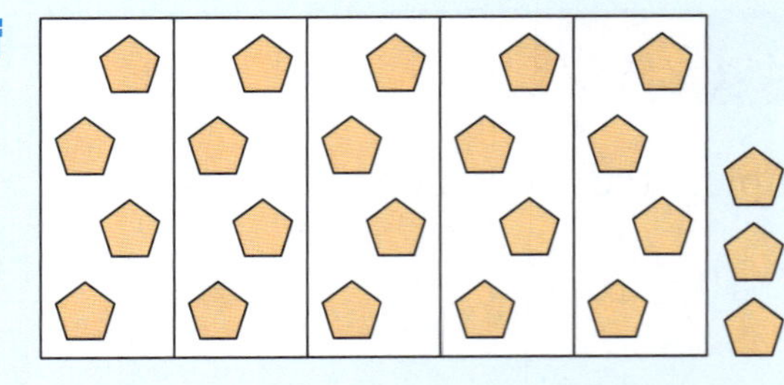

____

Exercise 15

Use a diagram to help answer each division:

a $7 \div 2 =$ ______ r ______

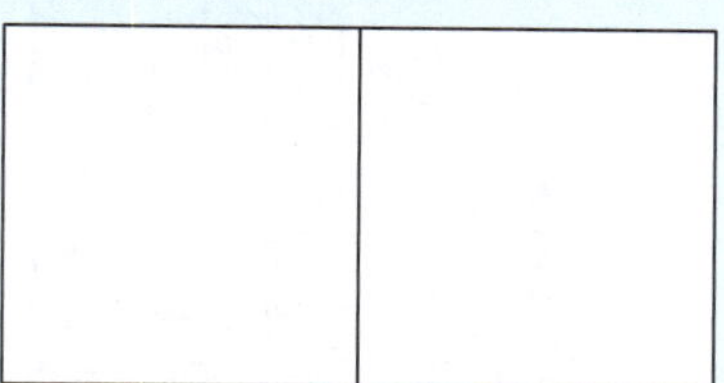

b $10 \div 3 =$ ____________

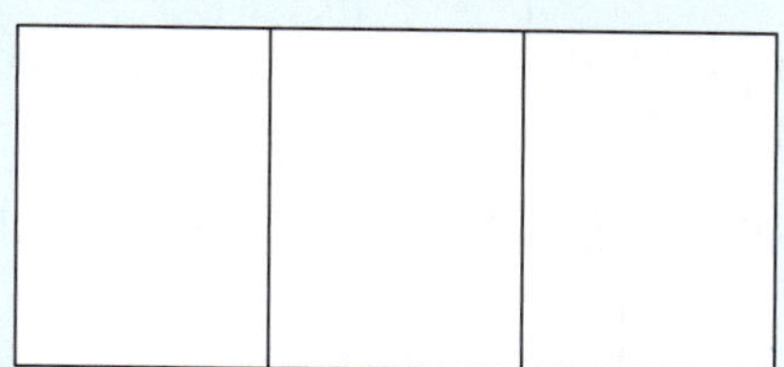

c $16 \div 3 =$ ____________

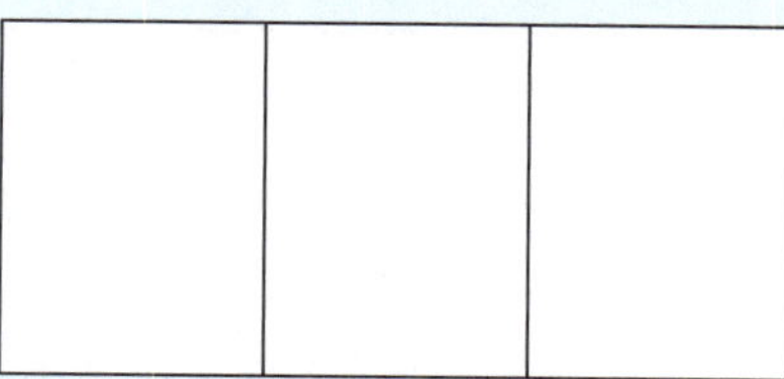

d $14 \div 4 =$ ____________

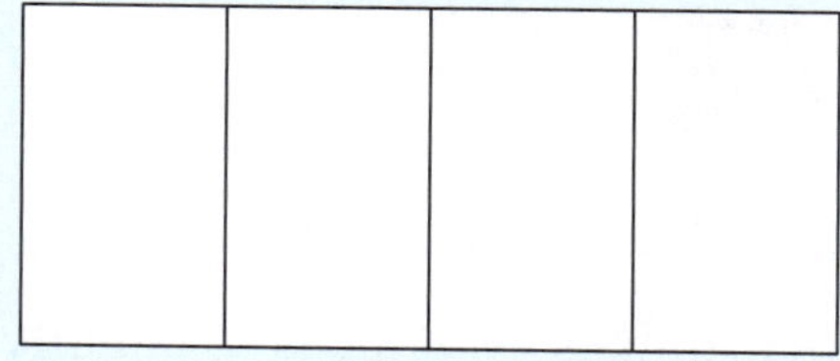

e $16 \div 5 =$ ____________

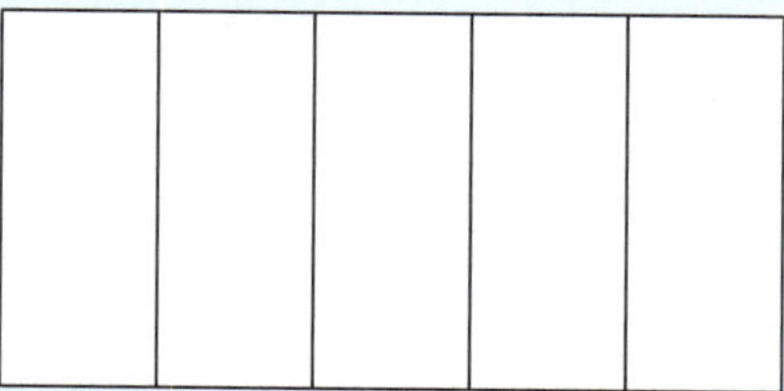

f $17 \div 2 =$ ____________

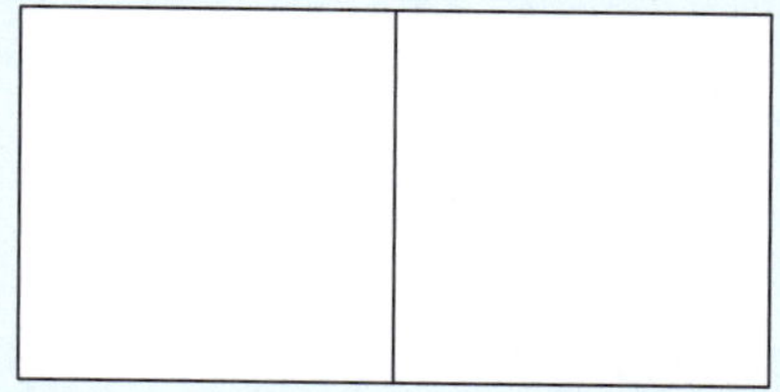

g $19 \div 3 =$ ____________

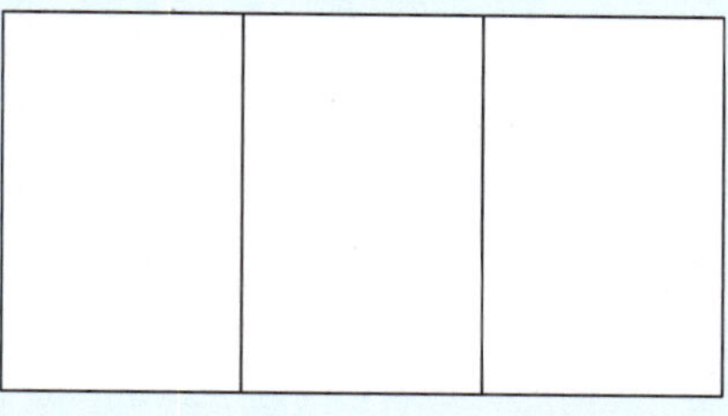

h $23 \div 4 =$ ____________

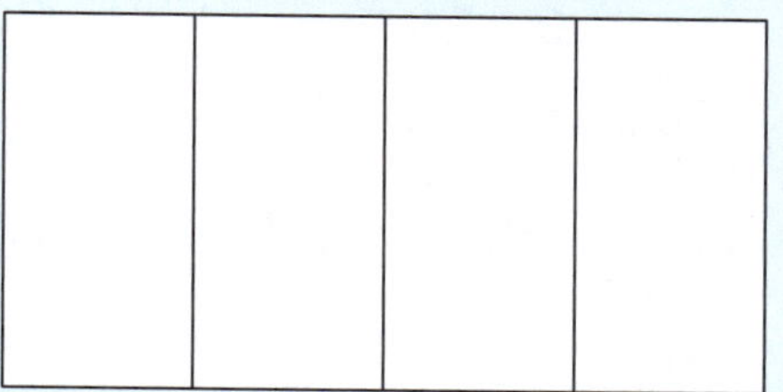

Exercise 16

Use counters to help answer each division:

a $9 \div 2 =$ ____________

b $11 \div 3 =$ ____________

c $13 \div 4 =$ ____________

d $15 \div 2 =$ ____________

e $20 \div 3 =$ ____________

f $17 \div 4 =$ ____________

g $17 \div 5 =$ ____________

h $25 \div 3 =$ ____________

i $25 \div 4 =$ ____________

j $29 \div 6 =$ ____________

k $25 \div 8 =$ ____________

l $29 \div 5 =$ ____________

Exercise 17

Solve each problem using a division:

a Brendan planted 24 cabbages in rows of 6.

Brendan planted ☐ rows of cabbages.

b Ellie stacked 35 bales of hay in a stack 5 bales high.

Ellie stacked ☐ bales on each level.

Exercise 18

Solve each problem using a division:

a Yiu Min arranges 16 cupcakes equally onto 3 plates.

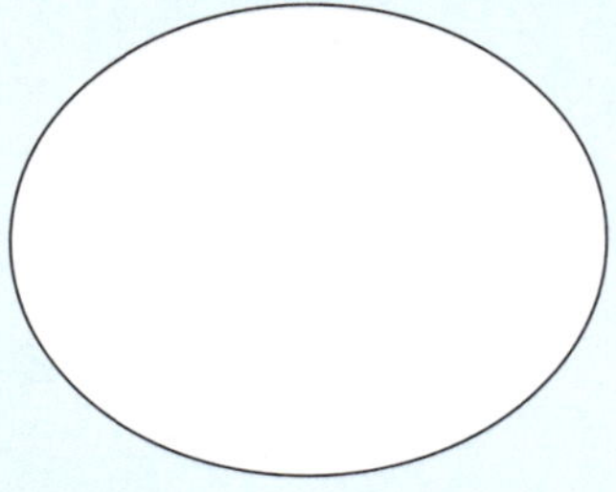 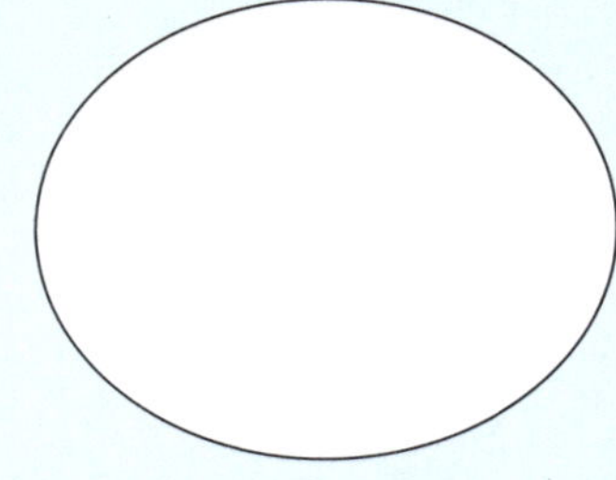 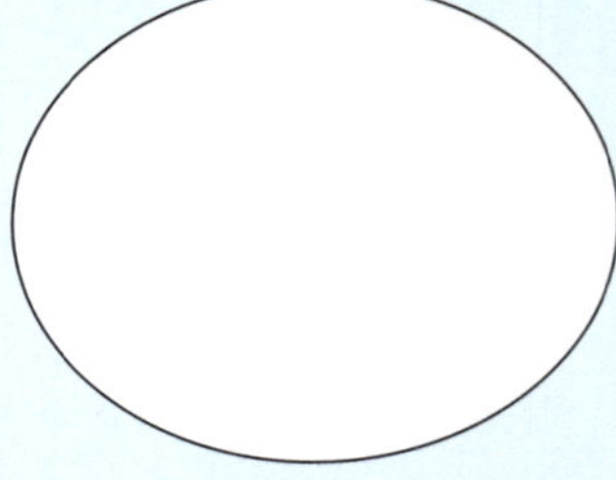

There are ______ cupcakes on each plate.

Yiu Min has ______ left over.

b Julia purchased a box of cookies. There were 29 cookies in the box.

Can Julia share them equally between 4 cookie jars? ______

Write a sentence to explain your answer.

....

....

Exercise 19

Solve each problem using a division:

a Georgia milks 23 litres of milk from her cows. She pours it into 2 litre bottles.

Georgia fills ☐ 2 litre bottles.

There is ☐ litre of milk remaining.

b Chef Colin needs 8 potatoes for a pot of soup.
Colin has 68 potatoes.

Colin can make ☐ pots of soup.

There will be ☐ potatoes left over.

Puzzle

The zoo has 59 hopping mice. They are kept in cages with either 8 or 9 hopping mice in each cage.

☐ cages contain 9 hopping mice.

Numbers that *do not* leave a remainder when divided by 2 are called **even** numbers.

Numbers that *do* leave a remainder when divided by 2 are called **odd** numbers.

Exercise 20

Using counters if necessary, divide each number below by 2. Record whether the number is even or odd.

~~4~~ ~~3~~ 8 16 7 9 15 22 13

17 10 29 24 25 21 19 18 33

even	4,
odd	3,

Discuss with your class any patterns in your lists.

Even numbers always end with ____, ____, ____, ____, or ____.

Odd numbers always end with ____, ____, ____, ____, or ____.

Exercise 21

Circle the even numbers.

27	41	18	12	9	25
14	16	17	21	15	43
29	62	70	47	77	58
94	87	83	102	119	61

Activity — Even and odd numbers

1 Natasha says:

"If I add two *even* numbers, the answer is always *even*."

Test Natasha's statement by writing 6 additions of even + even.

For example: $2+4=6$ even ✓

1 ______________ **4** ______________

2 ______________ **5** ______________

3 ______________ **6** ______________

Do you think Natasha is correct? Yes / No

2 Eddy says:

"If I add two *odd* numbers, the answer is always *odd*."

Test Eddy's statement by writing 6 additions of odd + odd.

1 ______________ **4** ______________

2 ______________ **5** ______________

3 ______________ **6** ______________

Do you think Eddy is correct? Yes / No

Complete:

If I add two *odd* numbers, the answer is always ______________.

3 Finish each statement using 5 examples to support your answer:

a even + odd = ...

b even − even = ...

c odd − odd = ...

d even − odd = ...

e even × even = ...

f odd × odd = ...

g even × odd = ...

$6 \div 2 = 3$ $\quad$ $12 \div 2 = 6$

$60 \div 2 = 30$ $\quad$ $120 \div 2 = 60$

$600 \div 2 = 300$ $\quad$ $1200 \div 2 = 600$

Exercise 22

Complete these divisions.

a

$9 \div 3 =$ ______

$90 \div 3 =$ ______

$900 \div 3 =$ ______

b

$8 \div 4 =$ ______

$80 \div 4 =$ ______

$800 \div 4 =$ ______

c

$10 \div 5 =$ ______

$100 \div 5 =$ ______

$1000 \div 5 =$ ______

d

$14 \div 2 =$ ______

$140 \div 2 =$ ______

$1400 \div 2 =$ ______

e

$18 \div 3 =$ ______

$180 \div 3 =$ ______

$1800 \div 3 =$ ______

f

$28 \div 7 =$ ______

$280 \div 7 =$ ______

$2800 \div 7 =$ ______

g

$32 \div 4 =$ ______

$320 \div 4 =$ ______

$3200 \div 4 =$ ______

h

$42 \div 6 =$ ______

$420 \div 6 =$ ______

$4200 \div 6 =$ ______

i

$36 \div 9 =$ ______

$360 \div 9 =$ ______

$3600 \div 9 =$ ______

j

$72 \div 8 =$ ______

$720 \div 8 =$ ______

$7200 \div 8 =$ ______

To divide larger numbers, we set out the division like this:

$96 \div 3$

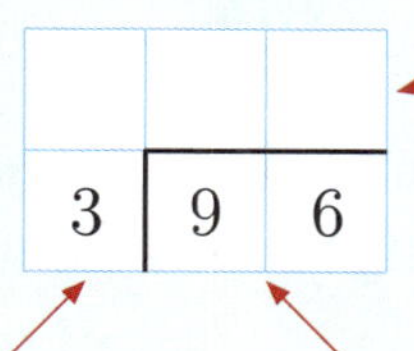

The answer is written here.

This is the number we are dividing by.

This is the number we are dividing.

To perform the division, we divide each place value in turn, starting with the highest.

	3	2
3	9	6

1 Set up your division.

2 There are 9 tens.

$9 \div 3 = 3$

We write 3 in the tens column above the line.

3 There are 6 units.

$6 \div 3 = 2$

We write 2 in the units column above the line.

So, $96 \div 3 = 32$.

Exercise 23

Complete each division:

a $24 \div 2$

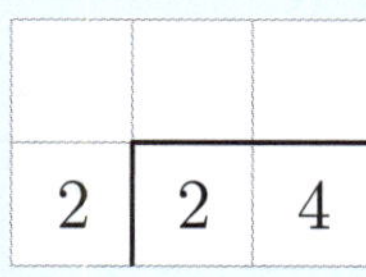

b $36 \div 3$

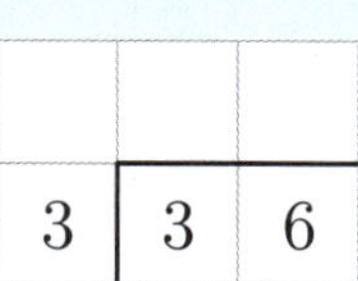

c $48 \div 4$

d $48 \div 2$

e $60 \div 3$

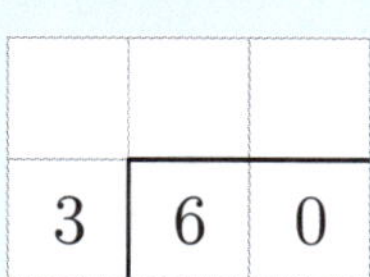

f $84 \div 4$

g $264 \div 2$

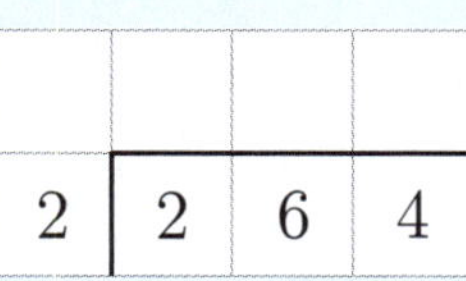

h $690 \div 3$

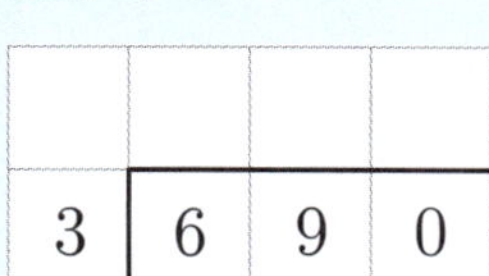

i $848 \div 4$

$72 \div 2$

1. Set up your division.
2. There are 7 tens.

   $7 \div 2 = 3$ r 1

   We write 3 in the tens column above the line.
   The remainder 1 ten is exchanged for 10 units.
3. We now have 12 units.

   $12 \div 2 = 6$

   We write 6 in the units column above the line.

So, $72 \div 2 = 36$.

Exercise 24

Complete each division:

a $56 \div 2$

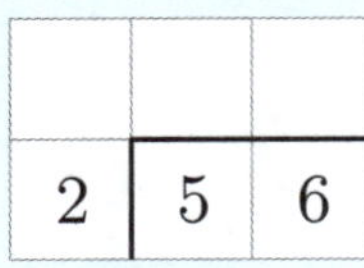

b $45 \div 3$

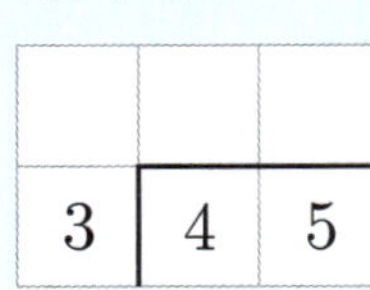

c $57 \div 3$

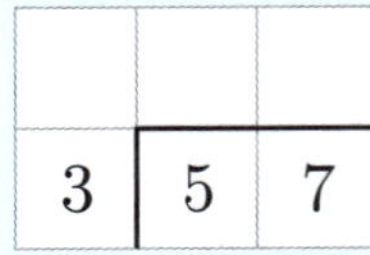

d $38 \div 2$

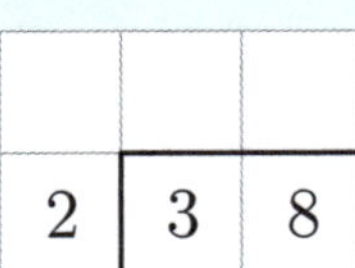

e $64 \div 4$

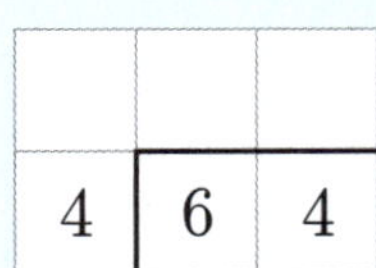

f $65 \div 5$

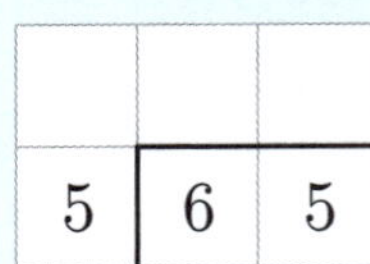

g $426 \div 3$

h $476 \div 4$

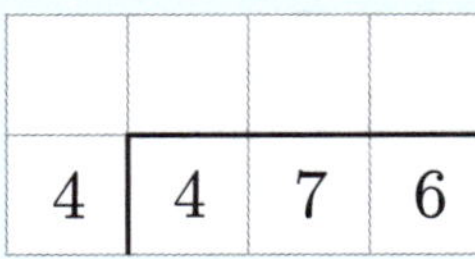

i $952 \div 7$

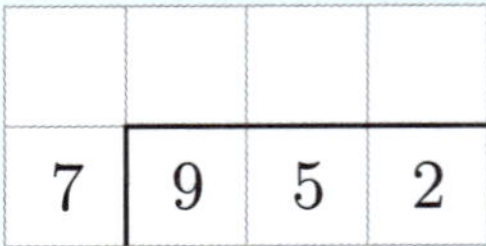

Exercise 25

Set out each division and find the answer:

a $54 \div 3$

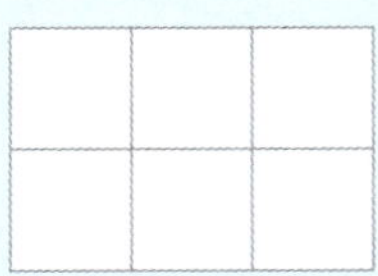

b $74 \div 2$

c $56 \div 4$

d $75 \div 5$

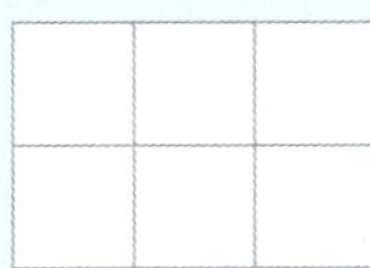

e $676 \div 4$

f $474 \div 3$

Exercise 26

Solve each problem using a division:

a Rachel went strawberry picking and collected 52 strawberries.

She divides them equally into 4 containers to take to work.

There are ______ strawberries in each container.

b Mrs Lewis buys 78 toy animals. She shares them equally between 2 kindergarten classes.

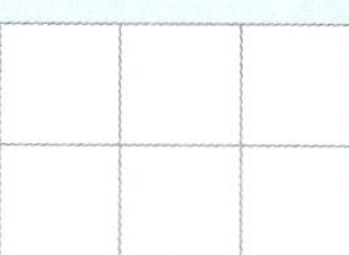

Each class will receive ______ toy animals.

$95 \div 2$

	4	7	r 1
2	9	[1]5	

1 Set up your division.

2 There are 9 tens.

$9 \div 2 = 4 \text{ r } 1$

We write 4 in the tens column above the line.
The remainder 1 ten is exchanged for 10 units.

3 We now have 15 units.

$15 \div 2 = 7 \text{ r } 1$

We write 7 in the units column above the line.
The remainder 1 is written as r 1 above the line on the end of the number.

So, $95 \div 2 = 47 \text{ r } 1$.

Exercise 27

Complete each division:

a $25 \div 2$

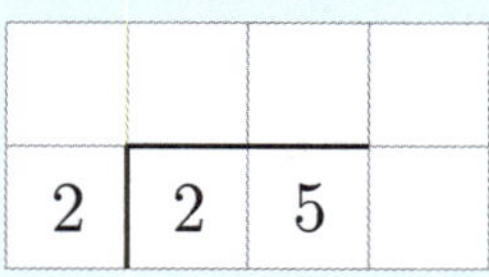

b $37 \div 3$

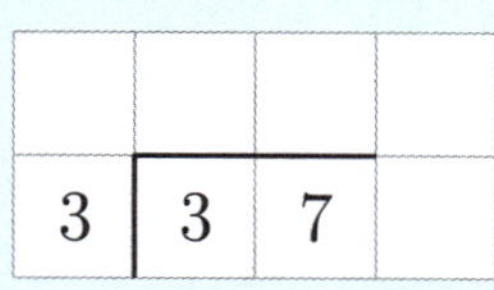

c $44 \div 3$

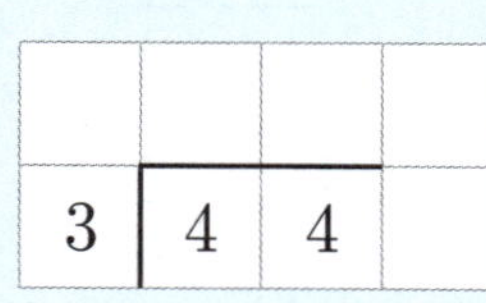

d $59 \div 2$

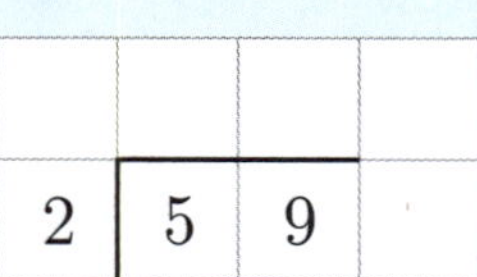

e $668 \div 5$

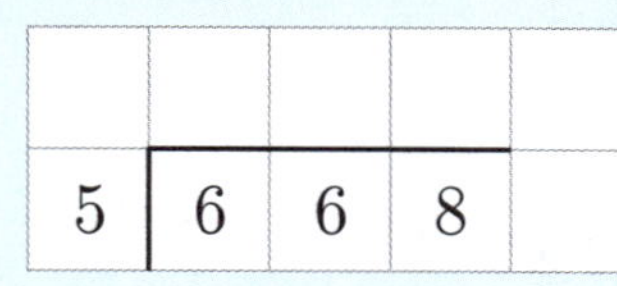

f $453 \div 4$

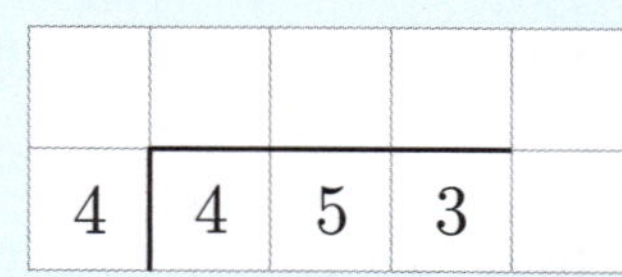

Exercise 28

Set out each division and find the answer:

a $37 \div 2$

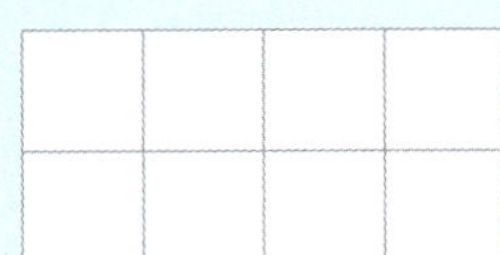

b $58 \div 3$

c $51 \div 4$

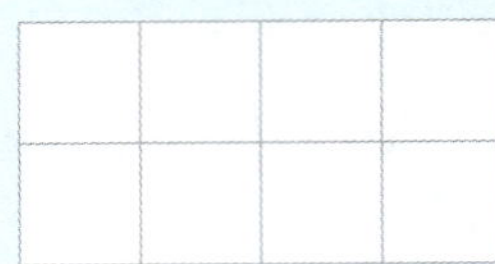

d $74 \div 6$

e $79 \div 5$

f $85 \div 6$

g $809 \div 7$

h $675 \div 4$

i $848 \div 5$

Exercise 29

Solve each problem using a division:

a Sukhi picks 77 apples from her tree and divides them equally between 3 boxes.

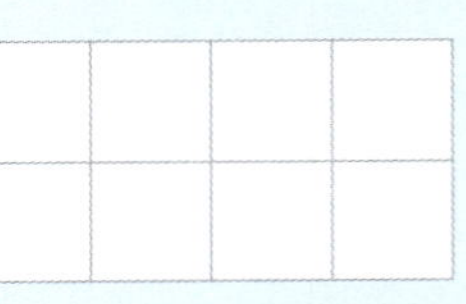

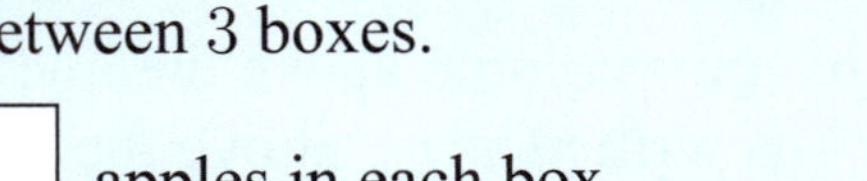

There are 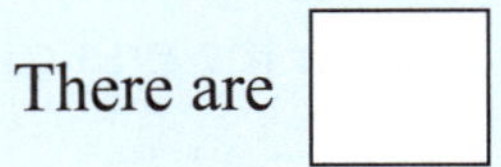 apples in each box.

There are 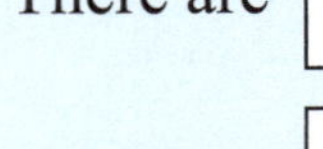 apples left over.

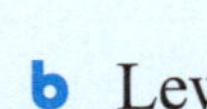

b Lewis picks 74 carrots from his garden and shares them equally between 5 bags.

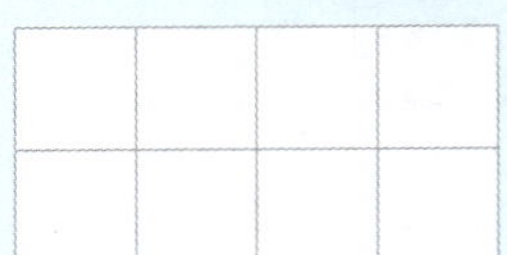

There are 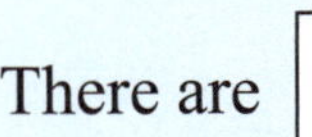 carrots in each bag.

There are carrots left over.

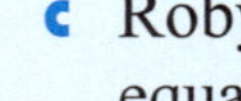

c Robyn has 97 sunflower seeds and plants them equally in 6 plant pots.

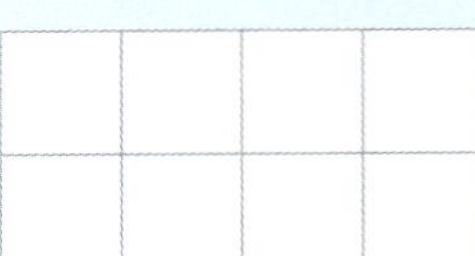

There are 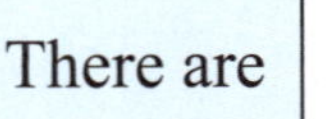 sunflower seeds in each pot.

There is 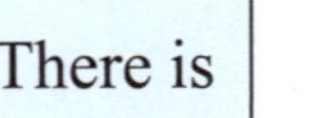 sunflower seed left over.

Exercise 30

Solve each problem using a division:

a Each tree in a plantation has 3 stakes.

There are 774 stakes in total.

There are ☐ trees in the plantation.

b There are 505 screws in a container.

Steve needs 4 screws for every plank in his deck.

Steve has enough screws for ☐ planks.

There will be ☐ screw left over.

Puzzle

In one week, Jim replaces tyres on 27 cars.

He replaces 1 tyre on 13 cars and 2 tyres on 5 cars.

If Jim replaces 55 tyres in total, how many cars had all of their tyres replaced?

☐ cars had all of their tyres replaced.

Revision

1 What division is shown in the diagram?

_______ ÷ _______ = _______

2 Use a diagram to show the division and find the answer.

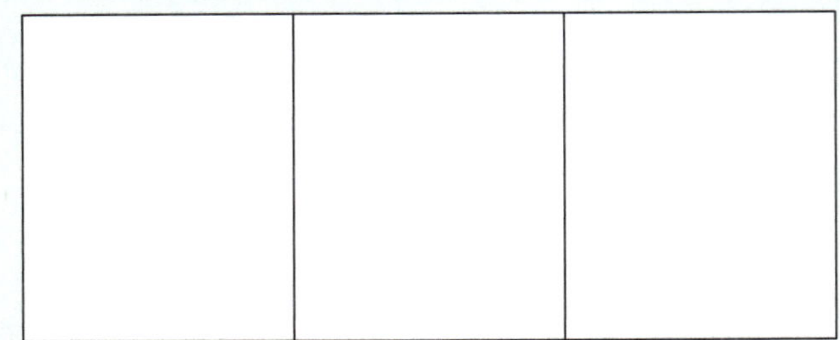

$18 \div 3 =$ _______

3 Complete these related multiplications and divisions.

a $5 \times 8 =$ _______

$8 \times$ _______ $=$ _______

_______ $\div 5 =$ _______

_______ $\div 8 =$ _______

b $8 \times$ _______ $= 72$

_______ $\times 8 =$ _______

_______ $\div 8 =$ _______

$72 \div$ _______ $=$ _______

c $7 \times 10 =$ _______

_______ $\times$ _______ $=$ _______

_______ $\div$ _______ $=$ _______

_______ $\div$ _______ $=$ _______

4 Complete each division:

a $15 \div 3 =$ _______ **b** $28 \div 7 =$ _______ **c** $64 \div 8 =$ _______

d $50 \div 10 =$ _______ **e** $42 \div 6 =$ _______ **f** $32 \div 4 =$ _______

g $20 \div$ _______ $= 4$ **h** _______ $\div 6 = 8$ **i** $56 \div$ _______ $= 7$

5 Complete each division:

a $160 \div 10 =$ _______ **b** $290 \div 10 =$ _______ **c** $6000 \div 10 =$ _______

d _______ $\div 10 = 54$ **e** _______ $\div 10 = 73$ **f** _______ $\div 10 = 940$

6 What division is shown by the diagram?

______ ÷ ______ = ______ r ______

7 Use a diagram to show the division and find the answer.

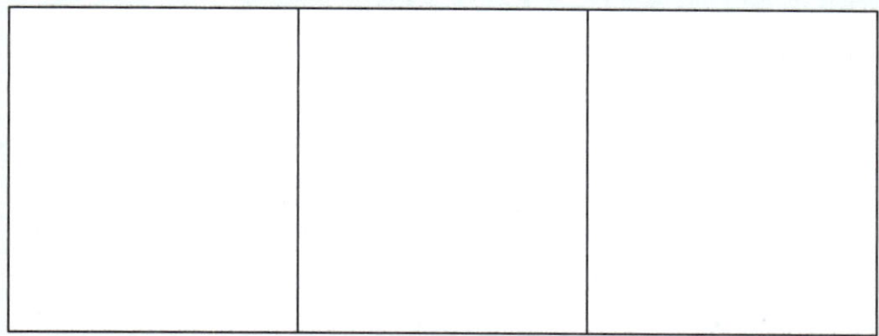

$17 \div 3 =$ ______ r ______

8 Answer each division:

a $19 \div 5 =$ ______ r ______

b $27 \div 8 =$ ______ r ______

c $32 \div 6 =$ ______ r ______

d $31 \div 7 =$ ______ r ______

9 Complete:

a Layla sorted 48 buttons into groups of 8.

There were ☐ buttons in each group.

b Edwin stacked 67 bricks in layers of 8.

There were ☐ complete layers.

There were ☐ bricks left over.

10 Circle the odd numbers:

42 14 86 17 35 68 73 59 91 40

11 Complete:

a $16 \div 2 =$ ______

$160 \div 2 =$ ______

$1600 \div 2 =$ ______

b $54 \div 9 =$ ______

$540 \div 9 =$ ______

$5400 \div 9 =$ ______

12 Complete each division:

a $39 \div 3$

3	3	9

b $85 \div 5$

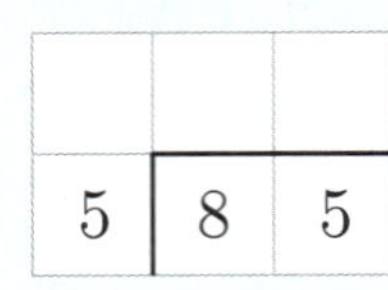

c $858 \div 3$

3	8	5	8

13 Set out each division and find the answer:

a $88 \div 4$

b $91 \div 7$

c $918 \div 6$

14 Laura had 68 biscuits.
She divided them equally onto 4 plates.
How many biscuits did Laura place onto each plate?

Laura placed ______ biscuits onto each plate.

15 Complete each division:

a $47 \div 4$

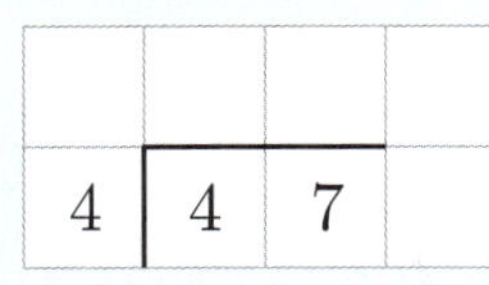

b $82 \div 7$

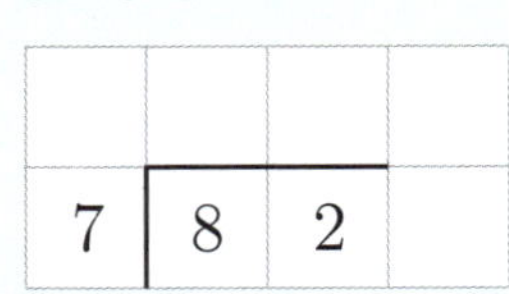

c $881 \div 6$

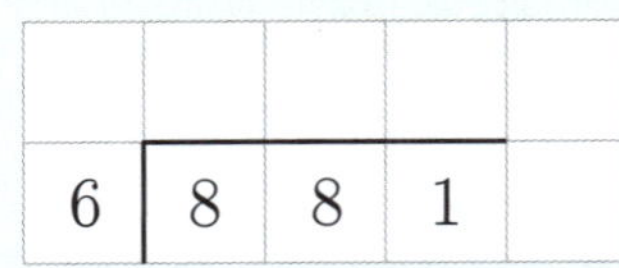

16 Set out each division and find the answer:

a $56 \div 5$

b $74 \div 3$

c $595 \div 4$

17 Anna pulled 79 carrots from her garden.
She shared them equally between her 4 horses.

Each horse received ☐ carrots.

There were ☐ carrots left over.

CHAPTER 7: FRACTIONS

When we divide a whole into *equally sized* pieces, we can write a **fraction**.

Scarlett cut her peach into 2 *equal pieces*.

Each piece is **one half** of the peach.

We write one half as the fraction $\frac{1}{2}$.

The whole peach has been divided into 2 **halves**.

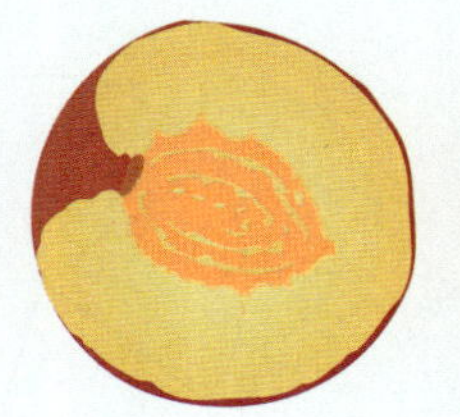

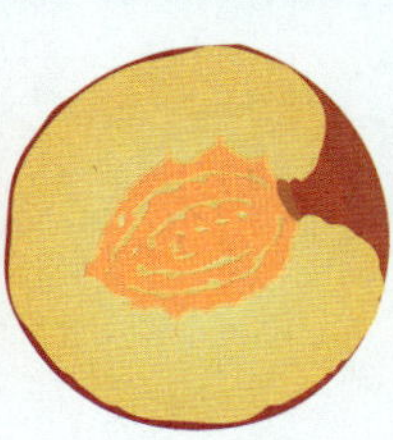

numerator → **1** ← We are looking at 1 part.
bar → —
denominator → **2** ← The whole is divided into 2 equal parts.

Exercise 1

The fraction $\frac{1}{3}$ is *one third*.

Explain why these figures do *not* show $\frac{1}{3}$.

a

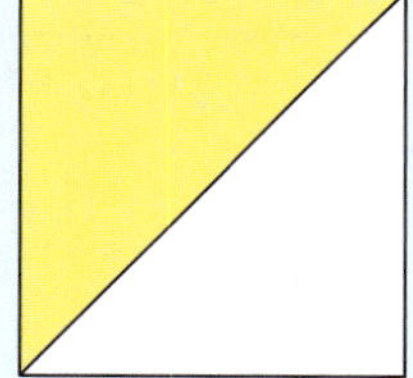

....

....

b

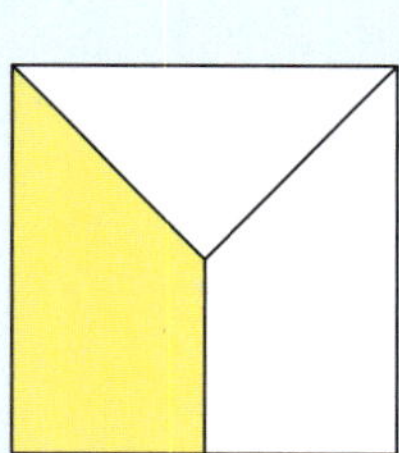

....

....

Exercise 2

The fraction alongside is *one quarter*.

Circle the numerator of the fraction.

$$\frac{1}{4}$$

Use this figure to show one quarter.

Exercise 3

The fraction alongside is *one fifth*. $\frac{1}{5}$

Circle the denominator of the fraction.

Circle the figure which corresponds to this fraction.

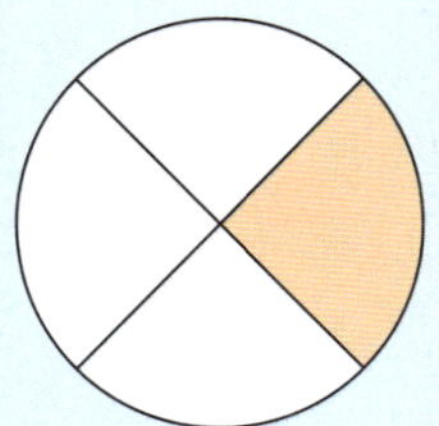
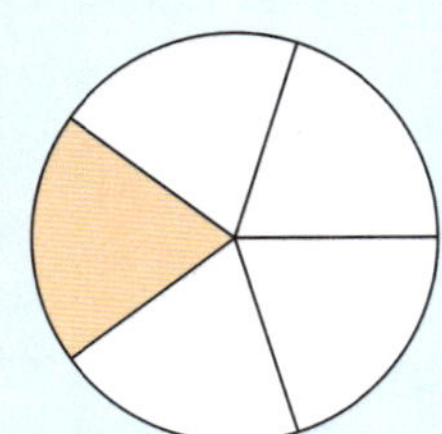
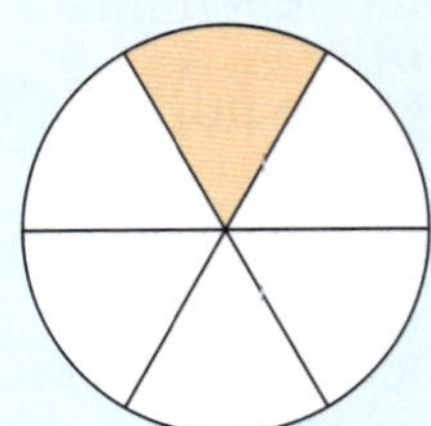

Discussion

What set of numbers are the fraction names similar to?

....

What fraction names are *different* from these numbers?

....

Exercise 4

Name the fraction.

a

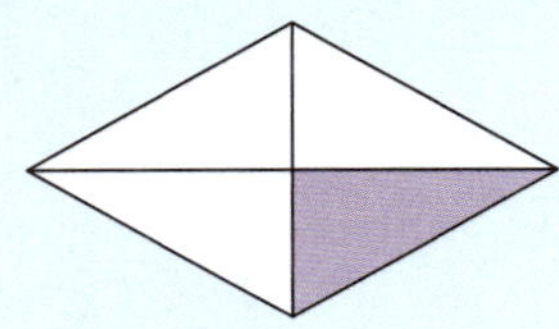
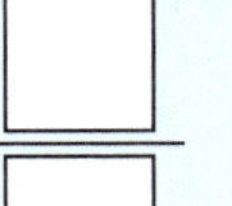
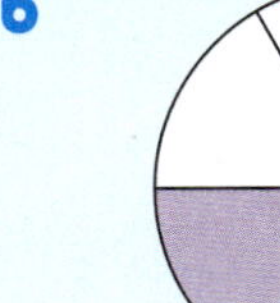

....

b

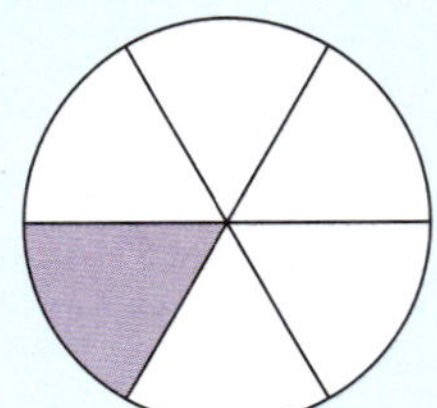
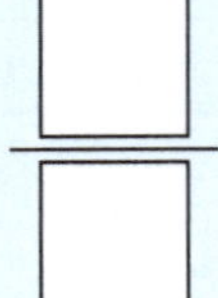

....

c

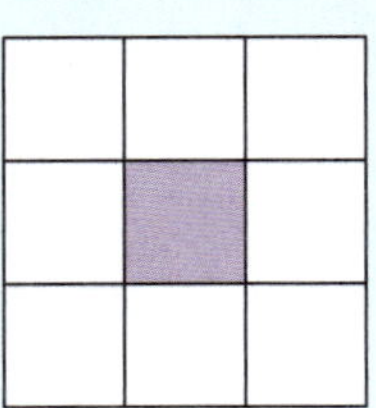
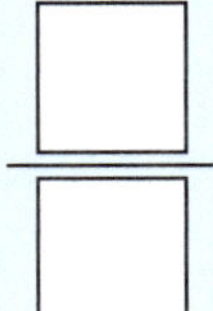

....

d

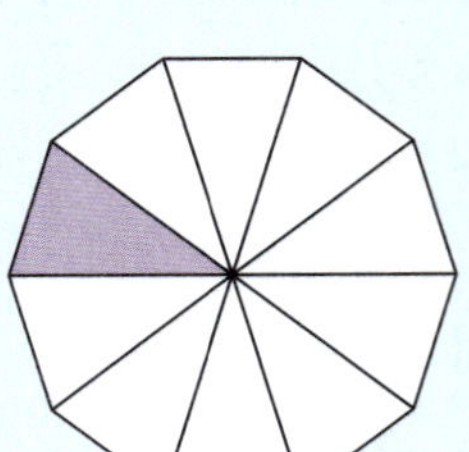
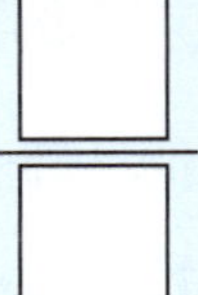

....

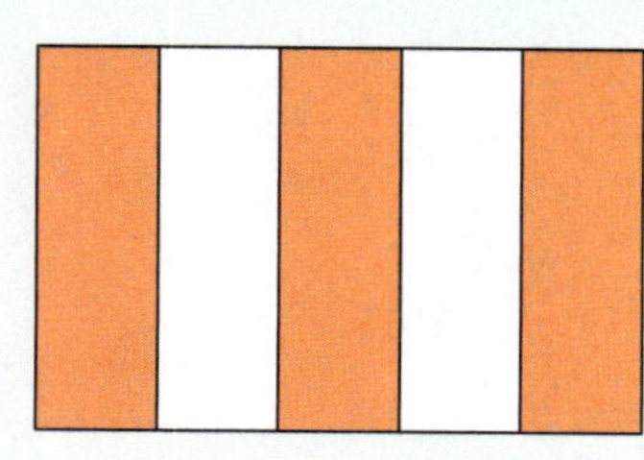

Three out of five equal pieces are shaded.

The fraction is $\frac{3}{5}$

or "three fifths".

Exercise 5

Write a fraction to show:

a two fifths ______

b three quarters ______

c four fifths ______

d three sixths ______

e two eighths ______

f five sixths ______

g seven tenths ______

h six eighths ______

i nine tenths ______

Exercise 6

Write each fraction in words:

a $\frac{2}{4}$ ______

b $\frac{3}{10}$ ______

c $\frac{3}{5}$ ______

d $\frac{6}{10}$ ______

e $\frac{3}{8}$ ______

f $\frac{4}{6}$ ______

g $\frac{9}{10}$ ______

h $\frac{2}{6}$ ______

Exercise 7

Shade the fraction given.

a $\frac{1}{3}$

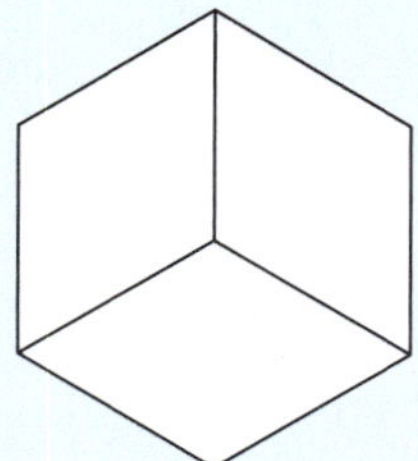

b $\frac{4}{5}$

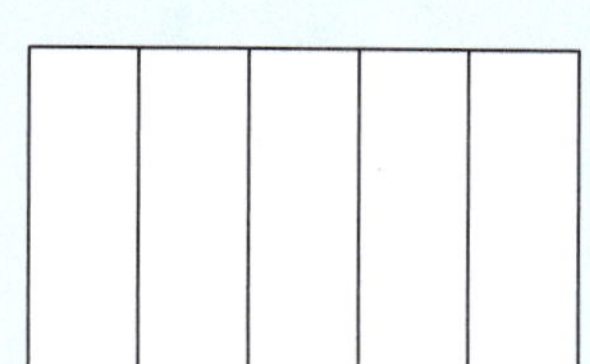

c $\frac{7}{10}$

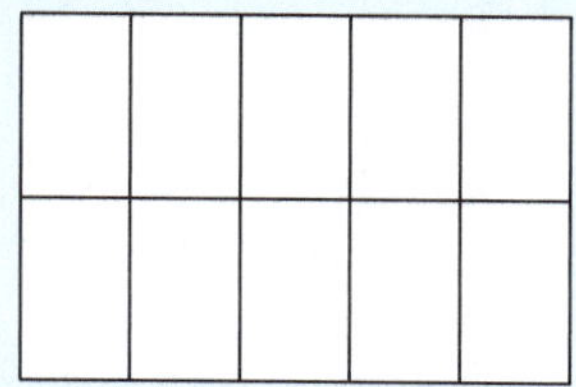

Shade the fraction given.

a three sixths

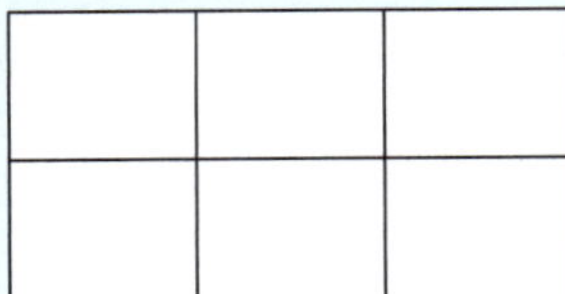

b nine tenths

c five eighths

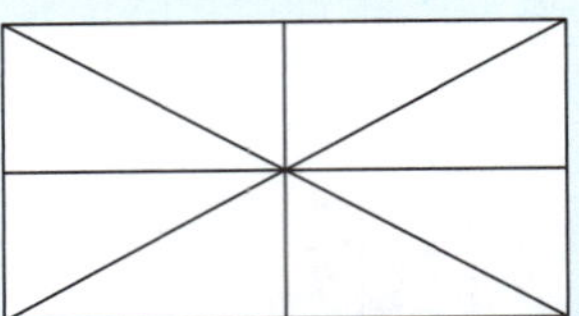

Exercise 9

What fraction of each shape is shaded?

a

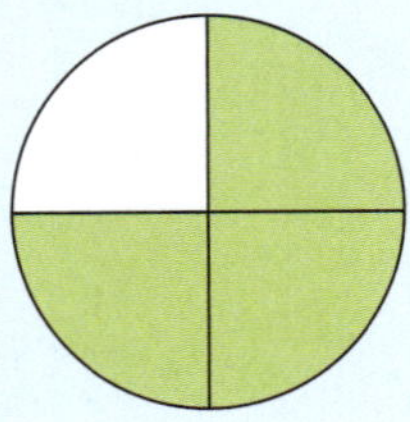

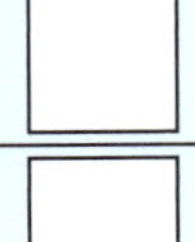

b

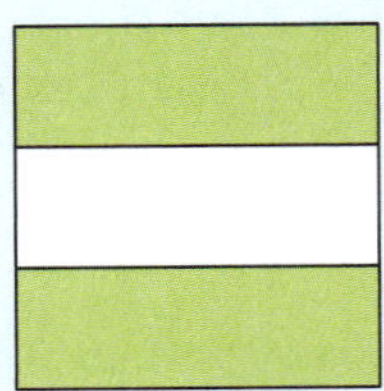

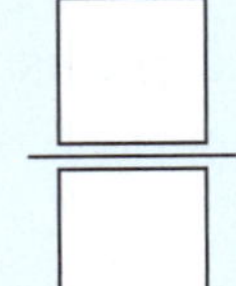

c

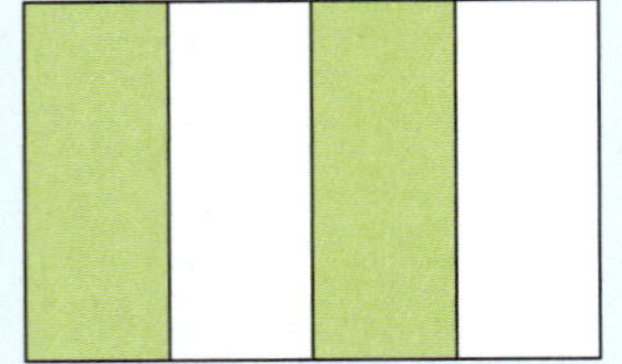

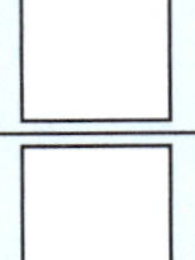

d

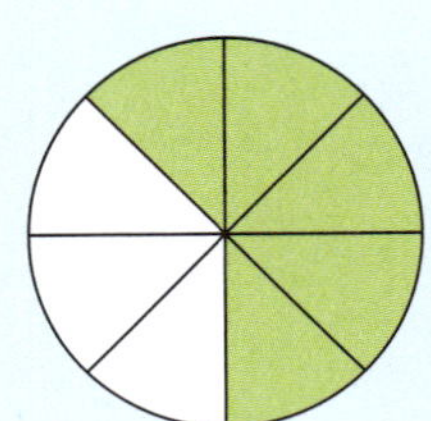

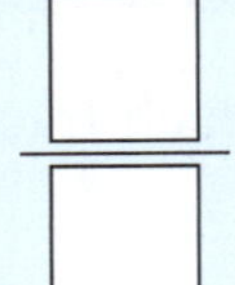

e

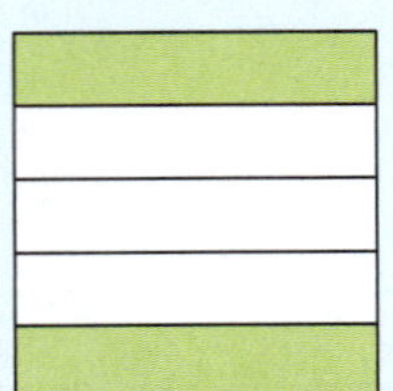

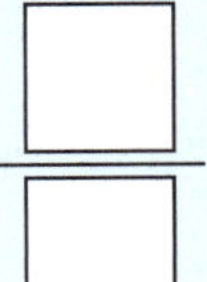

f

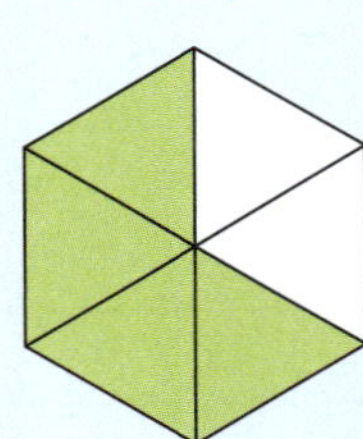

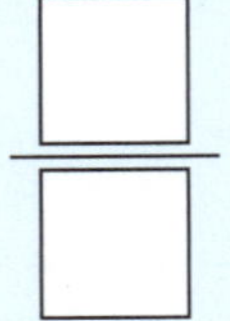

g

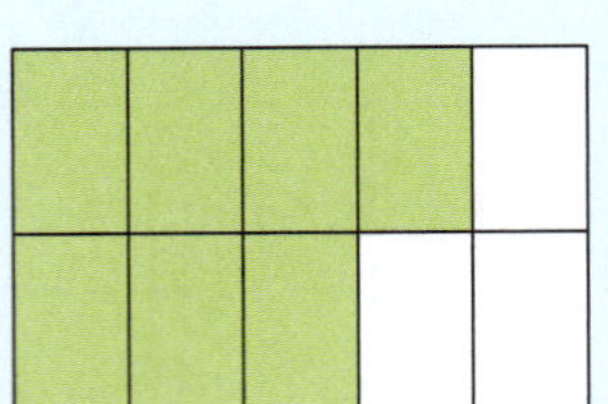

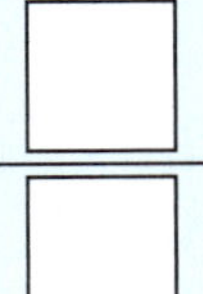

h

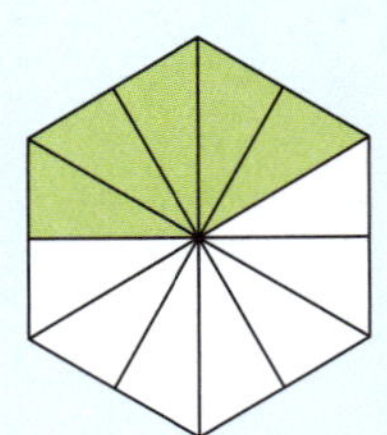

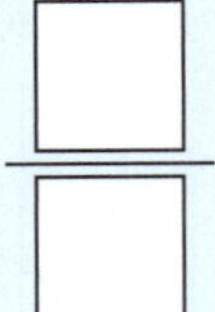

Exercise 10

Show each fraction:

a $\frac{2}{4}$

b $\frac{3}{8}$

c $\frac{5}{6}$

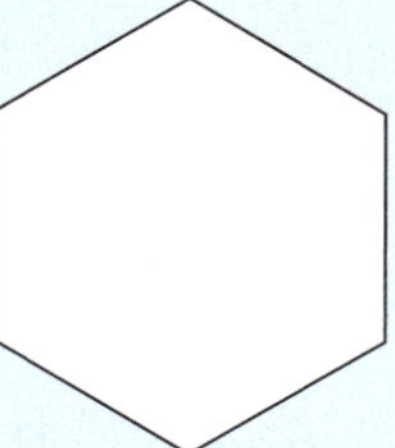

d $\frac{5}{9}$

e $\frac{2}{5}$

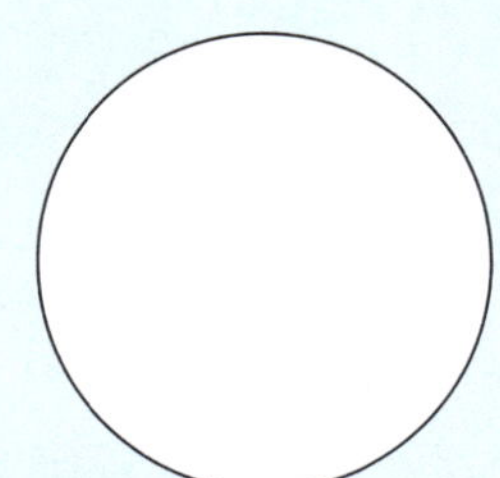

f $\frac{7}{12}$

Here are three fish.

2 out of 3 fish are circled.

$\frac{2}{3}$ are circled.

Exercise 11

What fraction of the fish in each group are circled?

a

b

c

d

e

f

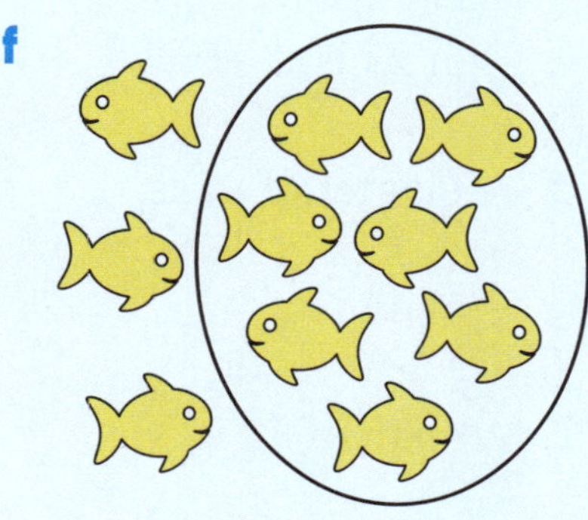

Exercise 12

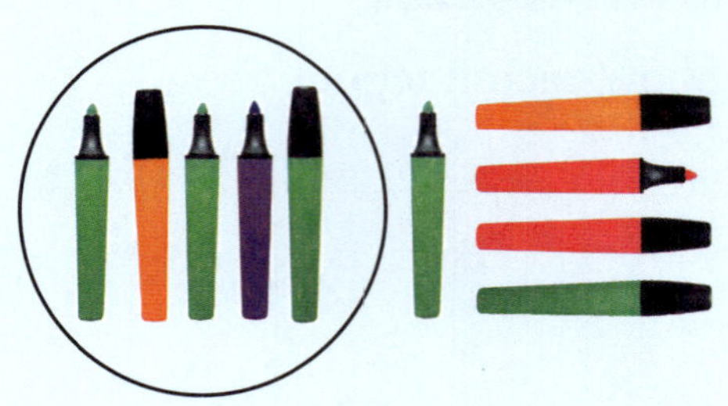

a How many pens are drawn? ______

b What fraction of the pens are pink? ______

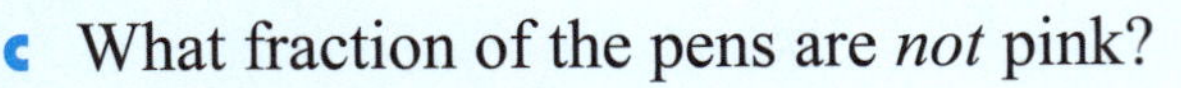

c What fraction of the pens are *not* pink? ______

d What fraction of the pens are inside the circle? ______

e Write two other fraction facts about the photograph.

____

____

When we **halve** a number, we divide it by 2.

The result is a **half** of the number.

$$10 \div 2 = 5$$

A half of 10 is 5.

Exercise 13

Complete:

a A half of 16 is $16 \div$ ______ $=$ ______.

b A half of 20 is ______ $\div$ ______ $=$ ______.

c $\frac{1}{2}$ of 80 is ______ $\div$ ______ $=$ ______.

d A third of 18 is ______ $\div 3 =$ ______.

e $\frac{1}{3}$ of 27 is ______ $\div$ ______ $=$ ______.

f A quarter of 32 is ______ $\div 4 =$ ______.

g $\frac{1}{4}$ of 44 is ______ $\div$ ______ $=$ ______.

h A sixth of 72 is ______ $\div$ ______ $=$ ______.

i $\frac{1}{6}$ of 42 is ______ $\div$ ______ $=$ ______.

Exercise 14

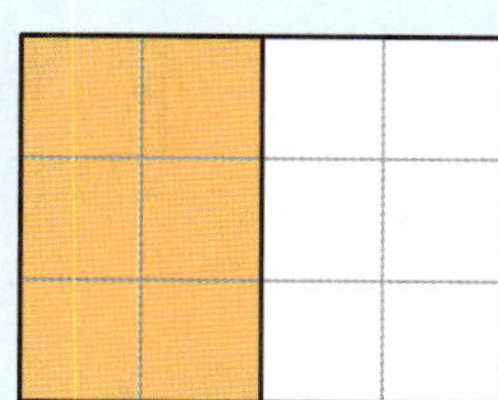

In each diagram there are $3 \times 4 = 12$ squares.

$\frac{1}{2}$ of 12 is 6.

By shading squares, complete each statement:

a 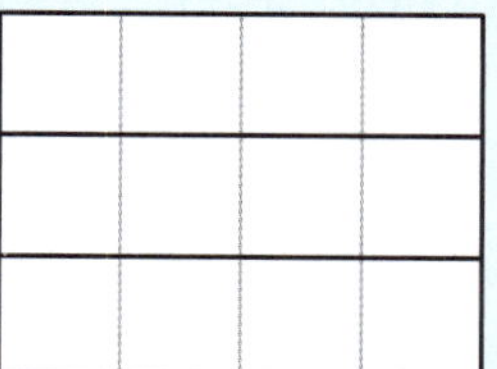

$\frac{1}{3}$ of 12 is ______

b 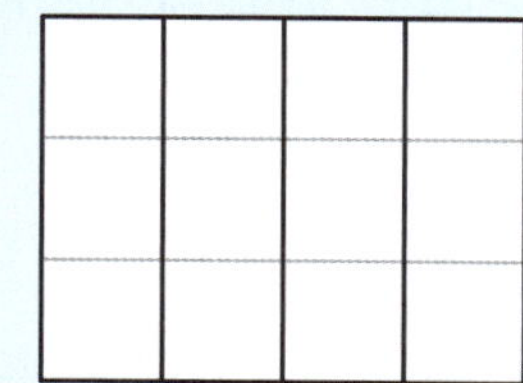

$\frac{3}{4}$ of 12 is ______

c 

$\frac{5}{6}$ of 12 is ______

Exercise 15

In each diagram there are $4 \times 6 =$ ______ squares.

By shading squares, complete each statement:

a 

$\frac{1}{6}$ of ______ is ______

b 

$\frac{2}{3}$ of ______ is ______

c 

$\frac{3}{8}$ of ______ is ______

Exercise 16

By shading squares, complete each statement:

a 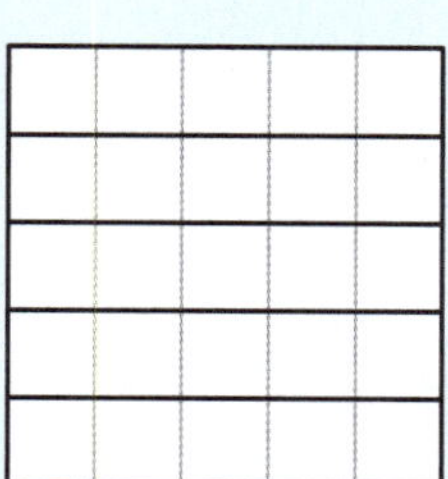

$\frac{2}{5}$ of 25 is ______

b 

$\frac{3}{4}$ of ______ is ______

c 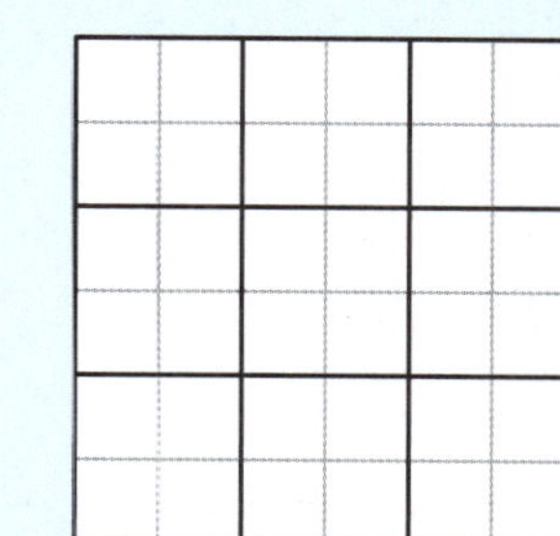

$\frac{2}{9}$ of ______ is ______

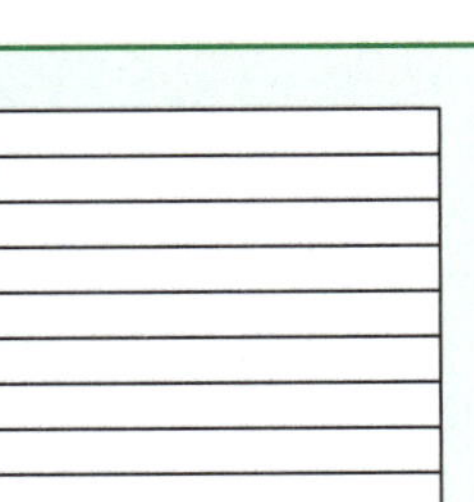

This square is made up of 10 equal pieces.

Each piece is one tenth of $\frac{1}{10}$ of the whole square.

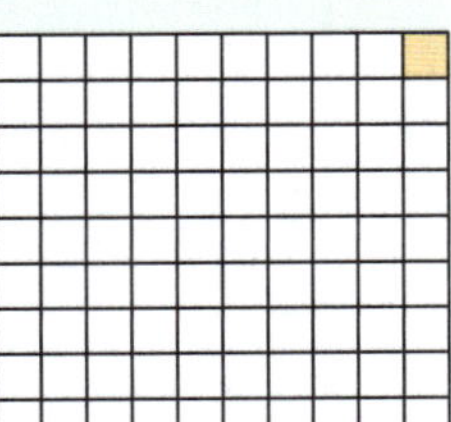

Each tenth can be divided into 10 equal pieces.

There are $10 \times 10 = 100$ little pieces in the whole square, so each piece is called **one hundredth**.

We write this as the fraction $\frac{1}{100}$.

Exercise 17

Write down the shaded fraction.

a 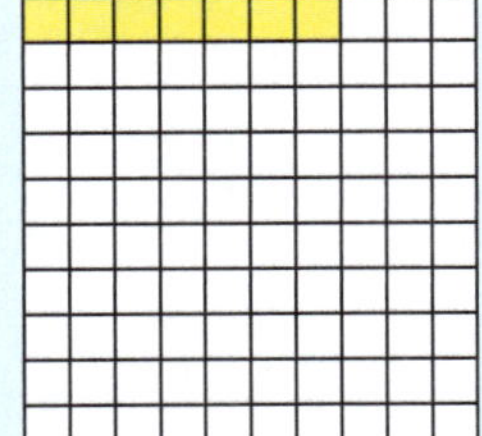 ....

b 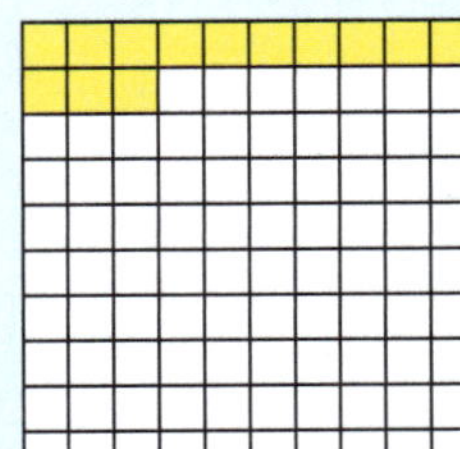 ....

c 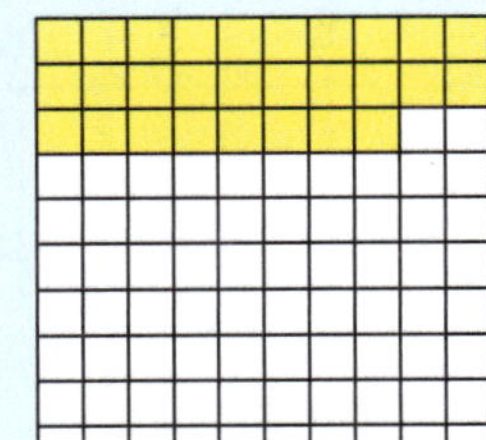 ....

d 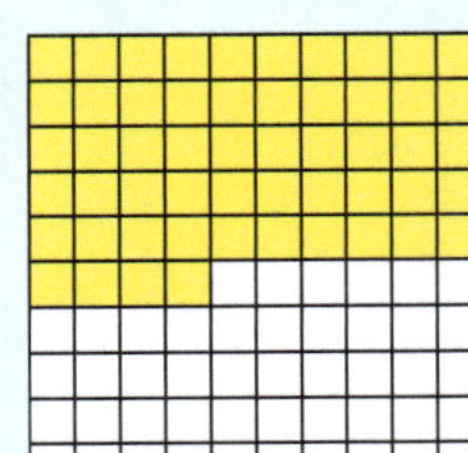 ....

Exercise 18

Shade each fraction:

a $\frac{3}{100}$

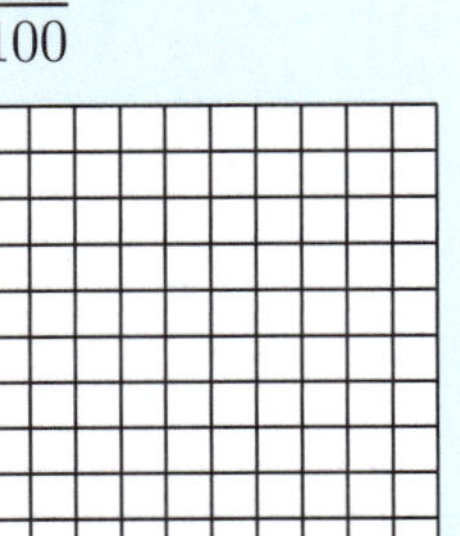

b $\frac{27}{100}$

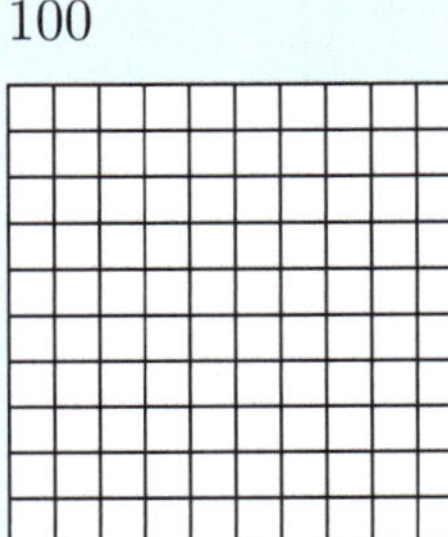

c $\frac{69}{100}$

Exercise 19

Use the squares to find each fraction:

a

$\frac{1}{10}$ of 10 = ______ $\frac{4}{10}$ of 10 = ______ $\frac{7}{10}$ of 10 = ______

$\frac{2}{10}$ of 10 = ______ $\frac{5}{10}$ of 10 = ______ $\frac{8}{10}$ of 10 = ______

$\frac{3}{10}$ of 10 = ______ $\frac{6}{10}$ of 10 = ______ $\frac{9}{10}$ of 10 = ______

b

$\frac{1}{2}$ of 10 = ______

Exercise 20

Use the squares to find each fraction:

a

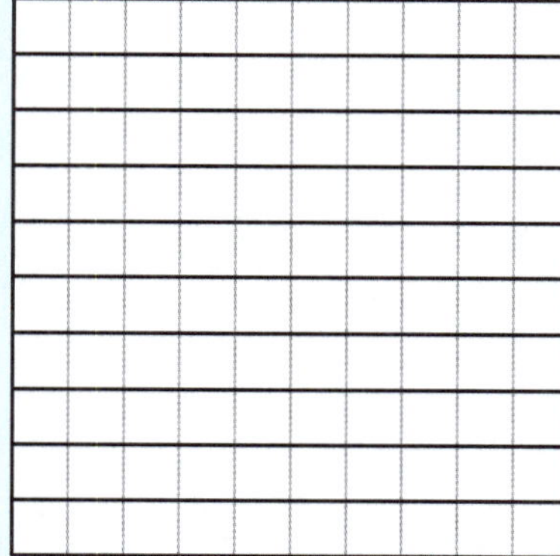

$\frac{1}{10}$ of 100 = ______ $\frac{6}{10}$ of 100 = ______

$\frac{2}{10}$ of 100 = ______ $\frac{7}{10}$ of 100 = ______

$\frac{3}{10}$ of 100 = ______ $\frac{8}{10}$ of 100 = ______

$\frac{4}{10}$ of 100 = ______ $\frac{9}{10}$ of 100 = ______

$\frac{5}{10}$ of 100 = ______

b

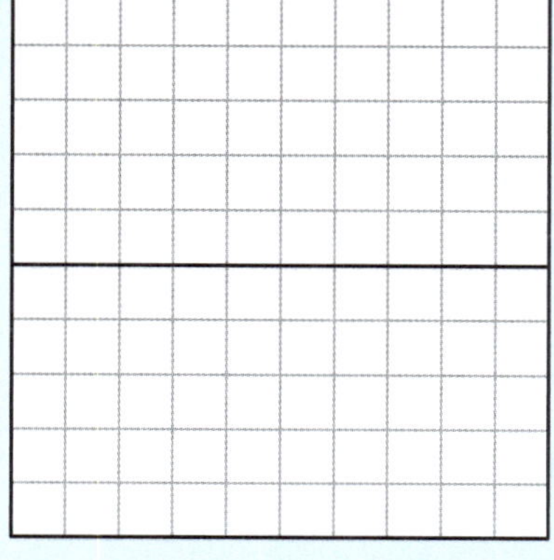

$\frac{1}{2}$ of 100 = ______

c

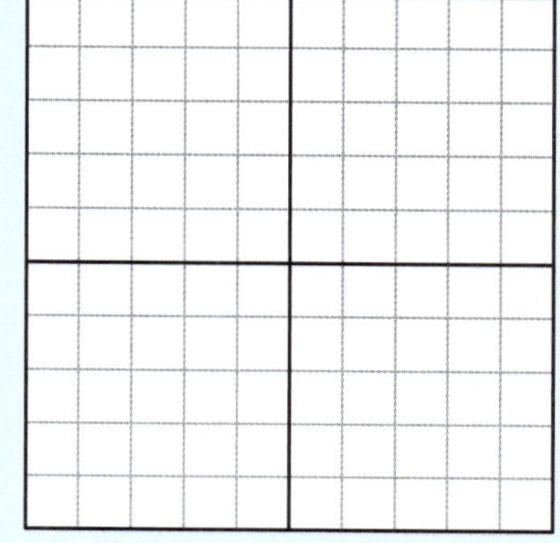

$\frac{1}{4}$ of 100 = ______

$\frac{2}{4}$ of 100 = ______

$\frac{3}{4}$ of 100 = ______

Exercise 21

Complete using a whole number.

a $\frac{4}{4} =$ ______ **b** $\frac{0}{5} =$ ______ **c** $\frac{9}{9} =$ ______ **d** $\frac{0}{10} =$ ______

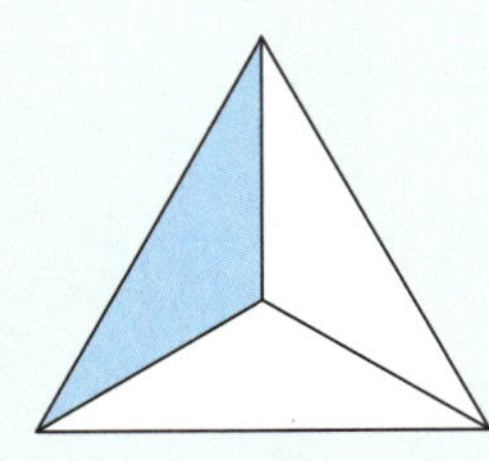

$\frac{1}{3}$ is shaded blue.

$\frac{2}{3}$ is not shaded.

$\frac{1}{3} + \frac{2}{3} = \frac{3}{3} = 1$

Exercise 22

Complete:

a

☐ is shaded.

☐ is not shaded.

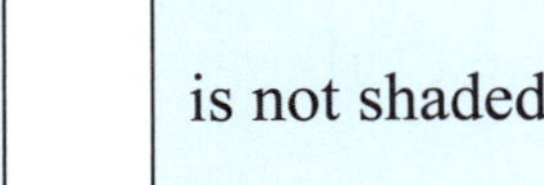

☐ + ☐ = ☐ = 1

b

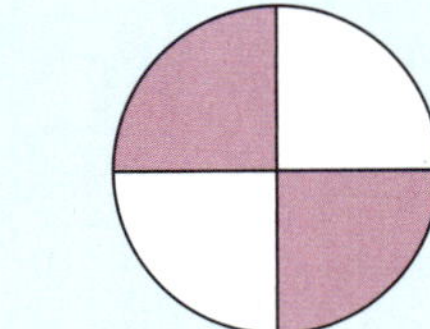

☐ is shaded.

☐ is not shaded.

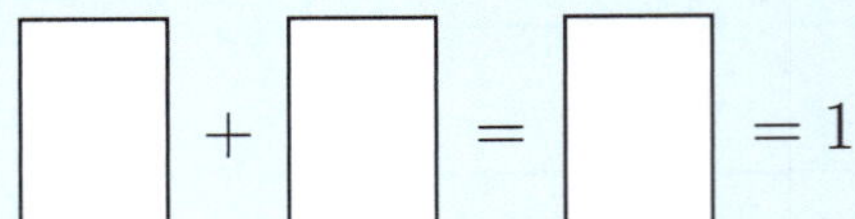

☐ + ☐ = ☐ = 1

c

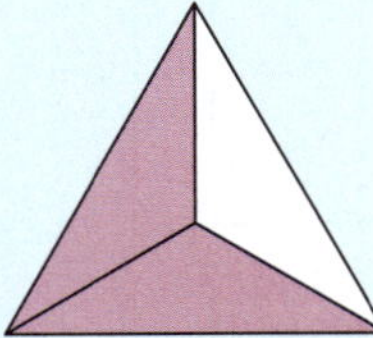

☐ is shaded.

☐ is not shaded.

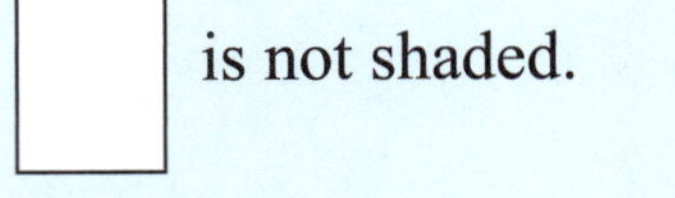

☐ + ☐ = ☐ = 1

d

☐ is shaded.

☐ is not shaded.

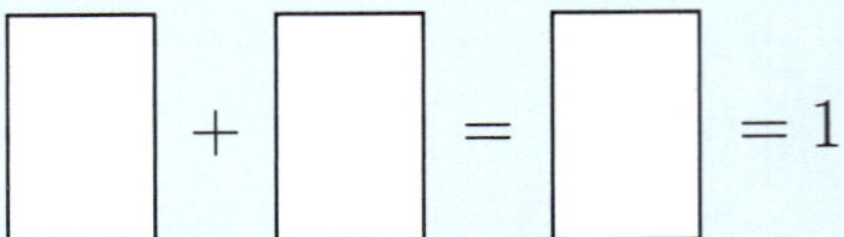

☐ + ☐ = ☐ = 1

e

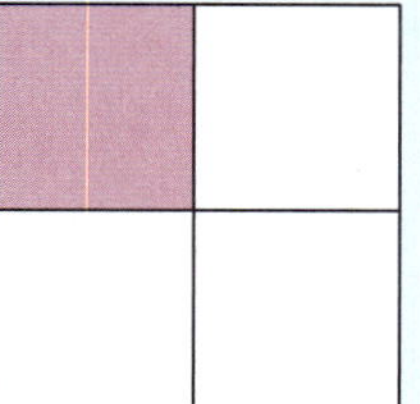

$\square$ is shaded.

$\square$ is not shaded.

$\square + \square = \square = 1$

f

$\square$ is shaded.

$\square$ is not shaded.

$\square + \square = \square = 1$

g

$\square$ is shaded.

$\square$ is not shaded.

$\square + \square = \square = 1$

h

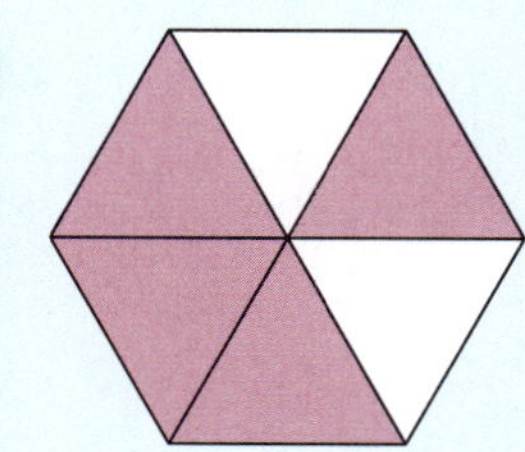

$\square$ is shaded.

$\square$ is not shaded.

$\square + \square = \square = 1$

i

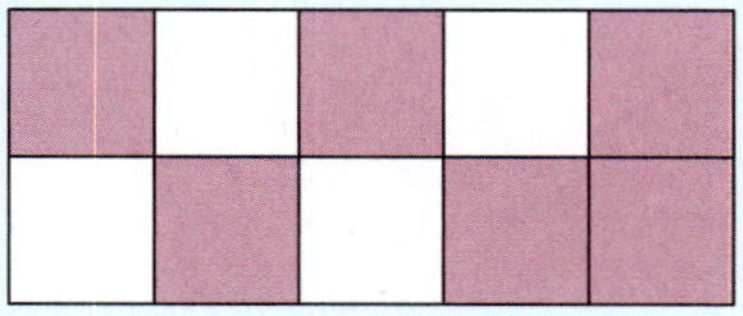

$\square$ is shaded.

$\square$ is not shaded.

$\square + \square = \square = 1$

Exercise 23

Complete:

a

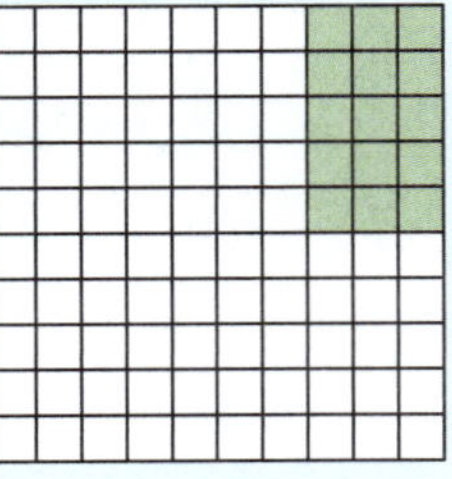

$\square$ is shaded.

$\square$ is not shaded.

$\square + \square = \square = 1$

b

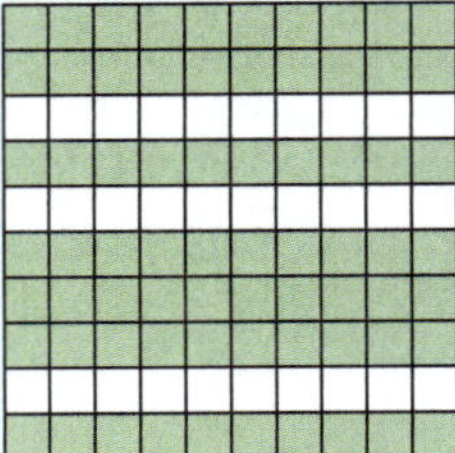

$\square$ is shaded.

$\square$ is not shaded.

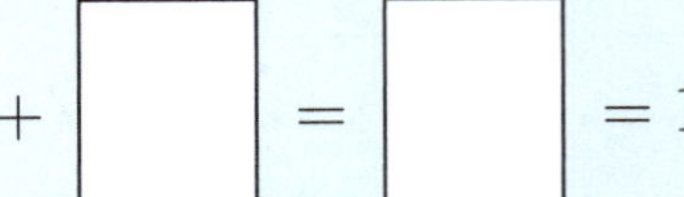

$\square + \square = \square = 1$

c

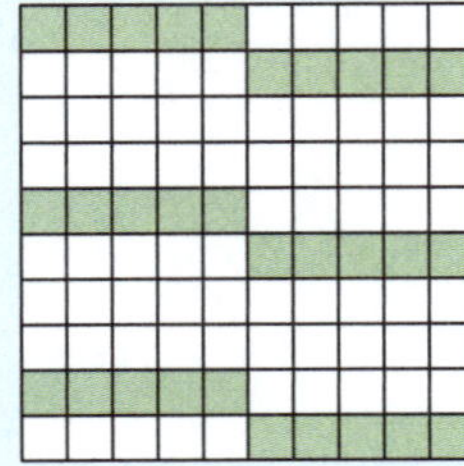

$\square$ is shaded.

$\square$ is not shaded.

$\square + \square = \square = 1$

d

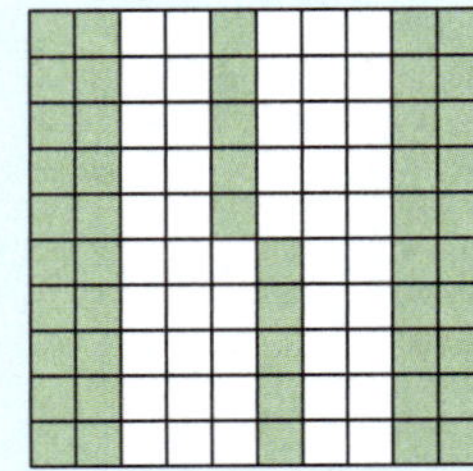

$\square$ is shaded.

$\square$ is not shaded.

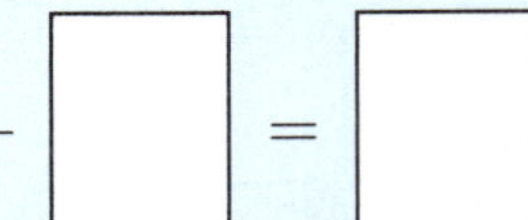

$\square + \square = \square = 1$

e

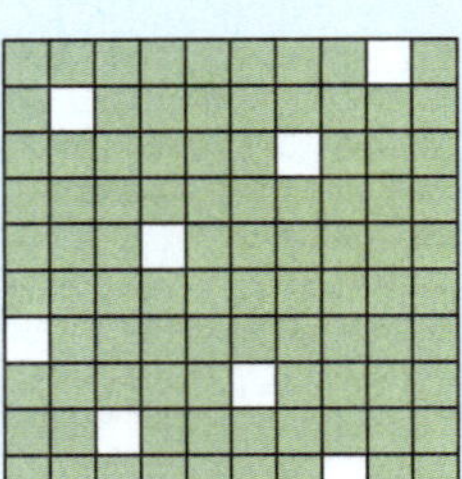

$\square$ is shaded.

$\square$ is not shaded.

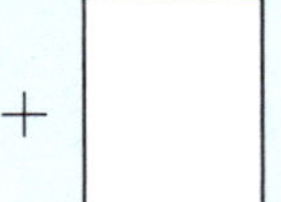

$\square + \square = \square = 1$

Exercise 24

Lewis ate $\frac{5}{6}$ of a sausage roll.

a Draw a rectangle and shade $\frac{5}{6}$.

b What fraction of the sausage roll does Lewis have left?

c Show how the two fractions add up to 1: $\square + \square = \square = 1$

Exercise 25

Tallulah ate $\frac{3}{8}$ of her ice block.

a Draw a rectangle and shade $\frac{3}{8}$.

b What fraction of the ice block does Tallulah have left?

c Show how the two fractions add up to 1: $\square + \square = \square = 1$

Puzzle

Leanne cut her apple pie into 6 equal pieces.

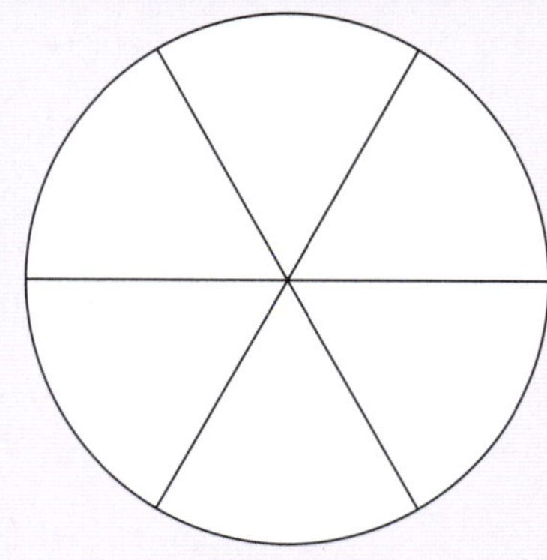

Her brother ate $\frac{1}{2}$ of the pie.

Shade her brother's portion blue.

Leanne ate $\frac{1}{3}$ of the pie.

Shade Leanne's portion red.

The fraction of the pie remaining is 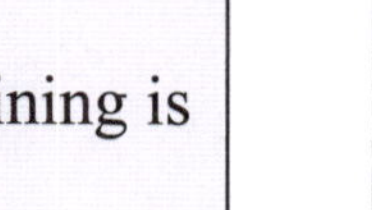.

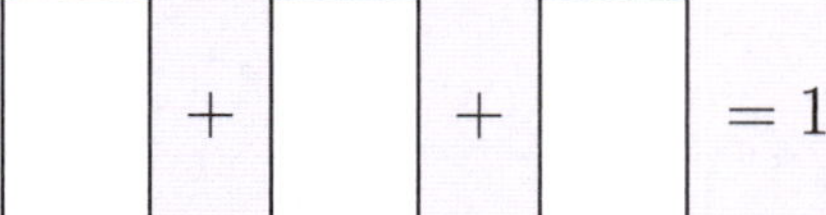

We can place whole numbers on a number line.

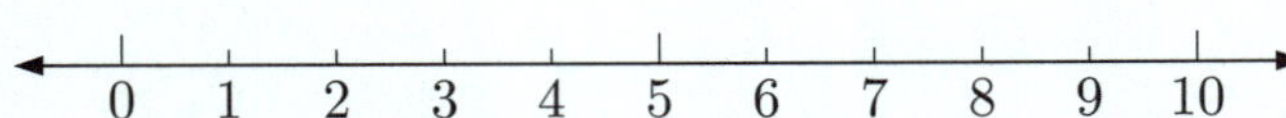

We can also place fractions on a number line by dividing the space between 0 and 1 into *equal parts*.

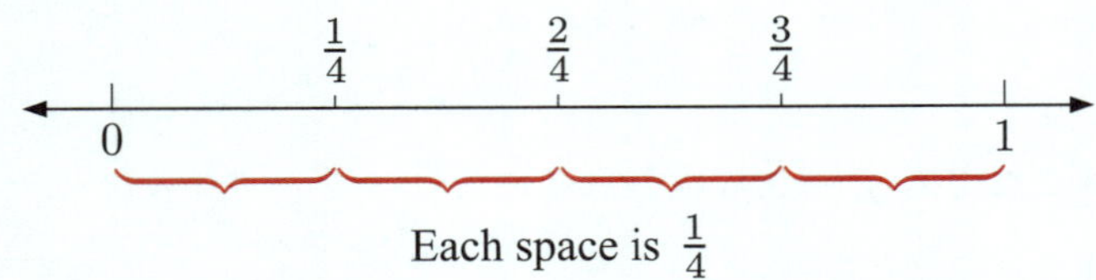

Exercise 26

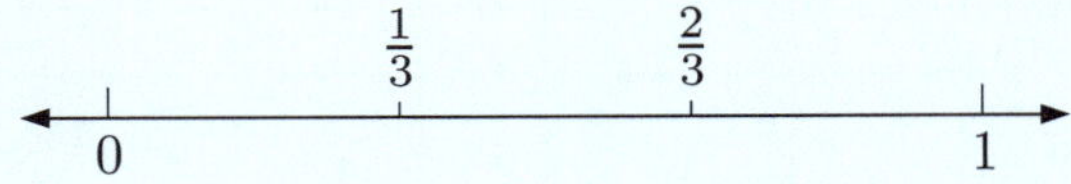

On this number line, the space between 0 and 1 is divided into ______ equal parts.

Each space between the marks is 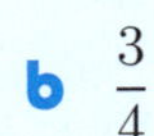.

Exercise 27

Place each fraction on the number line:

a $\frac{2}{3}$

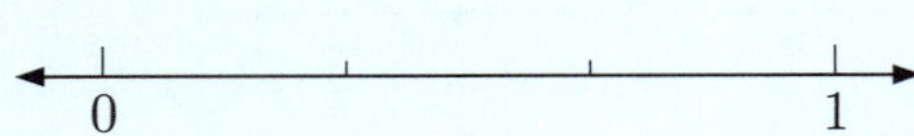

b $\frac{3}{4}$

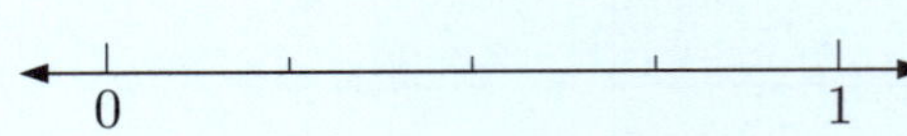

c $\frac{2}{5}$, $\frac{4}{5}$

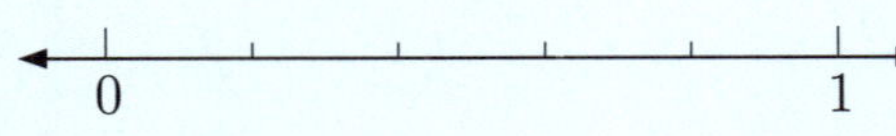

d $\frac{1}{6}$, $\frac{4}{6}$

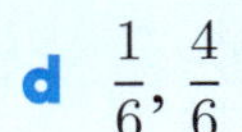

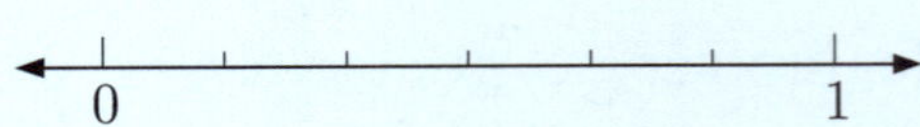

e $\frac{3}{5}$, $\frac{5}{5}$

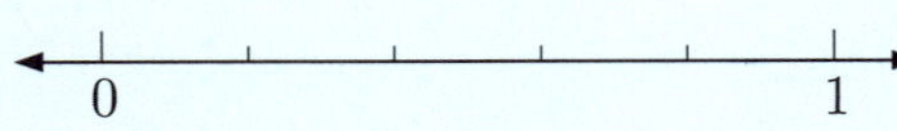

f $\frac{1}{8}$, $\frac{5}{8}$

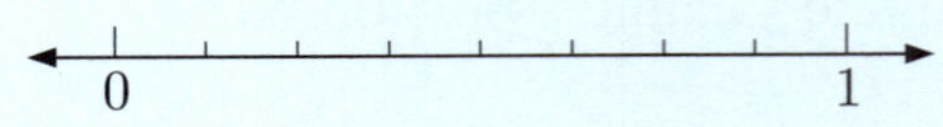

g $\frac{2}{8}$, $\frac{6}{8}$

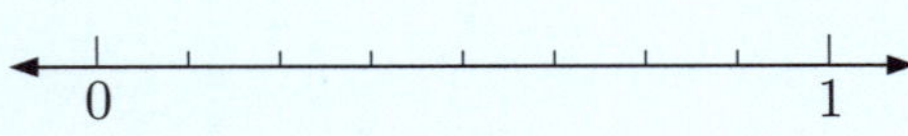

h $\frac{1}{10}$, $\frac{7}{10}$

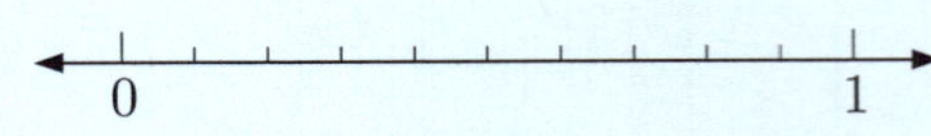

i $\frac{2}{5}, \frac{3}{5}, \frac{4}{5}$

j $\frac{1}{6}, \frac{3}{6}, \frac{5}{6}$

k $\frac{2}{8}, \frac{3}{8}, \frac{7}{8}$

l $\frac{2}{10}, \frac{5}{10}, \frac{6}{10}, \frac{9}{10}$

Exercise 28

Label each red dot with the fraction it represents:

a

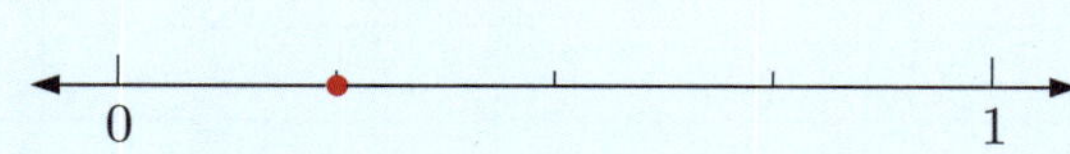

b

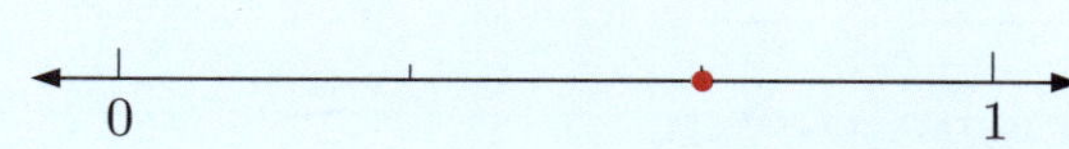

c

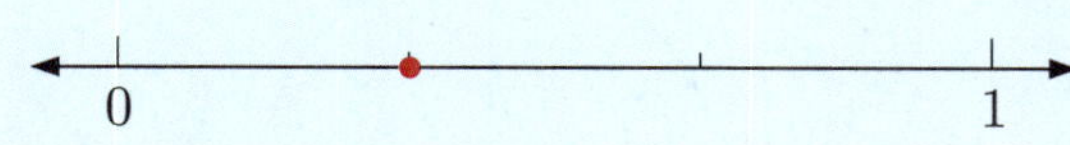

d

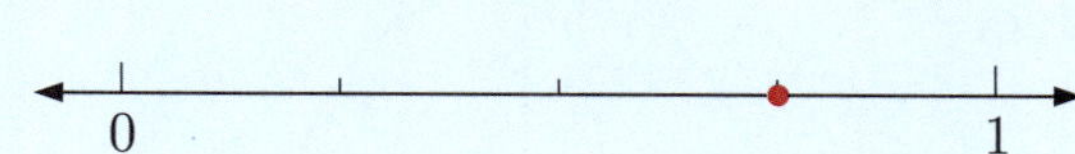

e

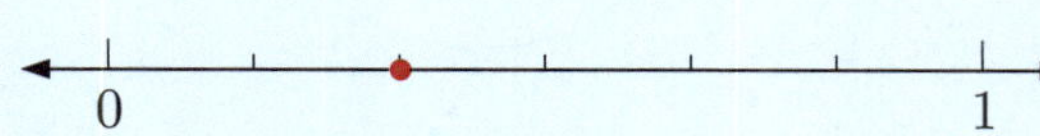

f

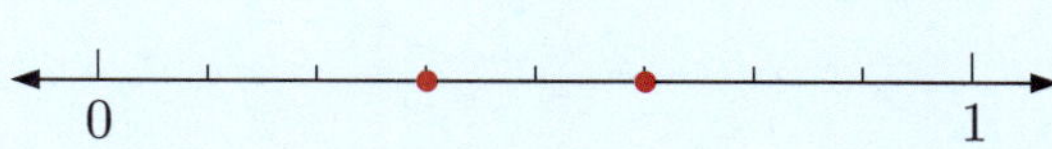

g

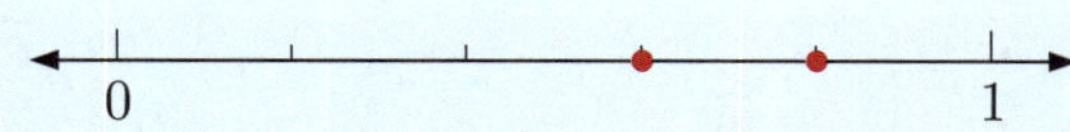

h

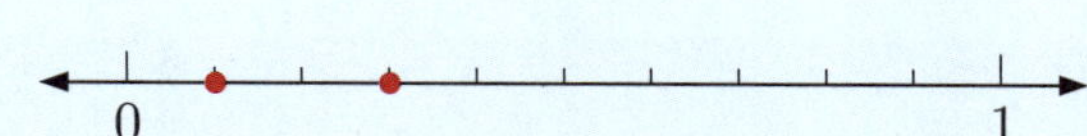

i

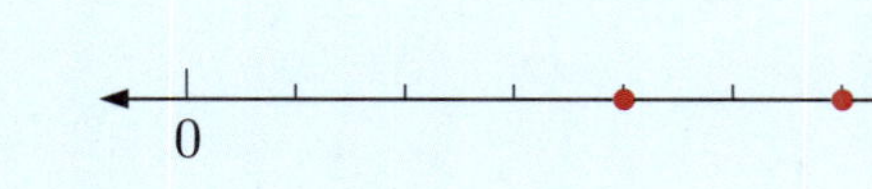

j

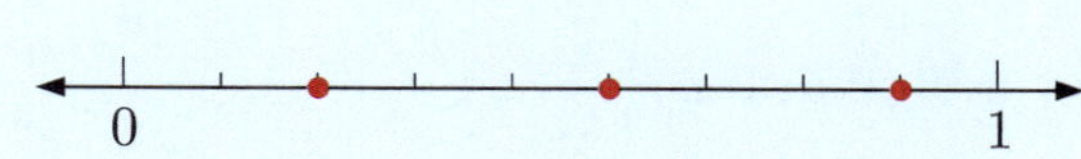

k

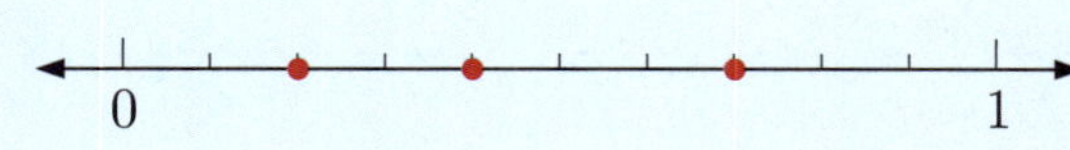

Two fractions are **equal** if they describe the same amount.

Exercise 29

Shade $\frac{1}{2}$.

Shade $\frac{2}{4}$.

$\frac{1}{2}$ is equal to $\frac{2}{4}$ because they describe ....

Discussion

Draw a red dot at $\frac{1}{2}$.

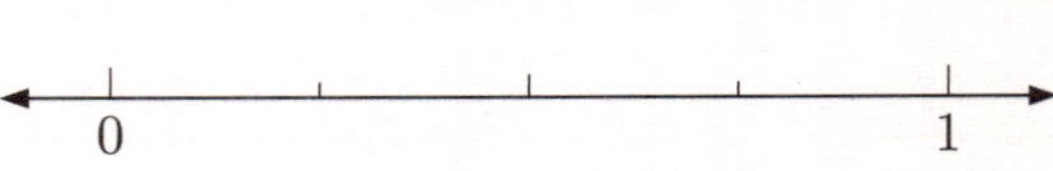

How can we tell from the number line that $\frac{1}{2}$ and $\frac{2}{4}$ are equal?

$\frac{1}{2}$ is equal to $\frac{2}{4}$ because they are at .... on the number line.

Exercise 30

Shade $\frac{1}{2}$.

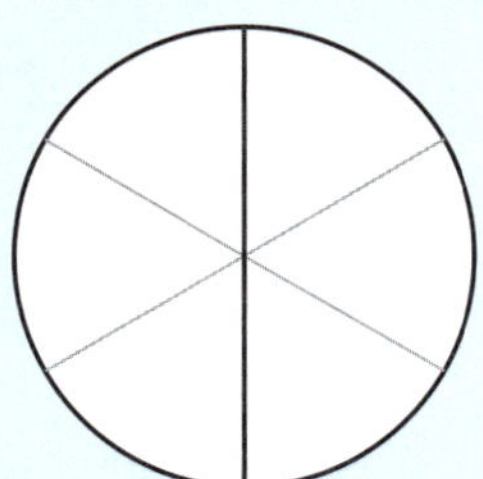

Shade $\frac{2}{3}$.

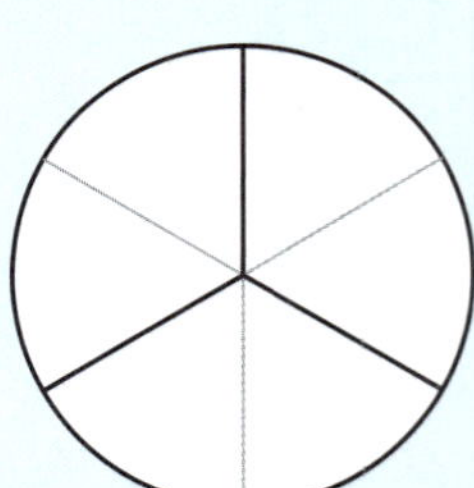

$\frac{1}{2}$ is not equal to $\frac{2}{3}$ because they describe .... amounts.

Discussion

Draw a red dot at $\frac{1}{2}$.

Draw a blue dot at $\frac{1}{3}$.

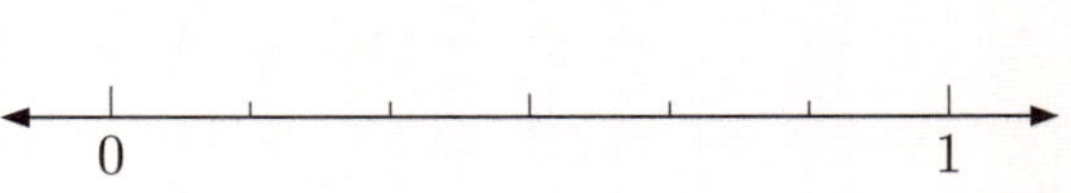

$\frac{1}{2}$ is not equal to $\frac{1}{3}$ because they are at .... on the number line.

Activity — Fraction wall

This diagram is called a **fraction wall**.

Each layer shows one whole.

Most of the layers have been divided into **fractions**.

1 whole							
$\frac{1}{2}$	$\frac{1}{2}$						
$\frac{1}{3}$	$\frac{1}{3}$	$\frac{1}{3}$					
$\frac{1}{4}$	$\frac{1}{4}$	$\frac{1}{4}$	$\frac{1}{4}$				
$\frac{1}{6}$	$\frac{1}{6}$	$\frac{1}{6}$	$\frac{1}{6}$	$\frac{1}{6}$	$\frac{1}{6}$		
$\frac{1}{8}$	$\frac{1}{8}$	$\frac{1}{8}$	$\frac{1}{8}$	$\frac{1}{8}$	$\frac{1}{8}$	$\frac{1}{8}$	$\frac{1}{8}$

You will need:

- scissors
- coloured pencils
- a print out of the fraction wall

What to do:

1 Colour each layer of the wall a different colour.

2 Use your scissors to carefully cut out each *layer* of the wall. *Do not cut out the individual bricks*.

3 Compare the halves with the quarters.

$\frac{1}{2}$	$\frac{1}{2}$		
$\frac{1}{4}$	$\frac{1}{4}$	$\frac{1}{4}$	$\frac{1}{4}$

We can see that $\frac{1}{2} = \frac{\square}{4}$ and $\frac{2}{2} = \frac{\square}{4}$.

4 By comparing layers of your fraction wall, complete:

a $\frac{1}{3} = \frac{\square}{6}$ **b** $\frac{1}{2} = \frac{\square}{6}$ **c** $\frac{2}{3} = \frac{\square}{6}$

d $\frac{1}{4} = \frac{\square}{8}$ **e** $\frac{2}{4} = \frac{\square}{8}$ **f** $\frac{3}{4} = \frac{\square}{8}$

g $\frac{4}{8} = \frac{\square}{2}$ **h** $\frac{4}{4} = \frac{\square}{6}$

Exercise 31

a Shade $\frac{1}{3}$.

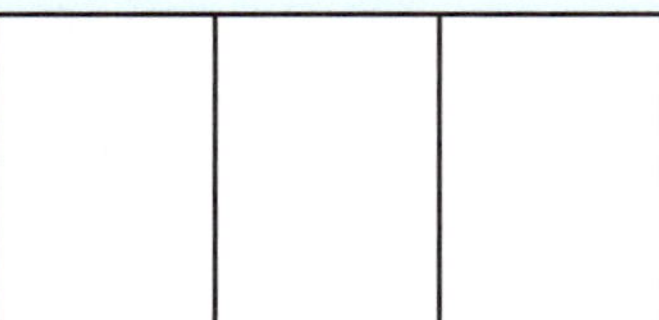

Shade $\frac{2}{6}$.

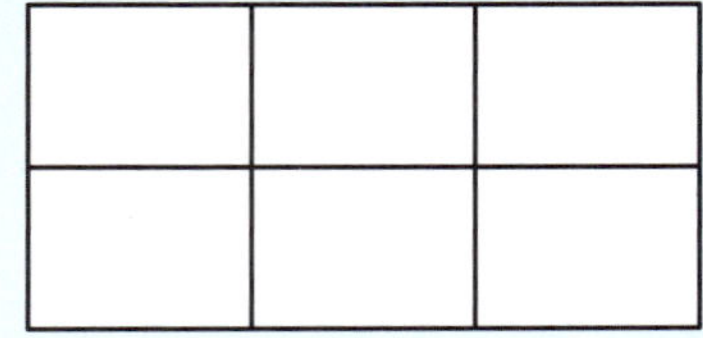

____ = ____

b Shade $\frac{2}{3}$.

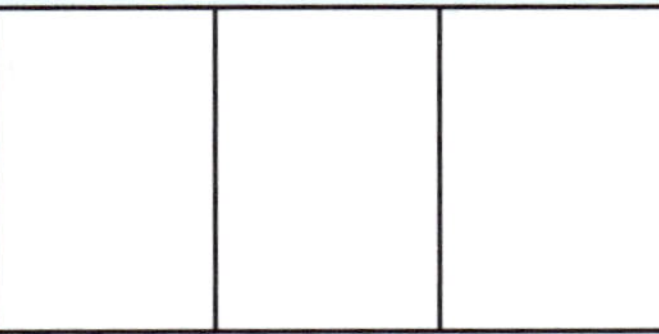

Shade $\frac{4}{6}$.

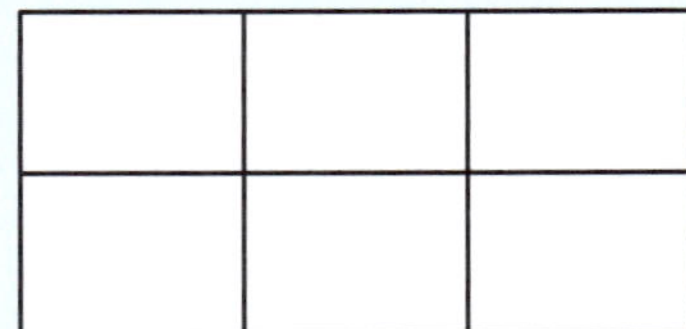

____ = ____

c Shade $\frac{1}{2}$.

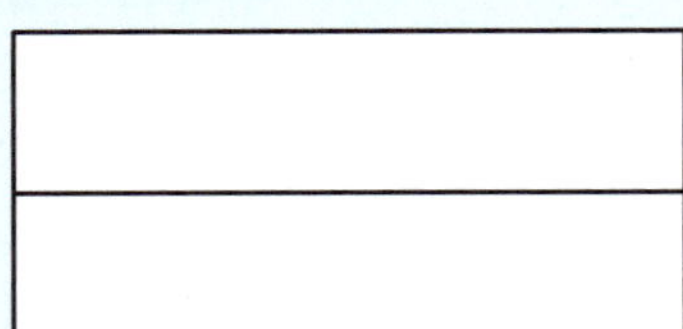

Shade $\frac{3}{6}$.

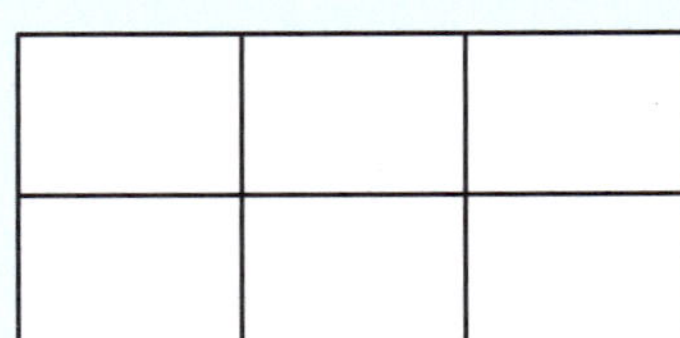

____ = ____

Mark the equal fractions on this number line.

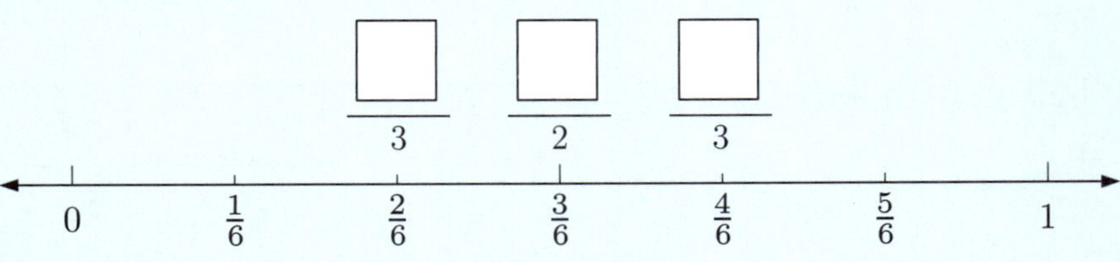

Discussion

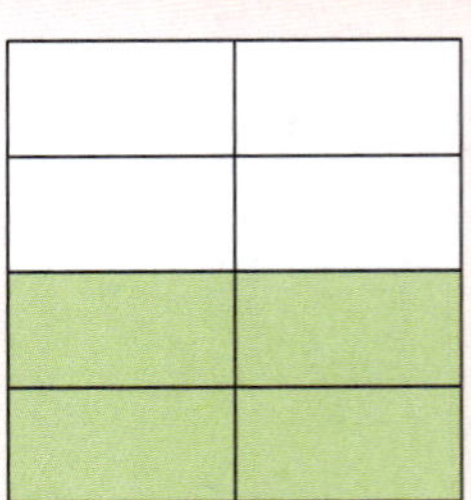

Do the fractions in these diagrams describe the same amount? ____

This means that ____ = ____ = ____.

Exercise 32

Write down the fraction shown in each diagram, then decide which fractions are equal:

a

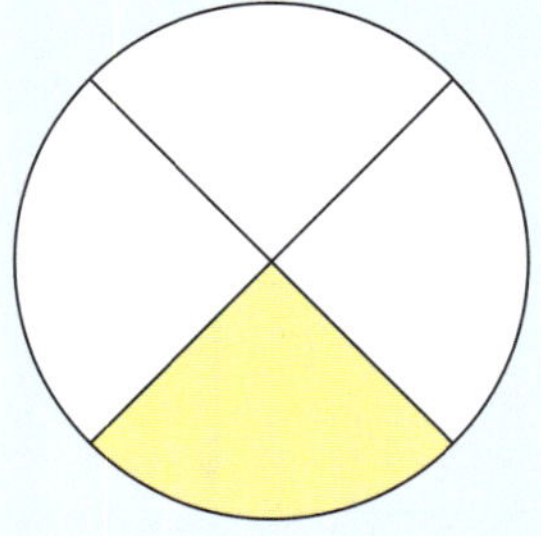

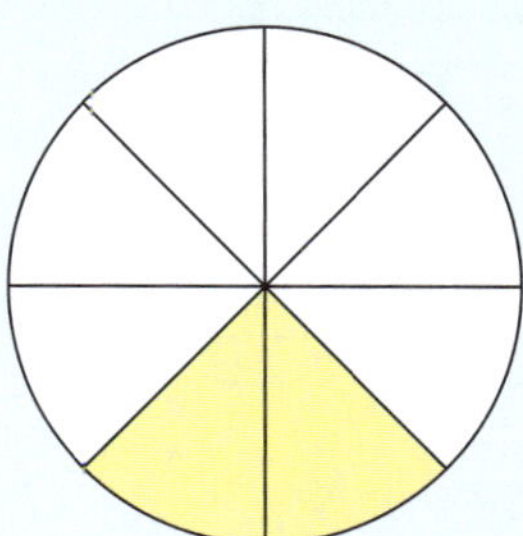

....

The fractions .... and .... are equal.

b

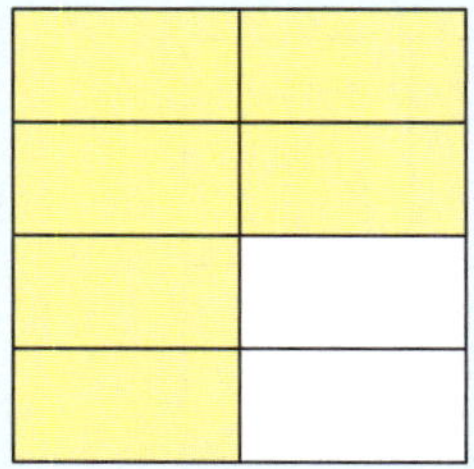

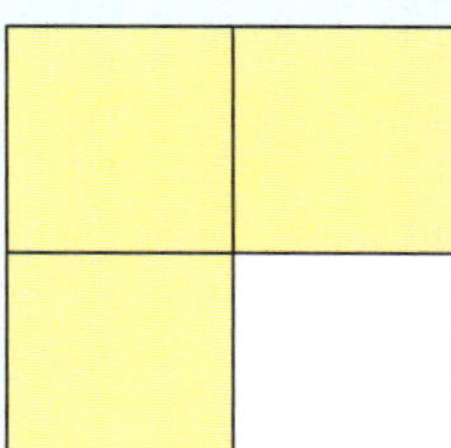

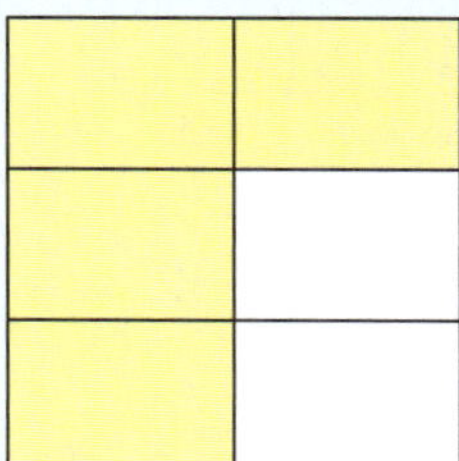

....

The fractions .... and .... are equal.

Discussion

What equal fractions can you write down from this diagram?

$$\frac{\square}{12} = \frac{\square}{6} = \frac{\square}{3}$$

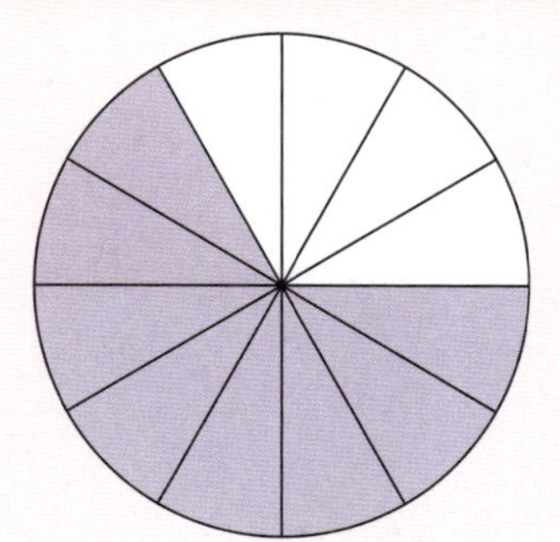

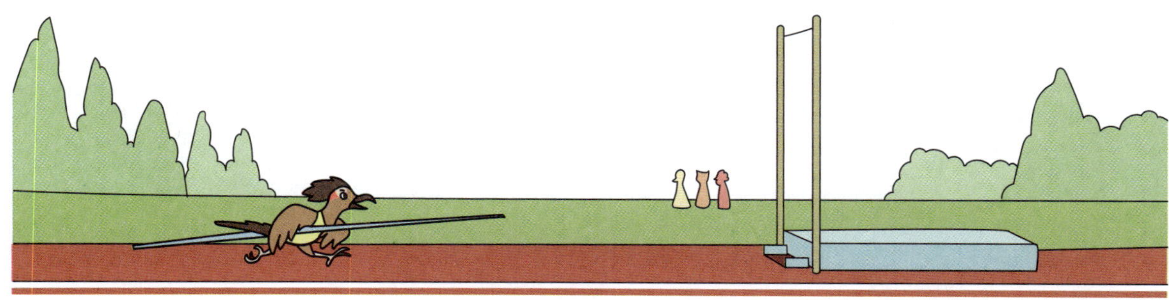

Exercise 33

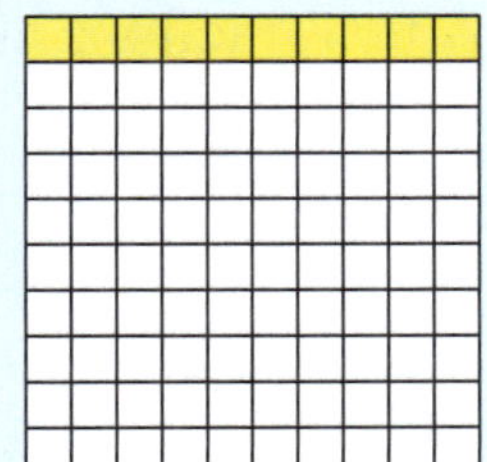

Complete:

$$\frac{1}{10} = \frac{10}{100}$$

$$\frac{2}{10} = \frac{\square}{100}$$

$$\frac{3}{10} = \frac{\square}{100}$$

$$\frac{\square}{10} = \frac{40}{100}$$

$$\frac{5}{10} = \frac{\square}{100}$$

$$\frac{\square}{10} = \frac{60}{100}$$

$$\frac{7}{10} = \frac{\square}{100}$$

$$\frac{\square}{10} = \frac{80}{100}$$

$$\frac{9}{10} = \frac{\square}{100}$$

$$\frac{\square}{10} = \frac{100}{100} = \square$$

Exercise 34

a Shade $\frac{1}{2}$.

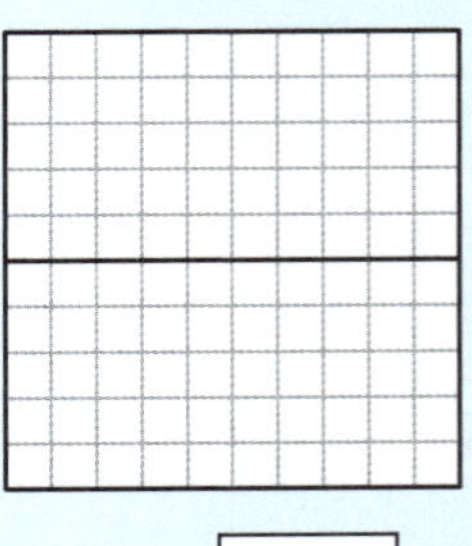

$$\frac{1}{2} = \frac{\square}{100}$$

b Shade $\frac{1}{4}$.

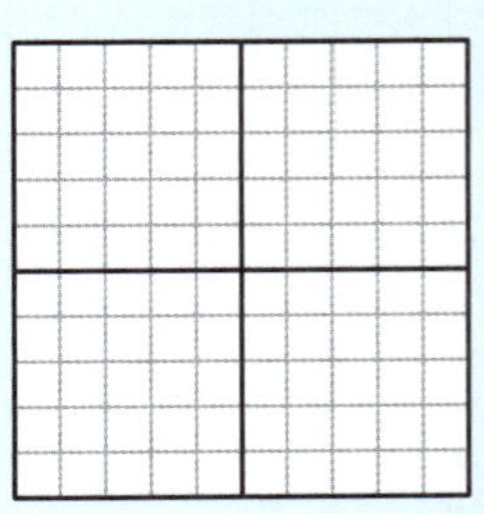

$$\frac{1}{4} = \frac{\square}{100}$$

c Shade $\frac{1}{5}$.

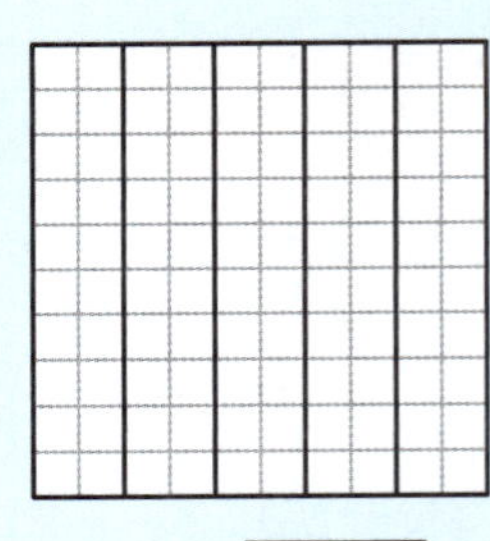

$$\frac{1}{5} = \frac{\square}{100}$$

Carys has 3 cupcakes. She cuts each cupcake in half.

In total, Carys has $3 \times 2 = 6$ halves.

$$\frac{6}{2} = 3$$

Since there is more than one whole, the numerator of $\frac{6}{2}$ is *greater* than the denominator.

A fraction which has numerator **greater** than its denominator is called an **improper fraction**.

Exercise 35

Complete each statement:

a

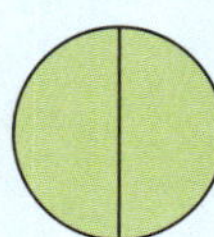
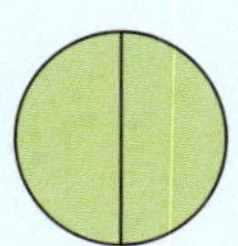

There are $3 \times 2 =$ ______ halves.

$3 = \frac{\square}{2}$

b

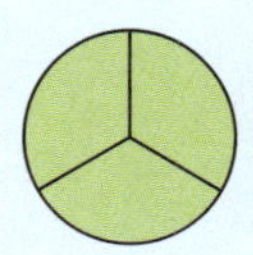
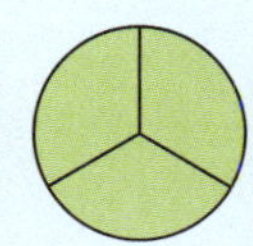
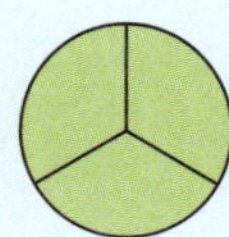
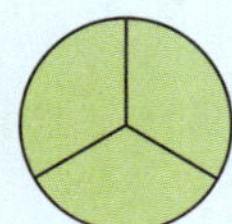

There are $4 \times$ ______ $=$ ______ thirds.

$4 = \frac{\square}{\square}$

c

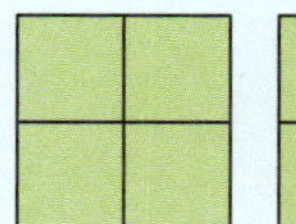
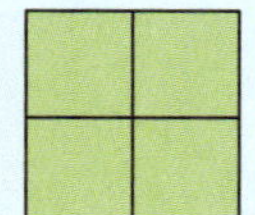

There are

$2 \times$ ______ $=$ ______ ________.

$2 = \frac{\square}{\square}$

d

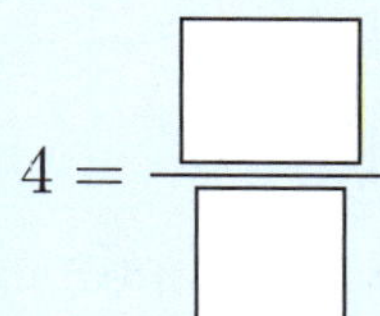
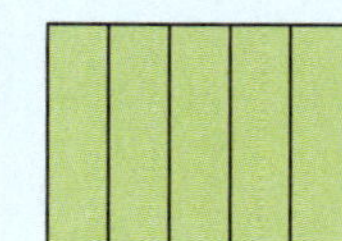
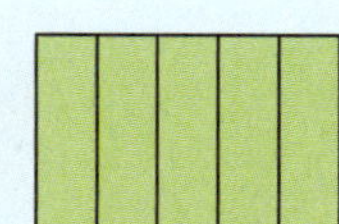
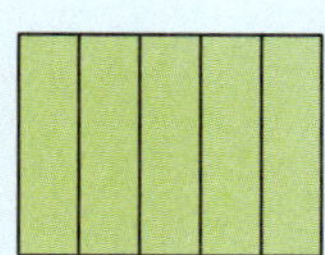

There are

$3 \times$ ______ $=$ ______ ________.

$3 = \frac{\square}{\square}$

e

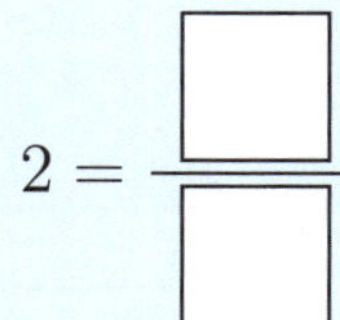
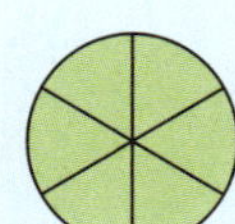
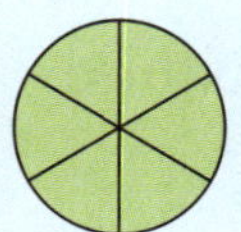
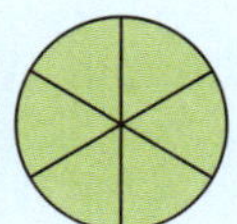

There are $4 \times$ ______ $=$ ______ ________.

$4 = \frac{\square}{\square}$

Exercise 36

Complete:

a In 2 wholes there are

$2 \times 3 = \square$ thirds.

$2 = \dfrac{\square}{3}$

b In 3 wholes there are

$3 \times \square = \square$ quarters.

$3 = \dfrac{\square}{4}$

Carys had 3 *whole* cupcakes, which is 6 *halves*.

She ate one half of a cupcake, so she has 5 halves left.

4 halves is the same as 2 wholes, so she has "two and a half" cupcakes left.

We write this as $2\frac{1}{2}$.

$2\frac{1}{2} = \frac{5}{2}$

$2\frac{1}{2}$ is a **mixed number** because it is made up of a whole number and a fraction.

Exercise 37

Draw a square around each improper fraction. Draw a circle around each mixed number.

$\frac{10}{2}$ $2\frac{3}{4}$ $\frac{8}{3}$ $4\frac{4}{5}$ $\frac{10}{4}$

$\frac{3}{5}$ $\frac{1}{6}$ 3 2 5

$3\frac{5}{10}$ $\frac{7}{3}$ $1\frac{1}{8}$ $\frac{6}{2}$ $\frac{9}{4}$

$\frac{2}{3}$ $\frac{3}{4}$

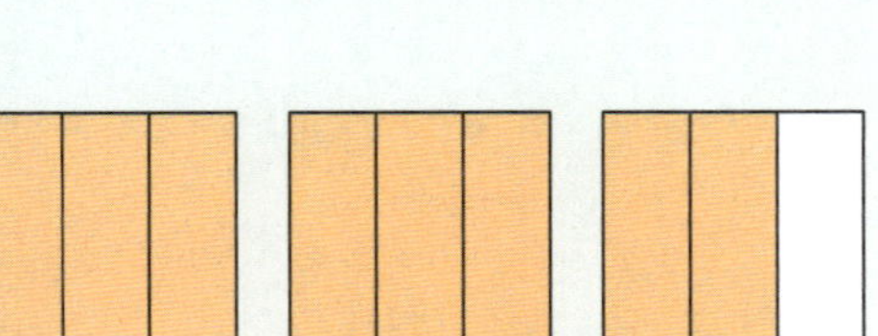

Each whole is divided into *thirds*.

Two wholes and two thirds are shaded. $2\frac{2}{3}$

Eight thirds are shaded. $\frac{8}{3}$

$2\frac{2}{3} = \frac{8}{3}$

Exercise 38

Write as a mixed number and as an improper fraction.

a 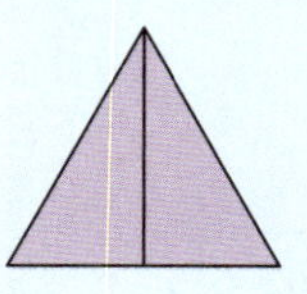 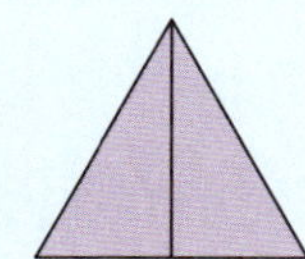 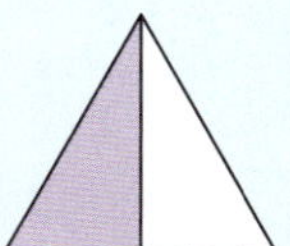

.... = ...

b 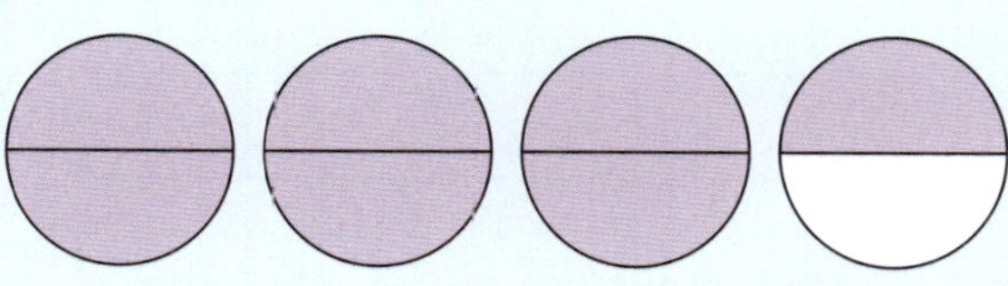

.... = ...

c 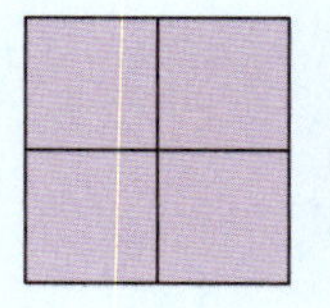 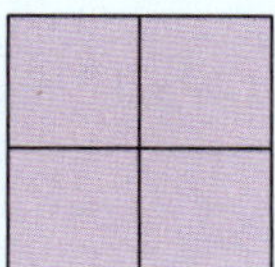 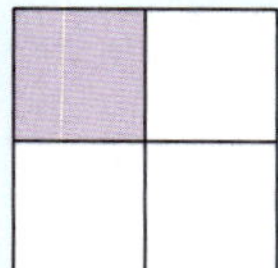

.... = ...

d 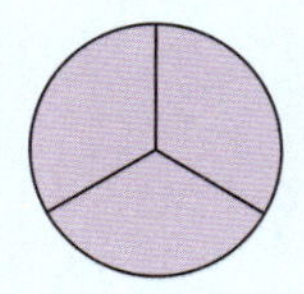 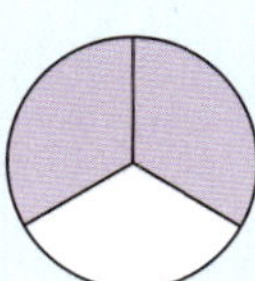

.... = ...

e 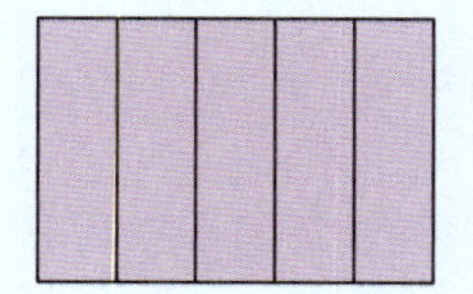 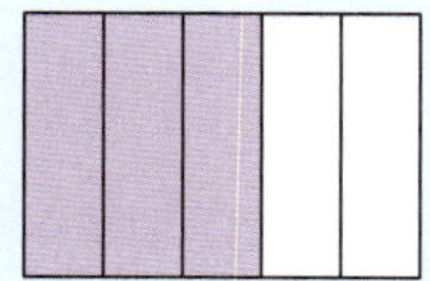

.... = ...

f 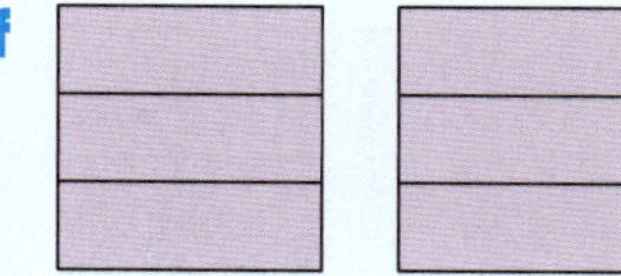 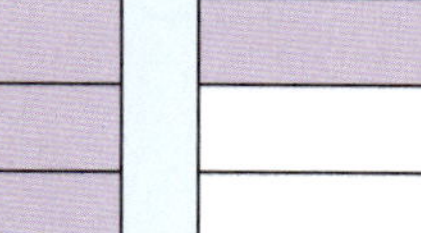

.... = ...

g 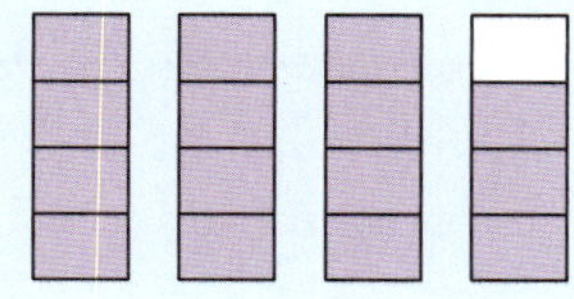

.... = ...

h 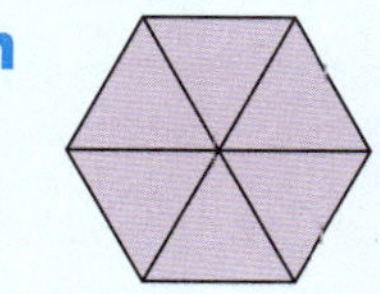 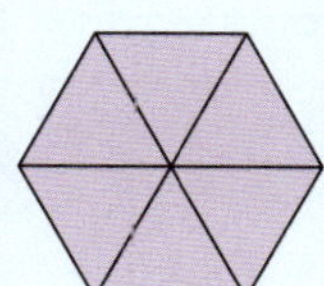 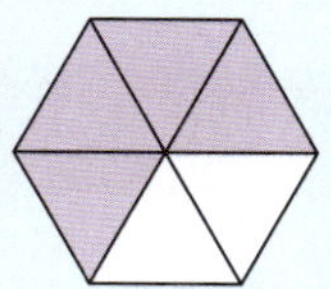

.... = ...

i 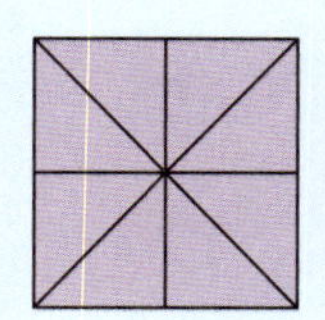 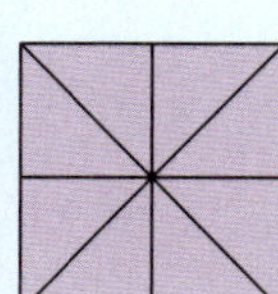 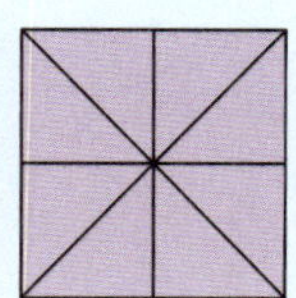 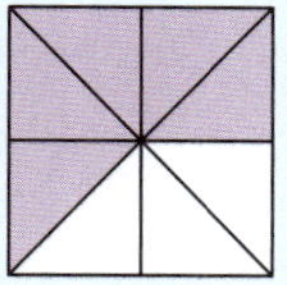

.... = ...

j 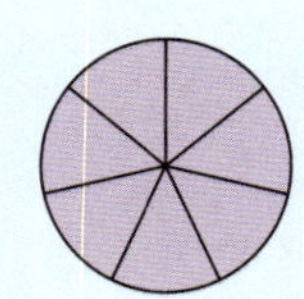 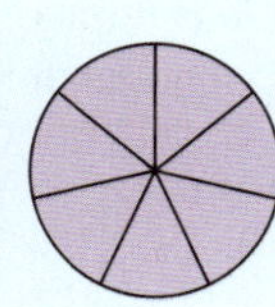 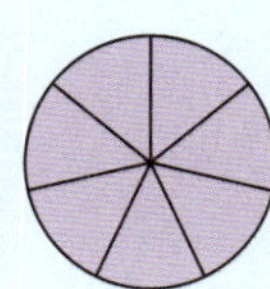 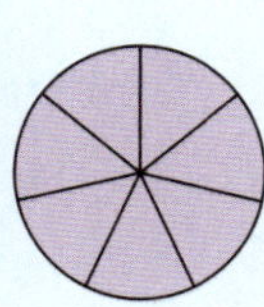 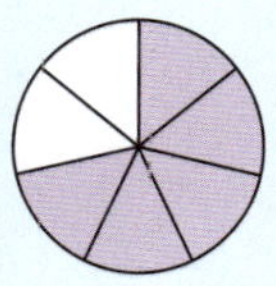

.... = ...

Exercise 39

Shade the mixed number.

a $2\frac{1}{4}$

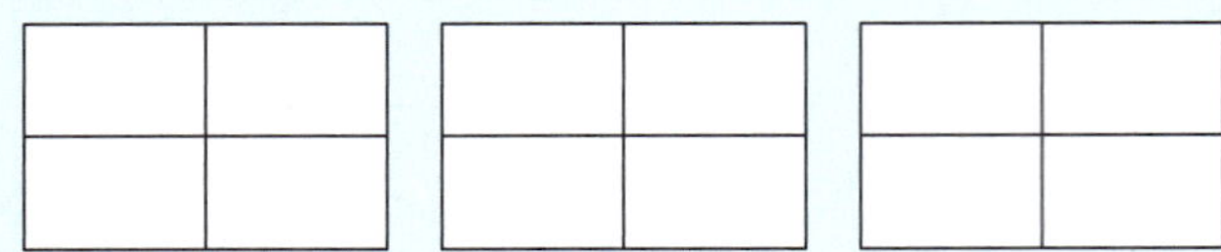

b $3\frac{1}{2}$

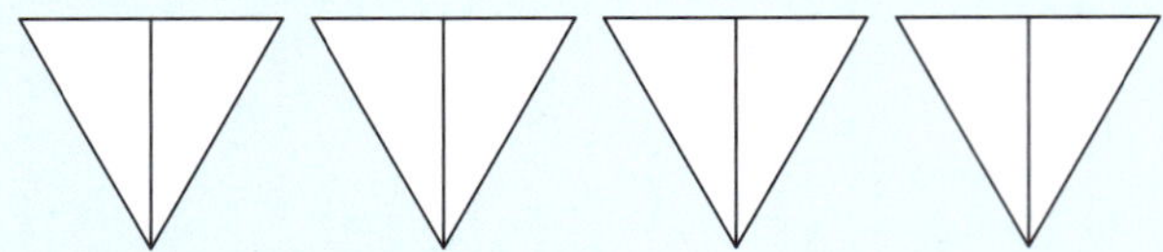

A mixed number is made up of a whole number and a fraction.

c $1\frac{1}{3}$

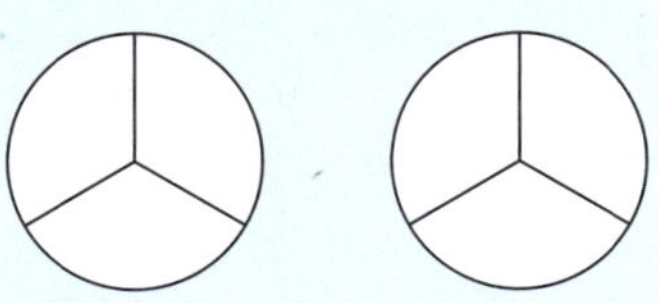

d $3\frac{1}{4}$

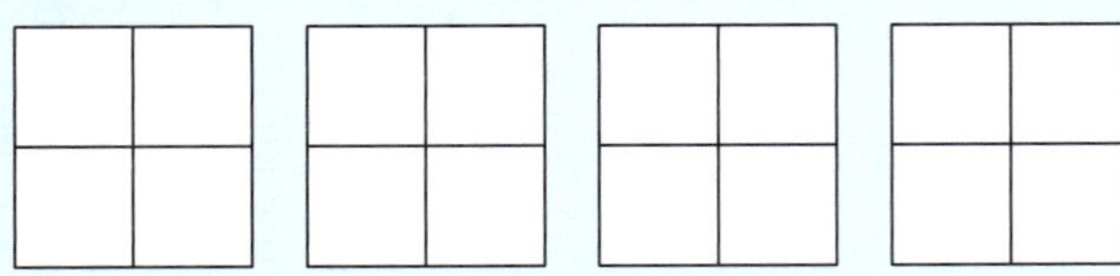

e $4\frac{2}{3}$

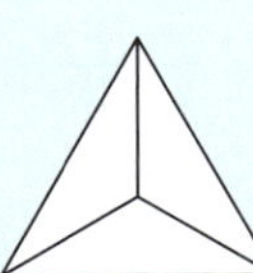

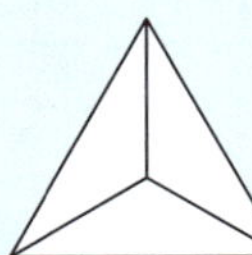
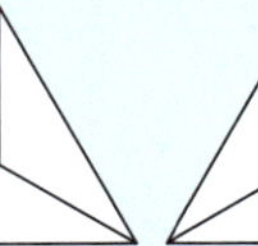
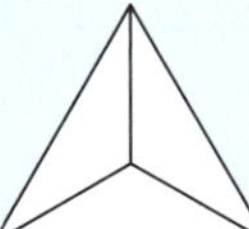
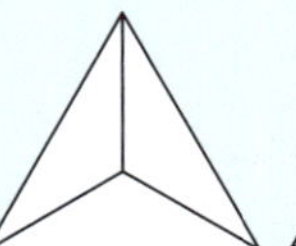
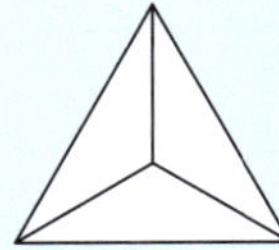

Exercise 40

Shade the improper fraction.

a $\frac{3}{2}$

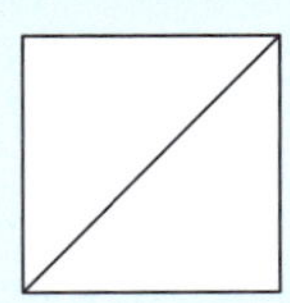
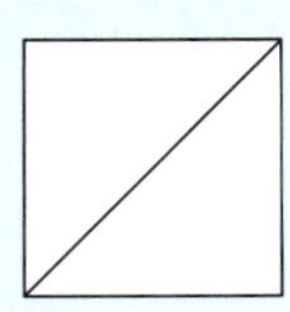

b $\frac{11}{4}$

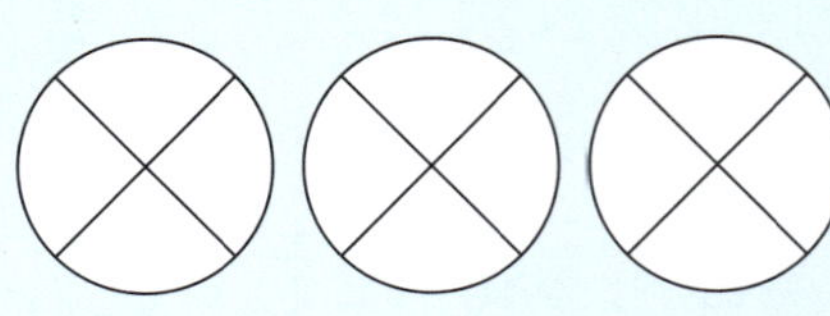

c $\frac{11}{6}$

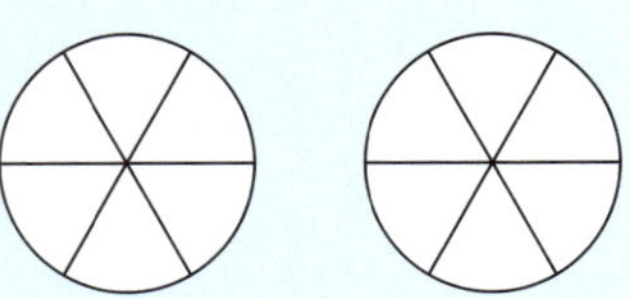

d $\frac{17}{4}$

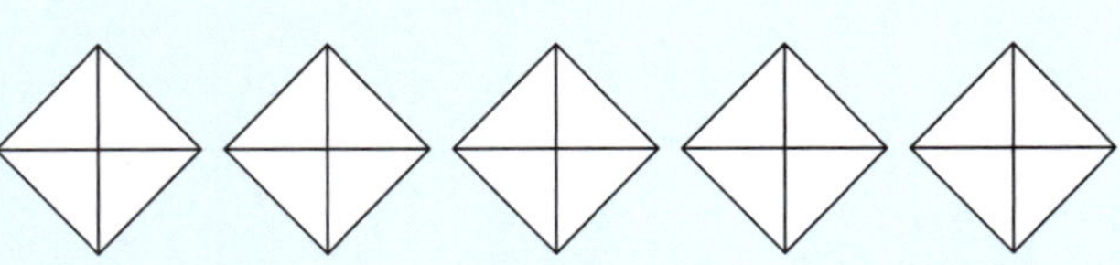

An improper fraction has numerator greater than its denominator.

e $\frac{14}{3}$

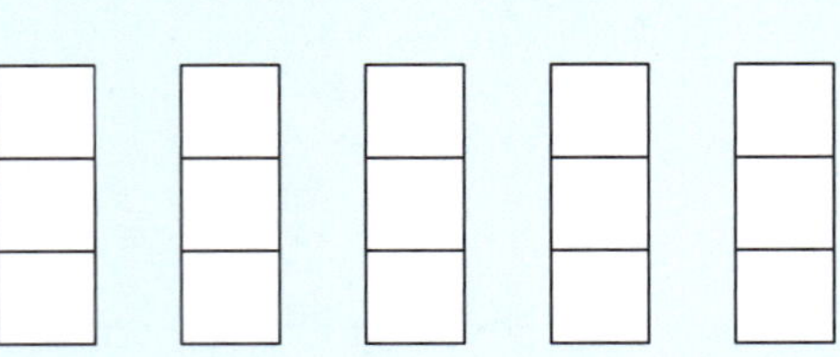

When working with tenths and hundredths, you can use blocks.

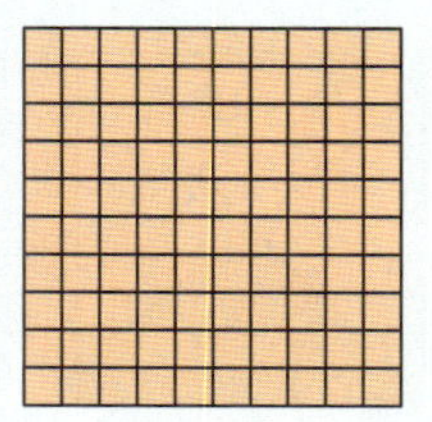

$1 = \frac{100}{100}$ $\frac{1}{10} = \frac{10}{100}$ $\frac{1}{100}$

Exercise 41

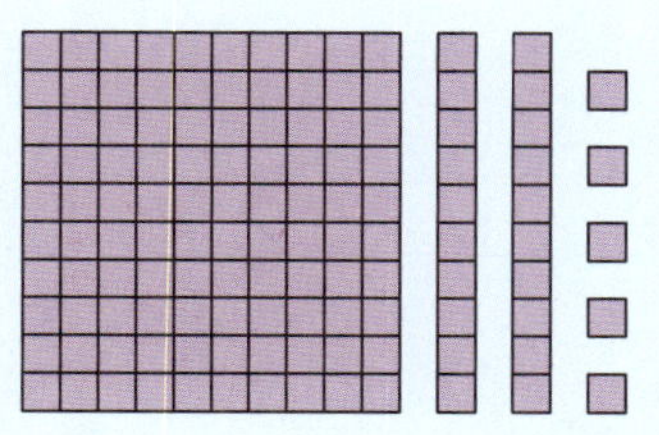

$1\frac{25}{100}$

Write as a mixed number.

a

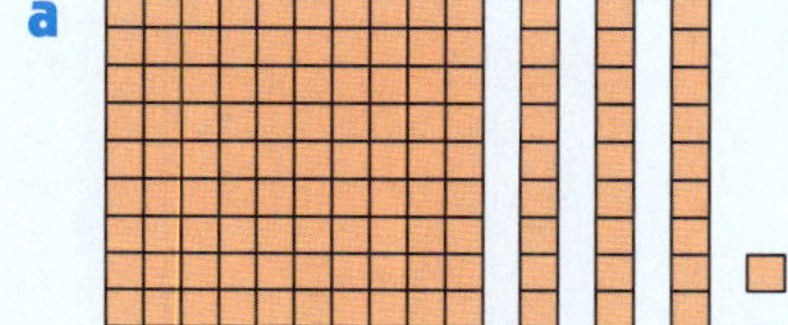

....

b

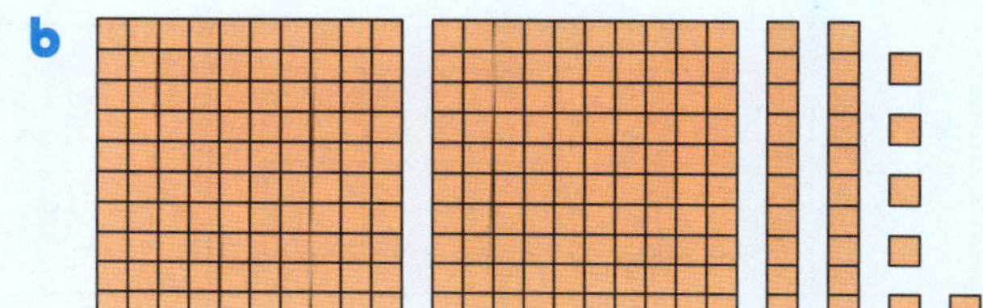

....

c

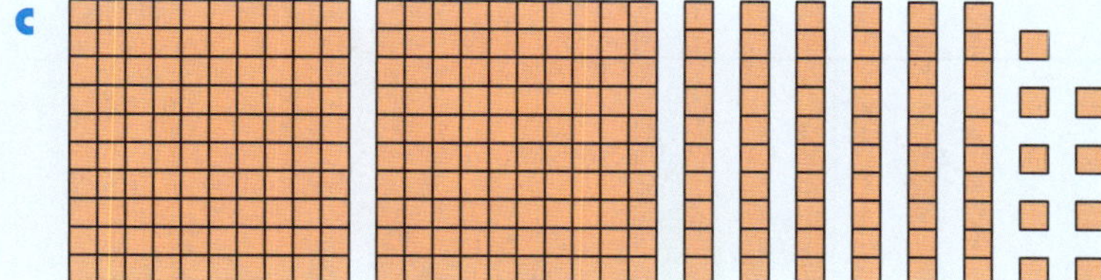

....

d

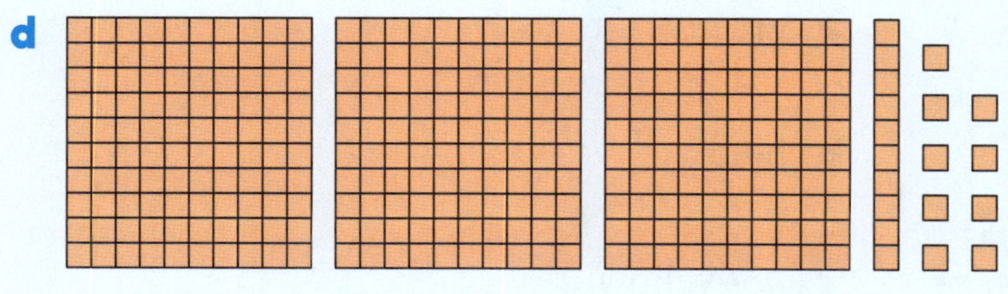

....

Revision

1 Explain why this figure does *not* show $\frac{1}{4}$.

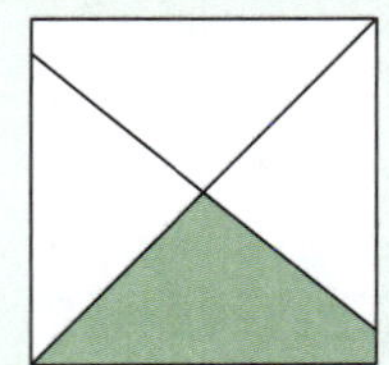

....

....

2 Complete the table of fractions.

Diagram	*Fraction*	*Words*
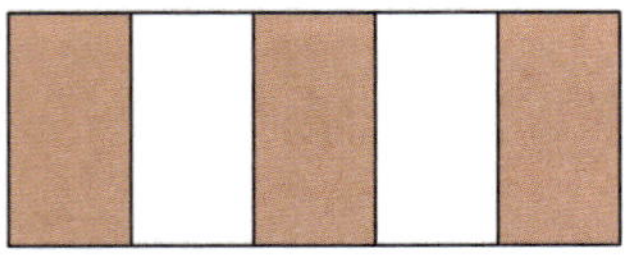	$\frac{3}{5}$	....
	....	one third
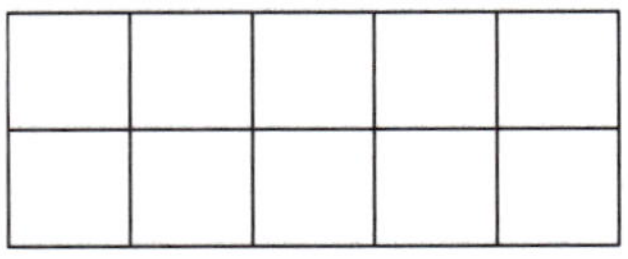	$\frac{4}{10}$	....
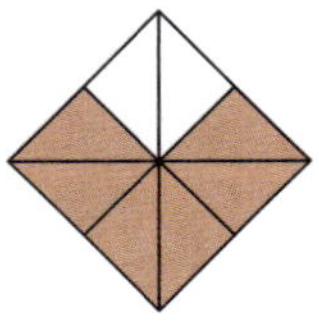	....	....
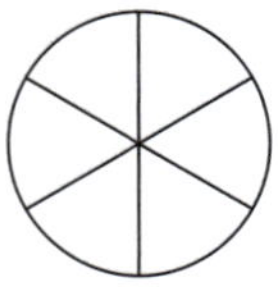	$\frac{2}{6}$	....

3

Write down the fraction of birds which are:

a green ....

b blue ....

c *not* yellow ....

4 **a** Write down the fraction.

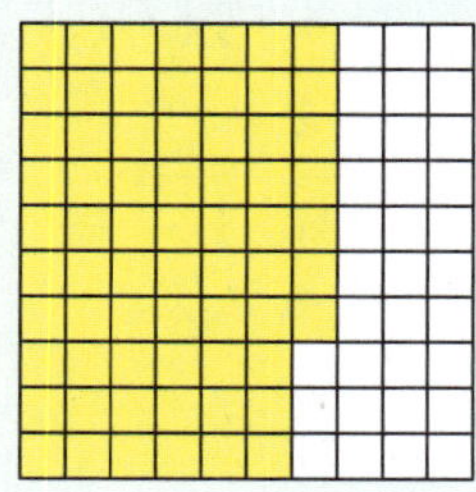

______

b Shade $\frac{43}{100}$.

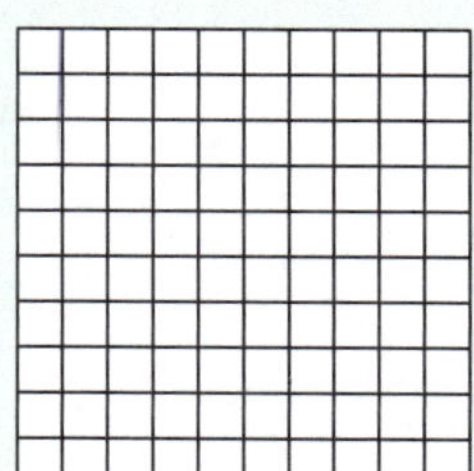

5 By shading squares, complete each statement:

a

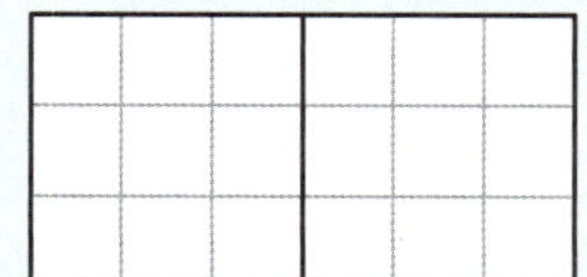

$\frac{1}{2}$ of 18 is ______

b

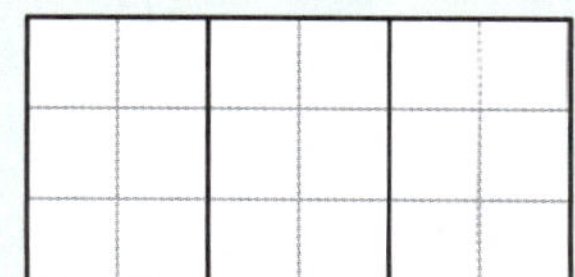

$\frac{2}{3}$ of 18 is ______

c

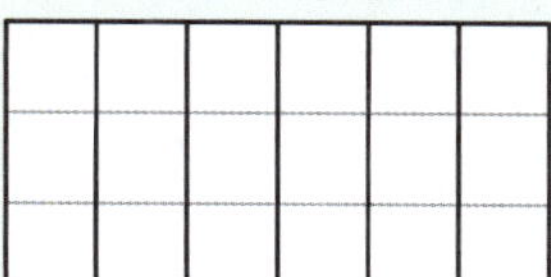

$\frac{5}{6}$ of 18 is ______

6 Complete:

a

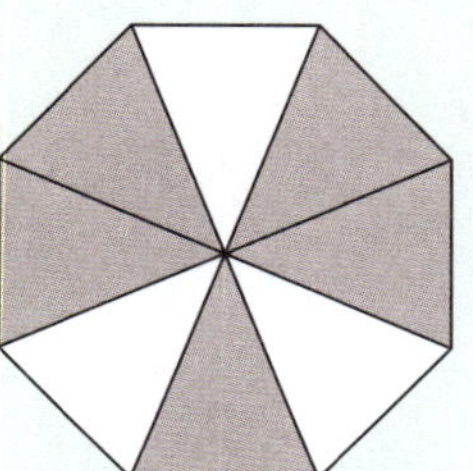

☐ is shaded.

☐ is not shaded.

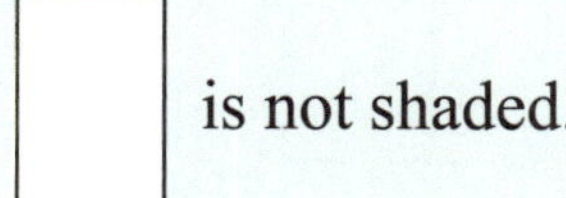

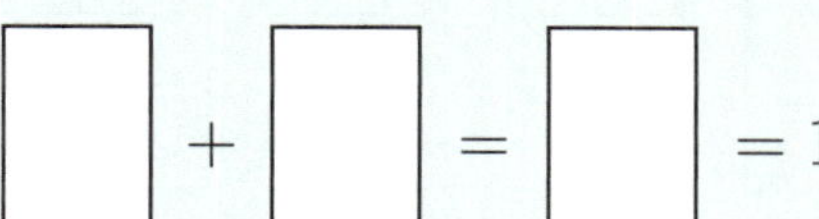

☐ + ☐ = ☐ = 1

b

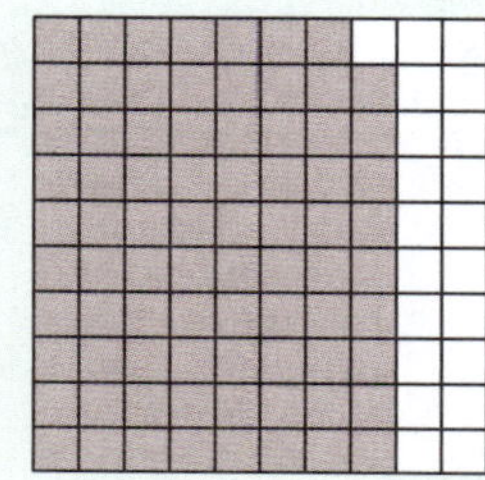

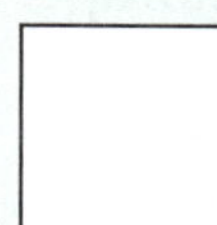

☐ is shaded.

☐ is not shaded.

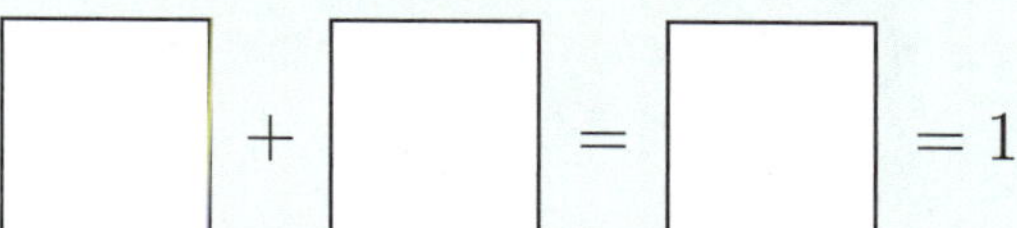

☐ + ☐ = ☐ = 1

7 Place each fraction on the number line:

a $\frac{2}{6}$, $\frac{5}{6}$

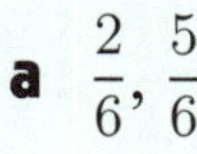

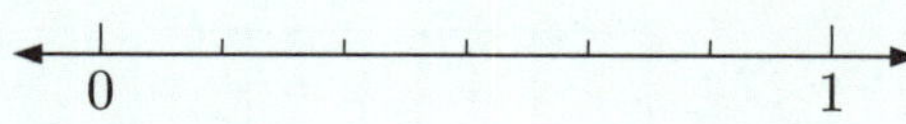

b $\frac{2}{10}$, $\frac{4}{10}$, $\frac{7}{10}$

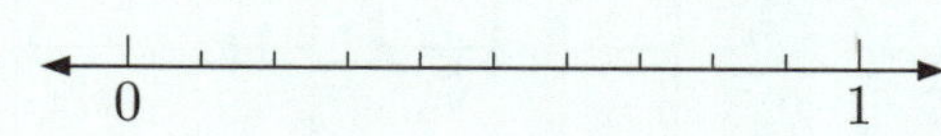

8 Label each red dot with the fraction it represents:

a

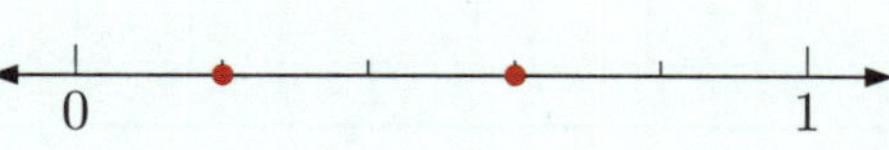

b

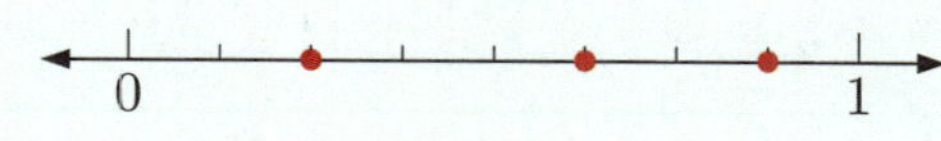

9 Shade $\frac{3}{4}$.

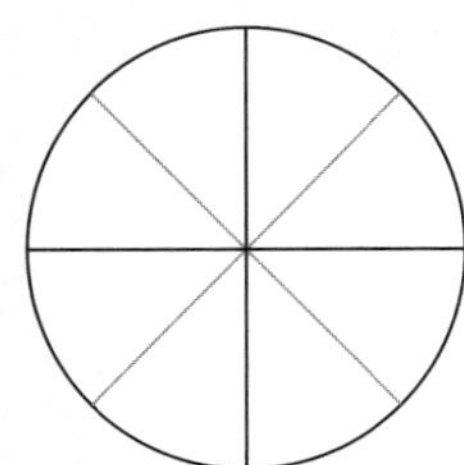

Shade $\frac{6}{8}$.

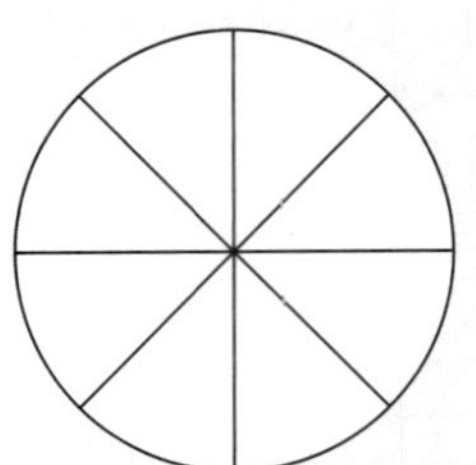

$\frac{3}{4}$ is equal to $\frac{6}{8}$ because they describe ________ ________ ______________.

10 Shade $\frac{2}{5}$.

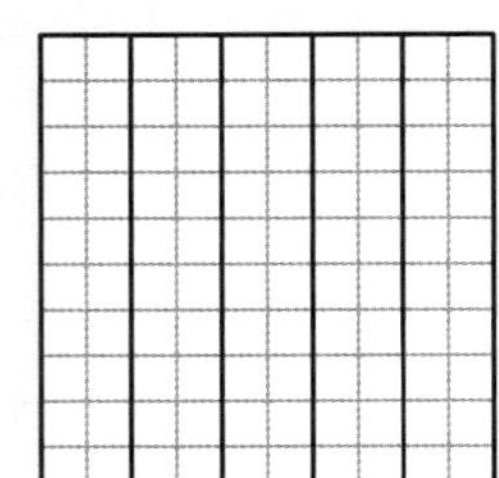

$\frac{2}{5} = \frac{\square}{100}$

11 Complete:

In 4 wholes there are $4 \times \square = \square$ halves.

$4 = \frac{\square}{2}$

12 Draw a square around each improper fraction. Draw a circle around each mixed number.

$3\frac{4}{8}$ $\quad 2 \quad$ $\frac{11}{6}$ $\quad$ $\frac{8}{5}$ $\quad$ $1\frac{1}{3}$ $\quad$ $\frac{21}{8}$ $\quad 4 \quad$ $3\frac{3}{10}$

13 Write as a mixed number and as an improper fraction.

a

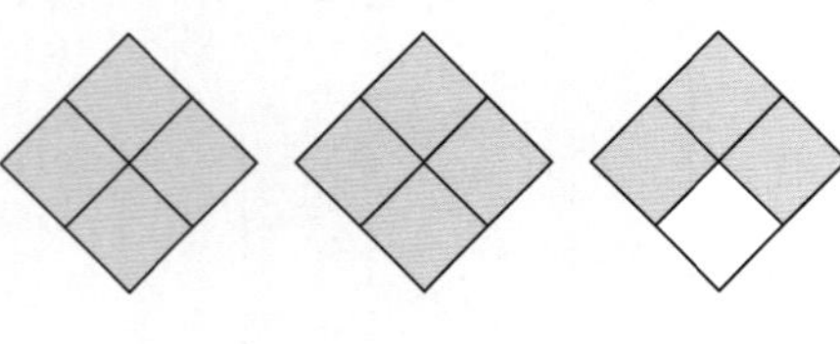

______ = ______

b

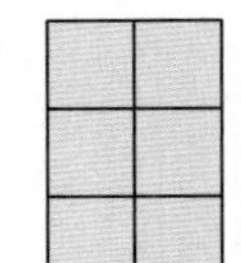 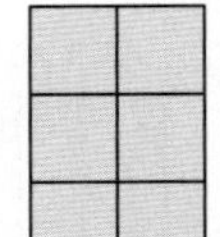 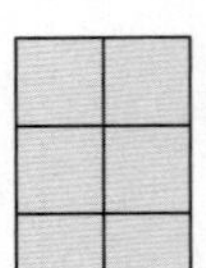 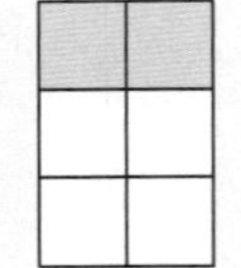

______ = ______

14 Shade the amount.

a $3\frac{3}{5}$

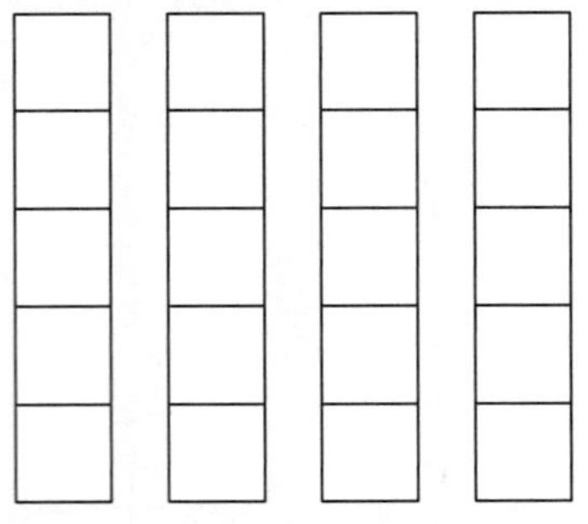

b $\frac{17}{8}$

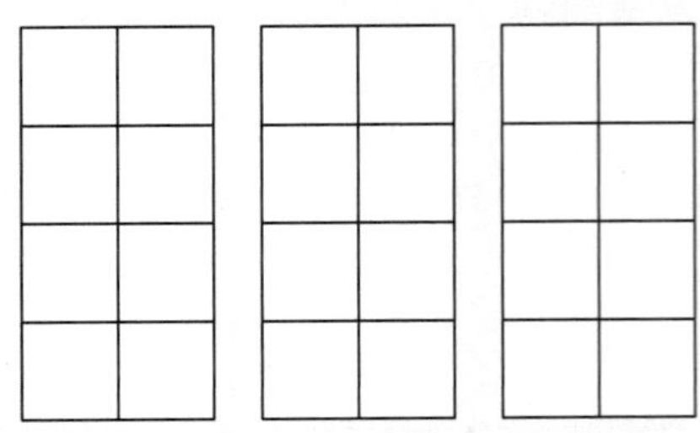

CHAPTER 8: TIME

Exercise 1

Complete each sentence using these words once each:

midday *sunset* *afternoon* *sunrise* *noon*
midnight *evening* *tonight* *morning*

Sunrise is in the ______________________ .

A word for "this evening" is ______________________ .

The day starts and ends at ______________________ .

Another name for noon is ______________________ .

Daytime is between ______________________ and ______________________ .

______________________ and ______________________ make up the pm times.

There are 12 hours between midnight and ______________________ .

Discussion

What do we mean by "the present"?

Exercise 2

Complete the diagram using *future*, *past*, and *present*.

yesterday today tomorrow

now

______________________ ______________________ ______________________

Analogue clocks and watches display time on faces like this.

The face shows 12 hours.

The **hands** tell us what the time is. They turn in a **clockwise** direction.

The **minute hand** is longer. It is pointing to 12.

The **hour hand** is shorter. It is pointing to 8.

The time is 8 o'clock.

Analogue time is written as *minutes to the hour*, or *minutes past the hour*.

There are 60 minutes in one hour.

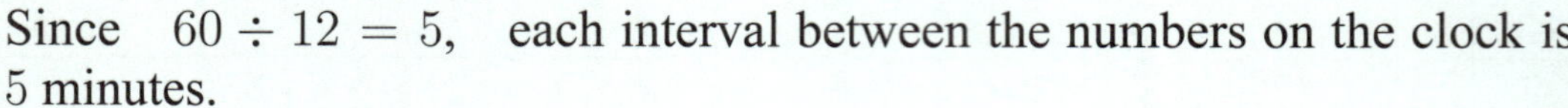

Since $60 \div 12 = 5$, each interval between the numbers on the clock is 5 minutes.

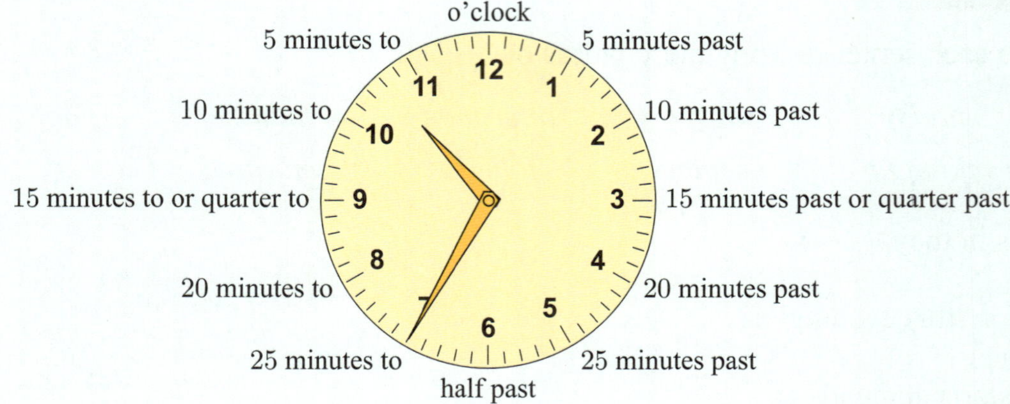

On this clock, the minute hand is showing 25 minutes to the hour.

The hour hand is between 10 and 11.

The time is "25 minutes to 11 o'clock" or "25 to 11".

Exercise 3

What time is shown on each clock?

a ______ o'clock

b half past ______

c ______________________

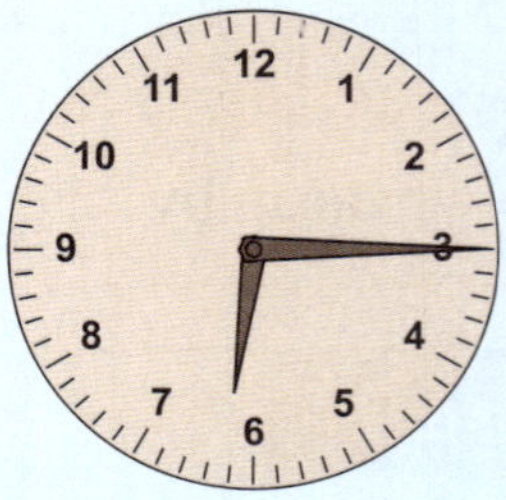

d ______________________

e ______________________

f ______________________

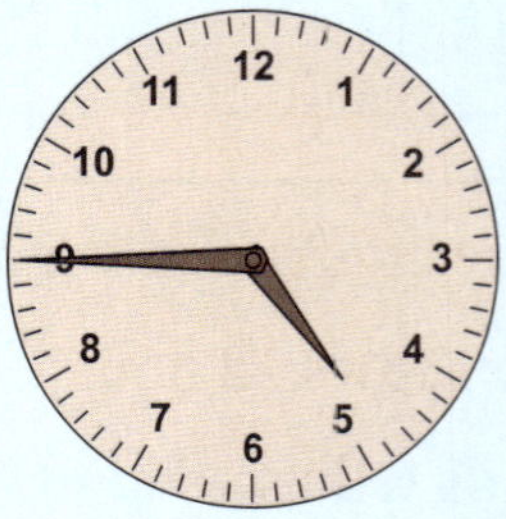

Exercise 4

What time is shown on each clock?

a ______ past ______

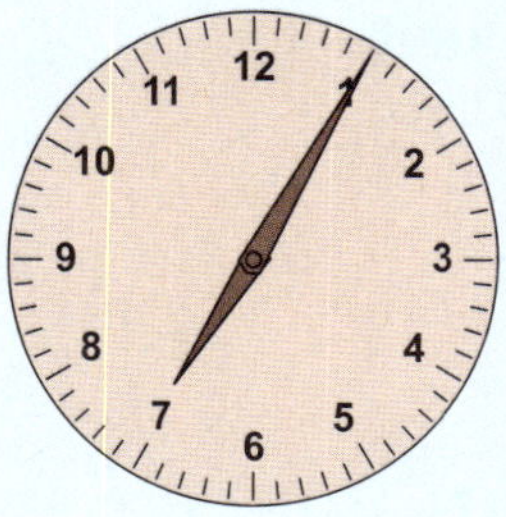

b ______ to ______

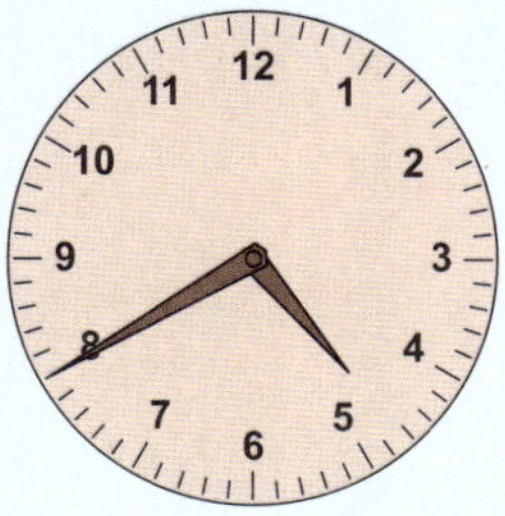

c ____________

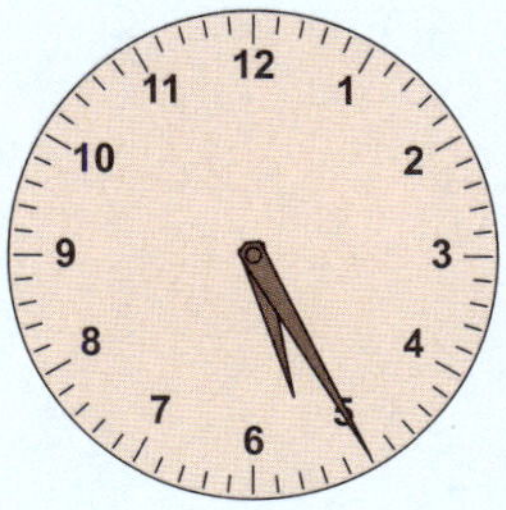

d ____________

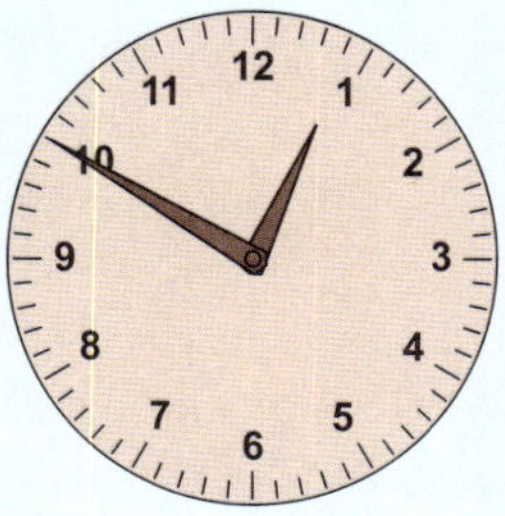

e ____________

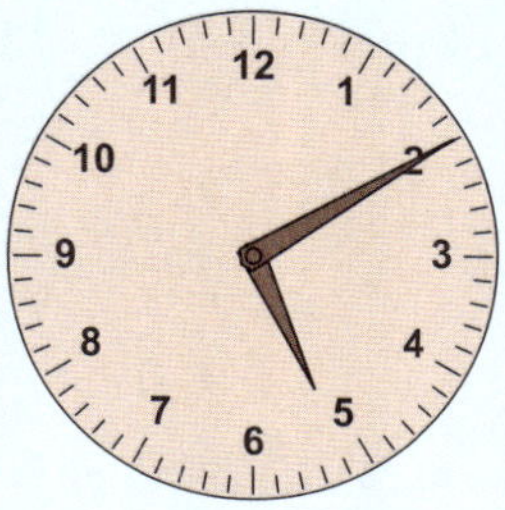

f ____________

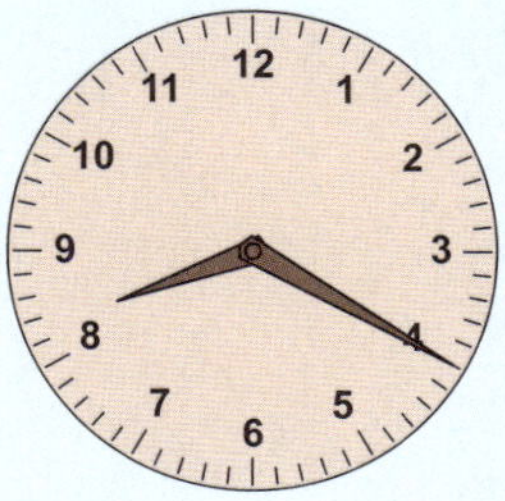

g ____________

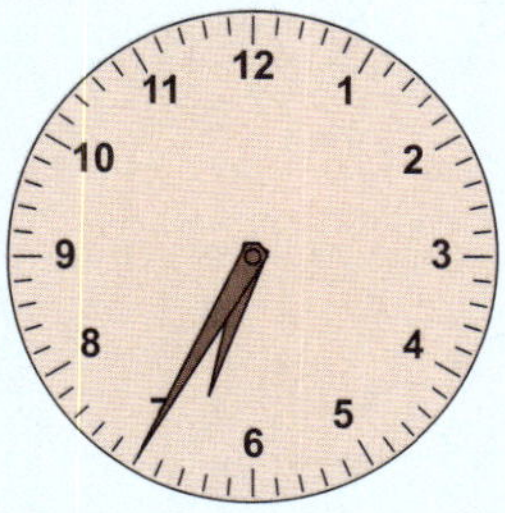

h ____________

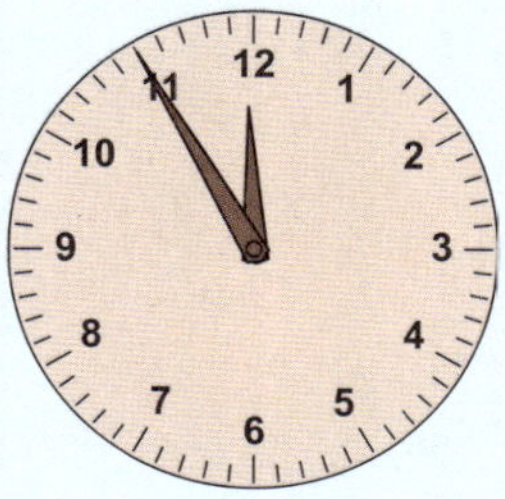

i ____________

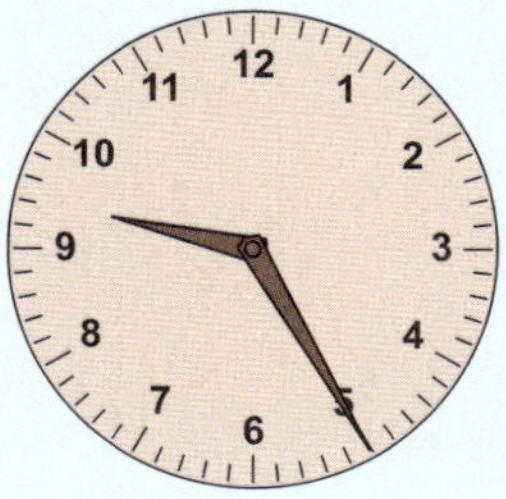

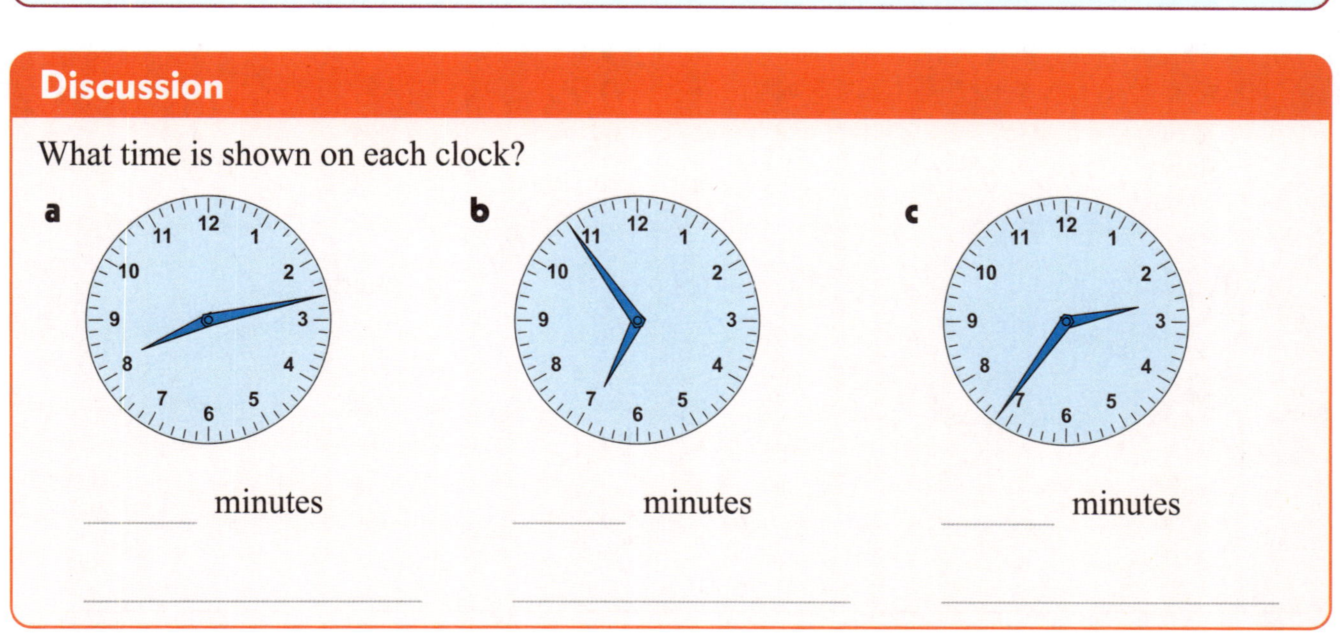

Exercise 5

Draw hands on each clock face to show the time:

a 6 o'clock

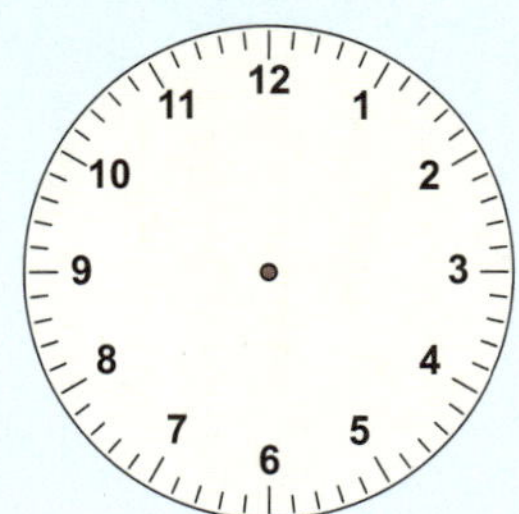

b quarter to 11

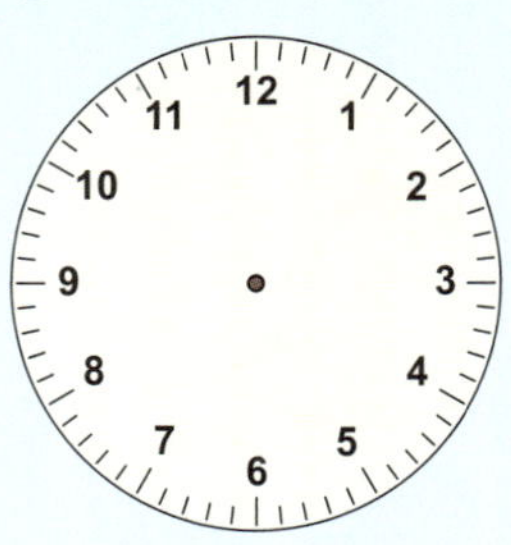

c 25 minutes to 5

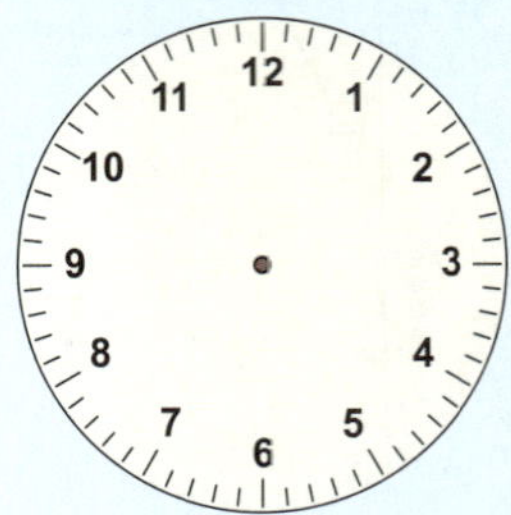

d 10 minutes past 7

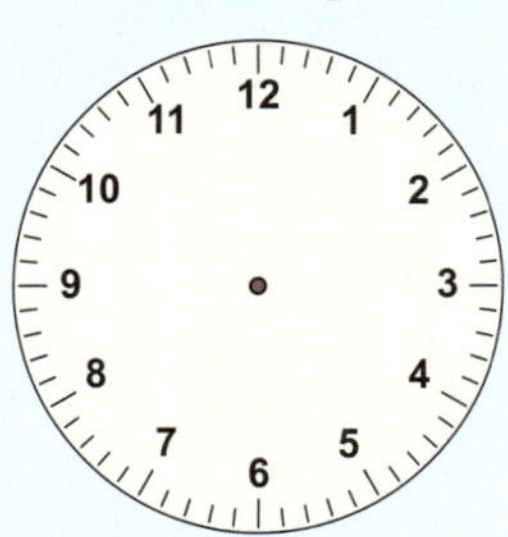

e half past 4

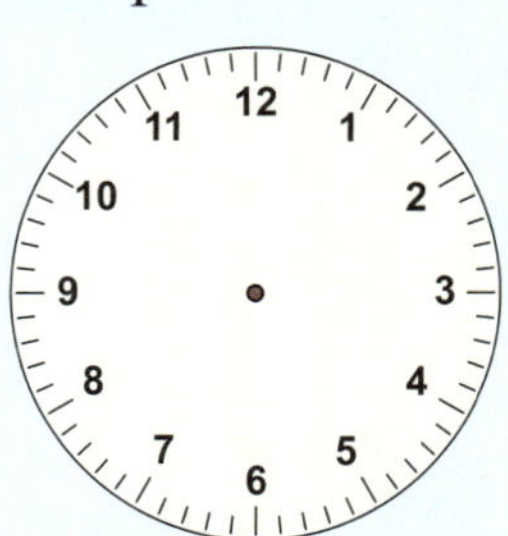

f quarter past 9

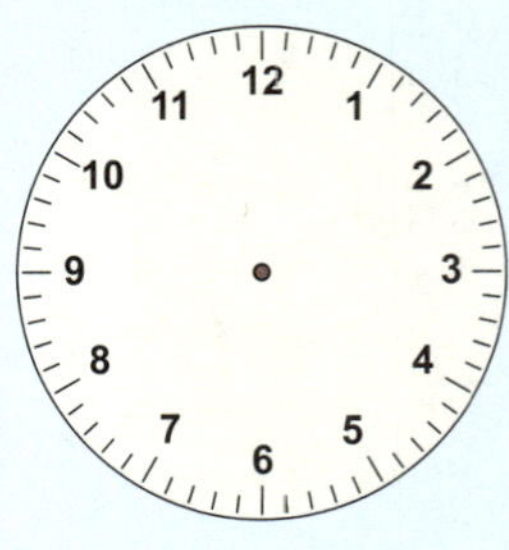

g 20 minutes to 2

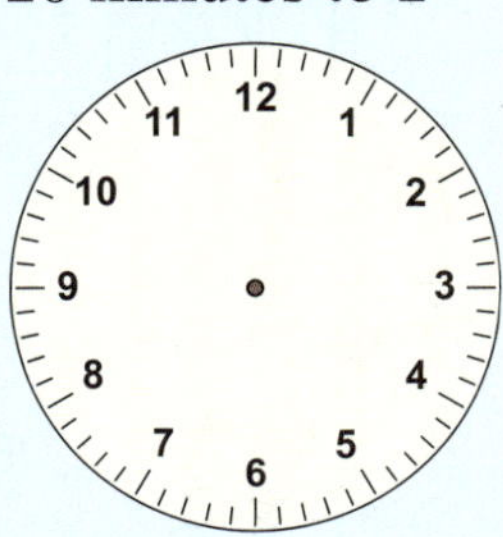

h 5 minutes past 3

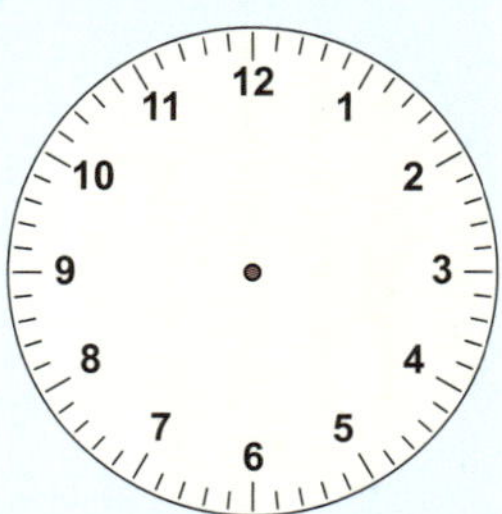

i quarter past 1

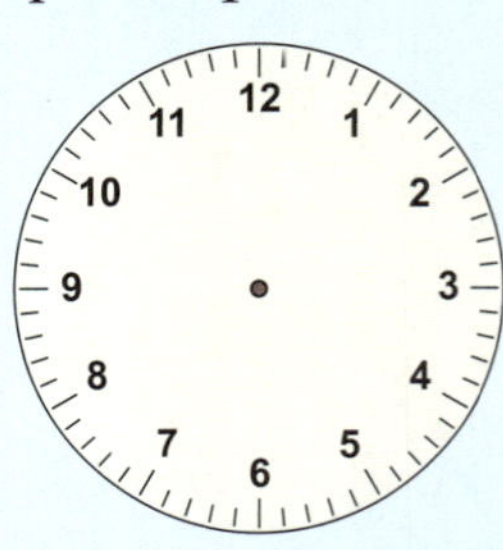

j 10 to 8

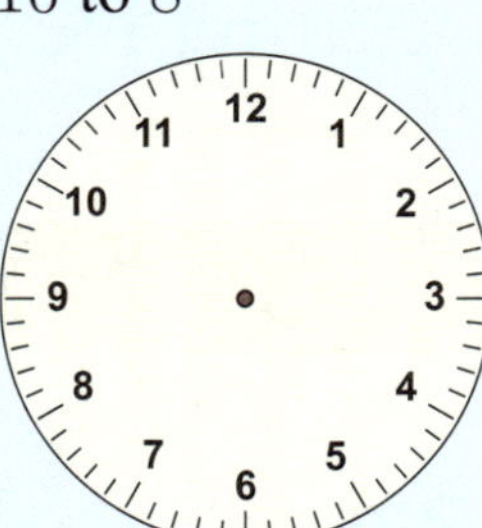

k 4 o'clock

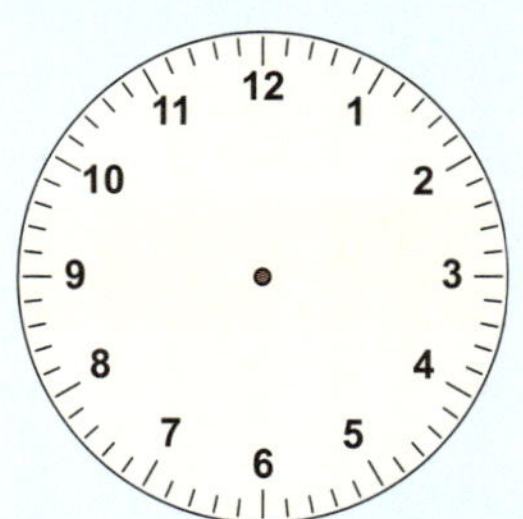

l 5 to 6

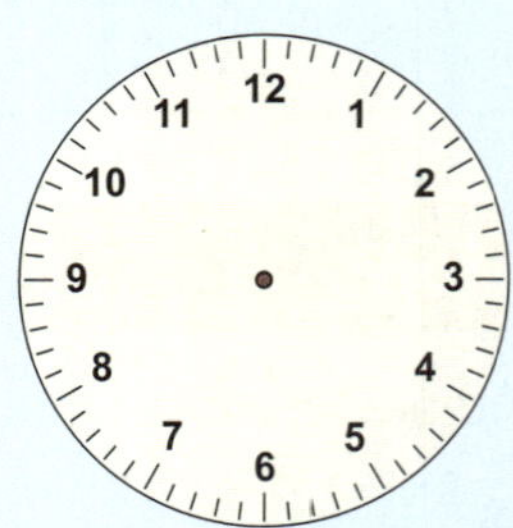

m 10 minutes past 11

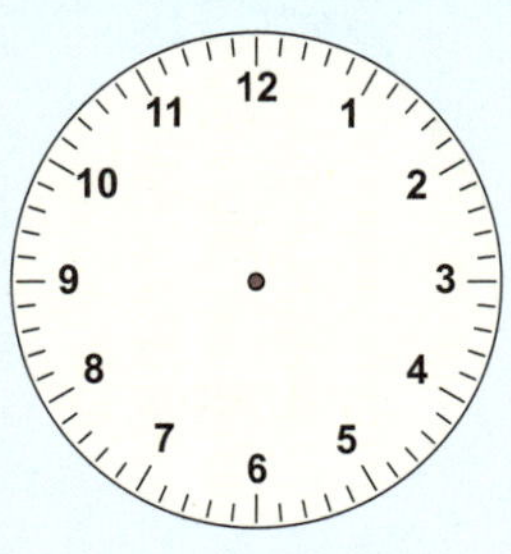

n 25 minutes past 10

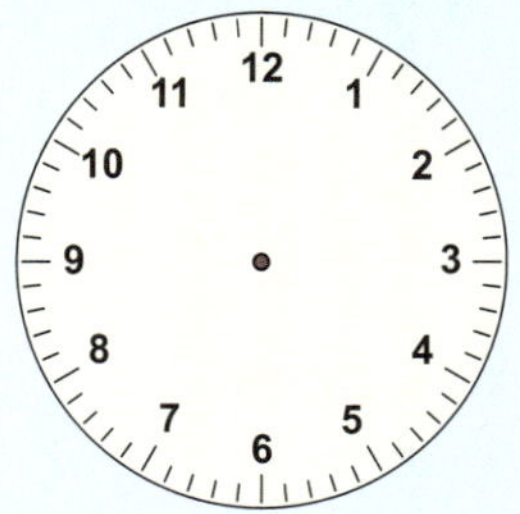

o quarter to 7

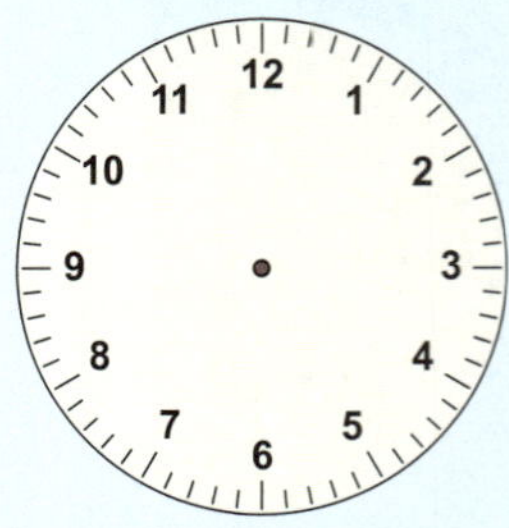

Exercise 6

What time is shown on each clock?

a

..........past..........

b

....

c

....

Exercise 7

Write how many minutes it takes for the minute hand to move from 12 to each number on the clock.

These numbers are the results of the .... times table.

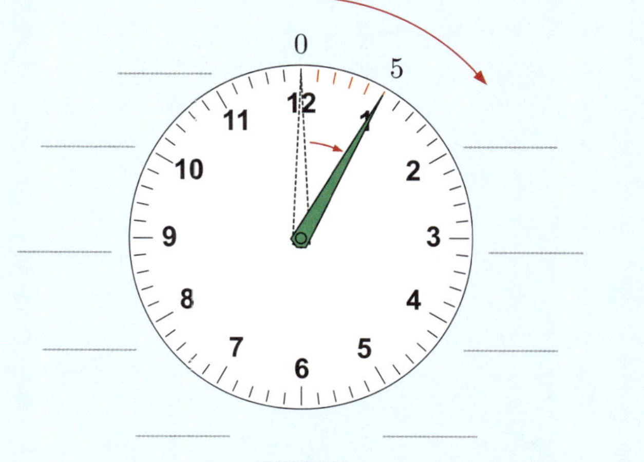

Digital clocks and watches display time on faces like this.

There are two numbers separated by a *colon* (:).

The number before the colon shows the **hour**.

The number after the colon shows the minutes **past the hour**.

The time is "three forty five".

45 minutes past 3 is the same as 15 minutes to 4.

So, the time is "15 minutes to 4" or "quarter to 4".

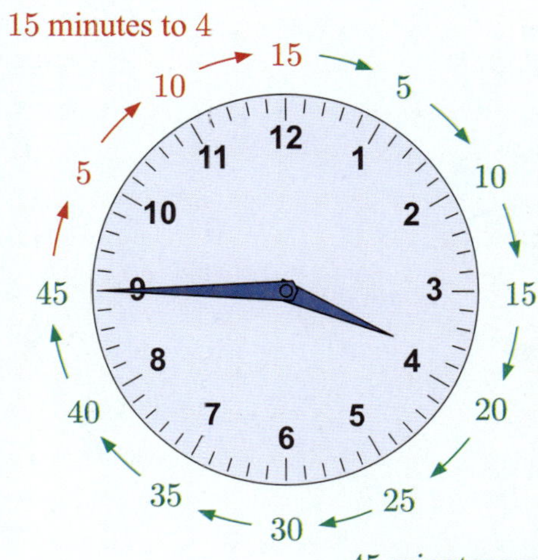

Exercise 8

Write as a digital time:

a

b

c

d

e

f

g

h

i

j

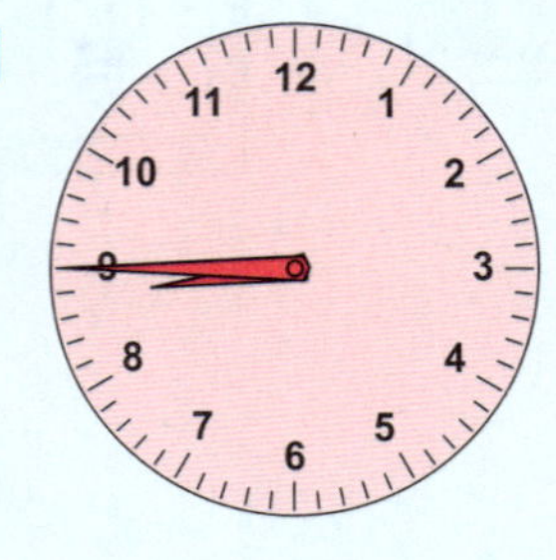

k

l

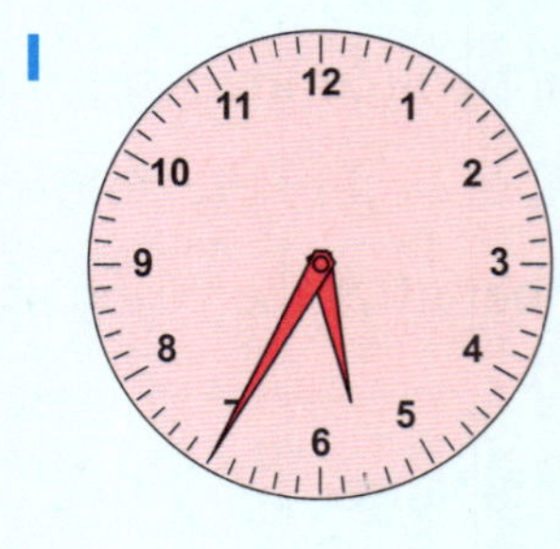

m

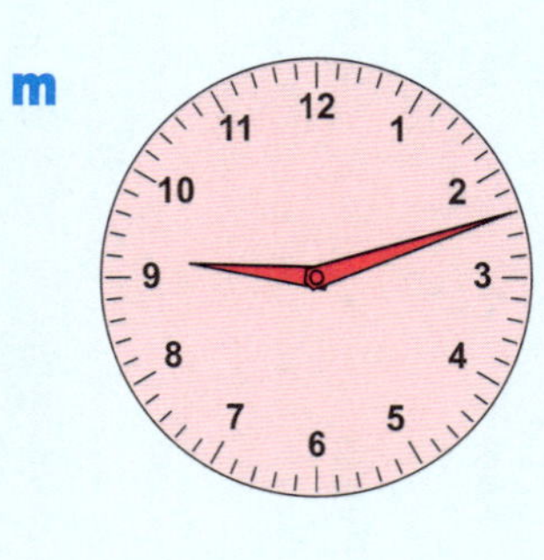

n

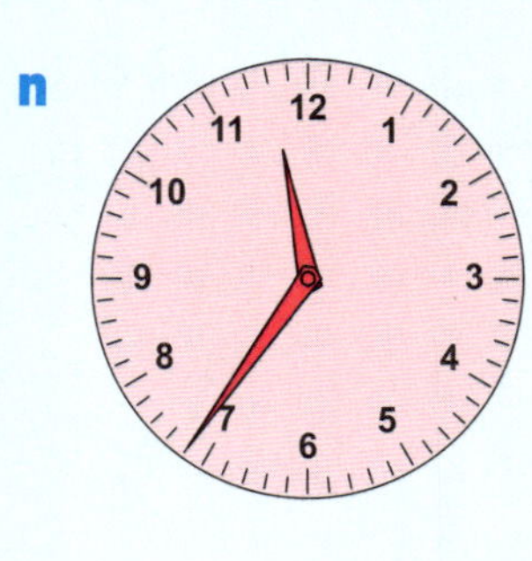

o

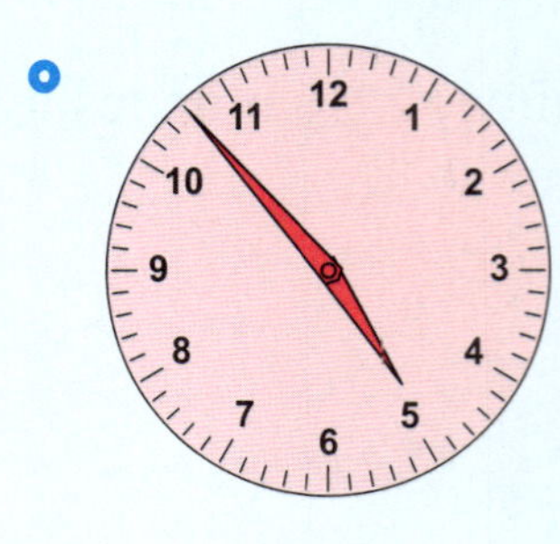

Exercise 9

Draw each time on the clock.

a 4:05

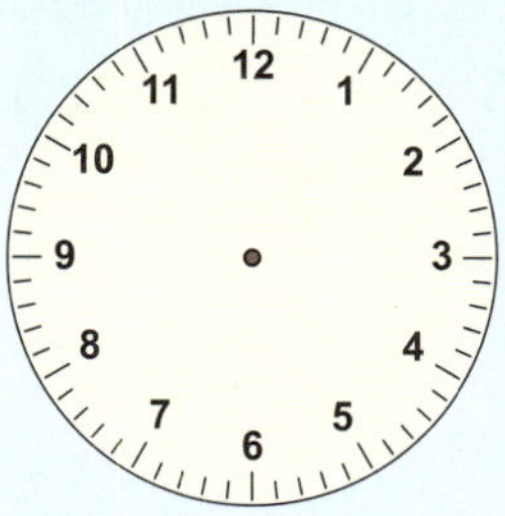

b 5:20

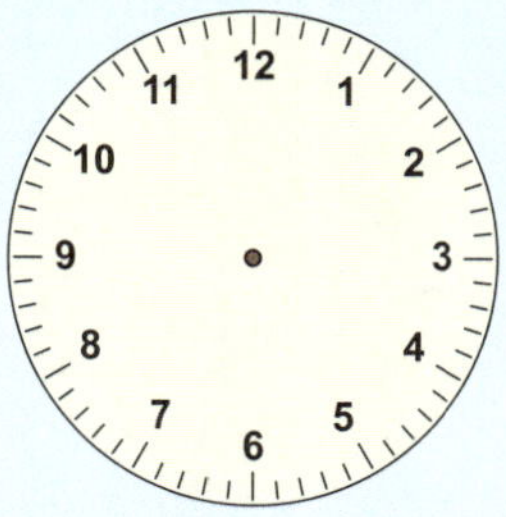

c 7:15

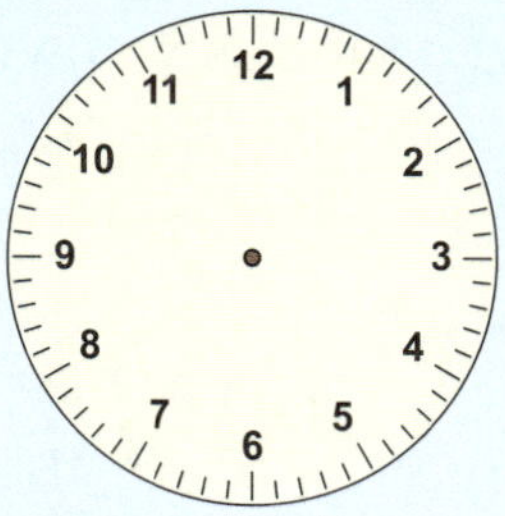

d 10:30

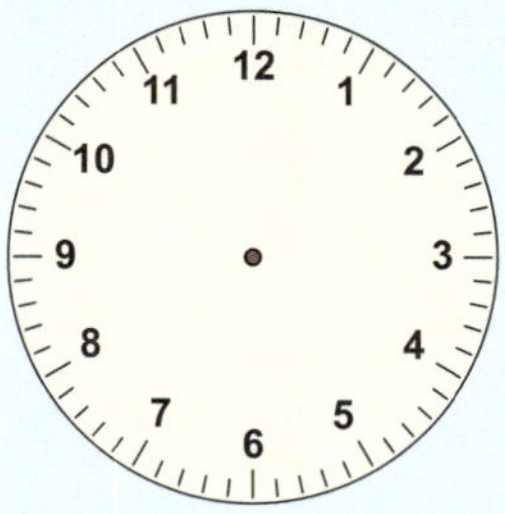

e 8:00

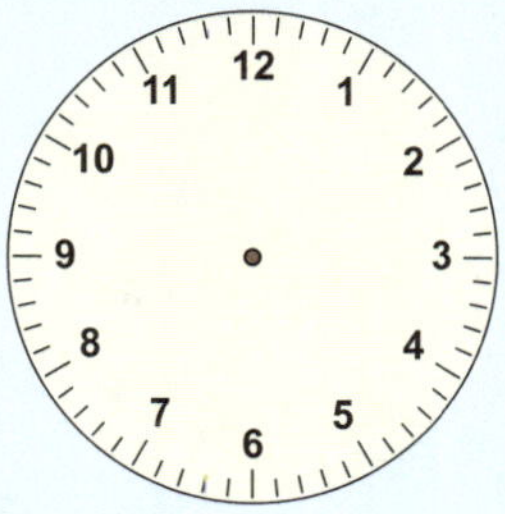

f 2:50

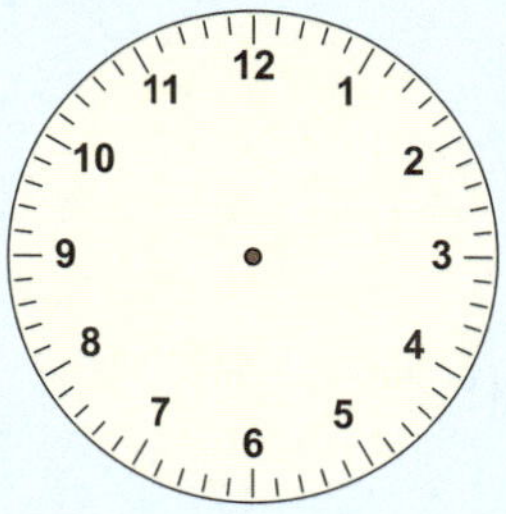

g 6:35

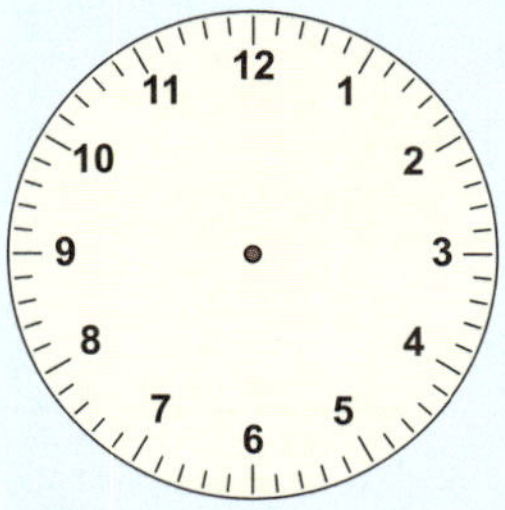

h 9:55

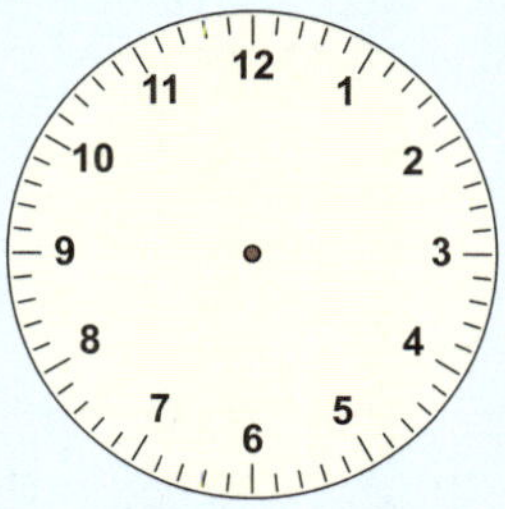

i 1:45

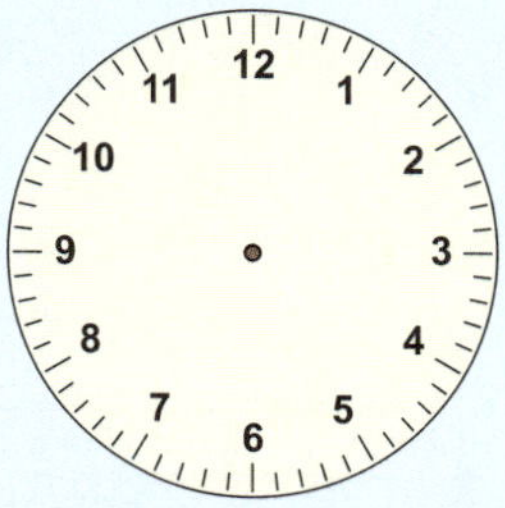

j 3:04

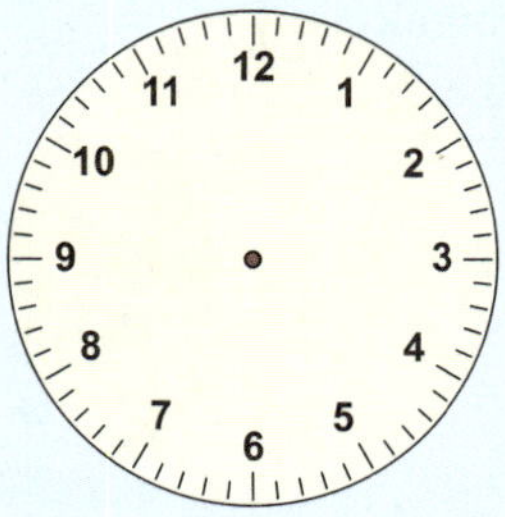

k 10:27

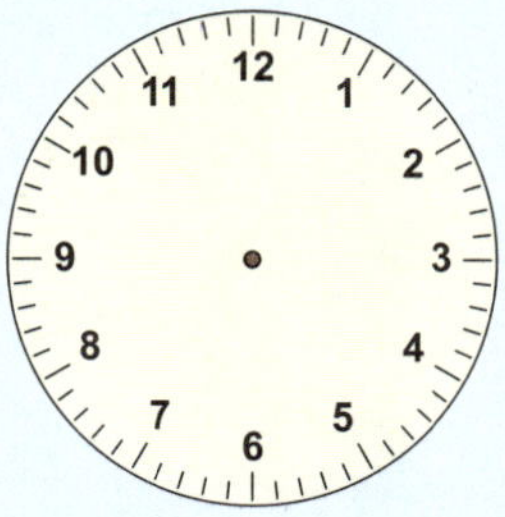

l 12:36

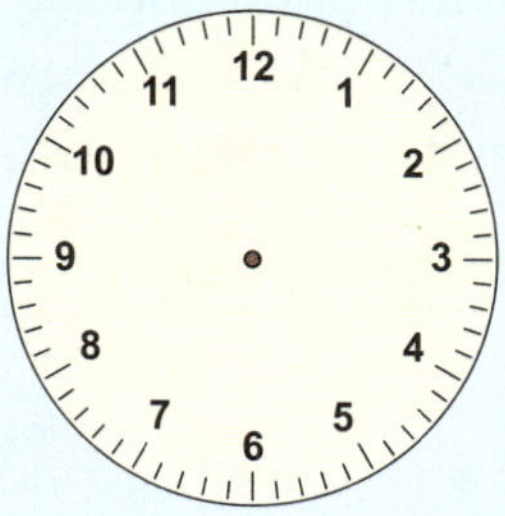

Discussion

How do you ask what the time is in other languages?

Does it always directly translate into English as "What is the time?"

Exercise 10

8:15 is read "eight fifteen".

12:00 is read "twelve o'clock".

Write each time as you would say it:

a 2:20 ....

b 6:35 ....

c 11:10 ....

d 9:55 ....

e 4:41 ....

f 5:23 ....

g 10:00 ....

h 12:18 ....

i 1:57 ....

j 7:05 ....

k 3:06 ....

2:05 is read "two 'o' five".

Discussion

With the analogue and digital times we have seen, do we know whether it is morning or afternoon?

What problems could this cause?

Let's meet at 8 o'clock!

Is this for breakfast or dinner?

To make sure there is no confusion, we use **am** and **pm** after the time.

We can see this on a **time line**.

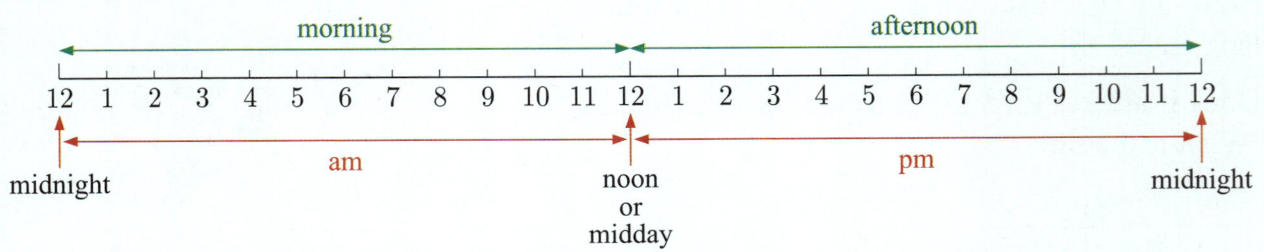

Exercise 11

Write each time using "quarter past", "half past", or "quarter to". Write whether it is morning, afternoon, or evening.

a 9:30 pm ______________ in the ______________.

b 1:15 pm ______________ in the ______________.

c 5:45 am ______________ in the ______________.

d 12:30 am ______________ in the ______________.

e 4:15 am ______________ in the ______________.

f 3:45 pm ______________ in the ______________.

g 11:30 pm ______________ in the ______________.

Exercise 12

Write each time using "to" or "past". Write whether it is morning, afternoon, or evening.

a 5:50 AM 10 to ______ in the ______________.

b 3:25 PM ______ past ______ in the ______________.

c 4:05 AM ______________ 4 in the ______________.

d 11:20 PM ______________________________

e 7:10 AM ______________________________

f 10:45 PM ______________________________

g 12:25 AM ______________________________

h 1:35 PM ______________________________

i 8:07 PM ______________________________

j 6:52 AM ______________________________

Exercise 13

Write as a digital time:

a 5 minutes past 2 in the morning ....

b 10 minutes to 3 in the afternoon ....

c 25 minutes past 6 in the morning ....

d 20 minutes to 9 in the evening ....

e 15 minutes past 7 in the morning ....

f 5 minutes to midnight ....

Activity

In this Activity you will use a clock to add on or take away time.

You can either use an analogue clock in your classroom, or click the icon alongside to use our clock software.

What to do:

1 Set the time to 2:00 pm.

a To find the time 1 hour *after* 2:00 pm, we *add on* 1 hour.

The time is ....

b Reset the time to 2:00 pm.
To find the time 30 minutes *after* 2:00 pm, we *add on* 30 minutes.

The time is ....

c The time 45 minutes *after* 2:00 pm is ....

d The time 2 hours and 10 minutes *after* 2:00 pm is ....

2 Set the time to 9:00 am.

a To find the time 1 hour *before* 9:00 am, we *take away* 1 hour.

The time is ....

b Reset the time to 9:00 am.
To find the time 10 minutes *before* 9:00 am, we *take away* 10 minutes.

The time is ....

c The time 55 minutes *before* 9:00 am is ....

d The time 2 hours and 15 minutes *before* 9:00 am is ....

3 Look at the time on the clock.
Write down what the time will be after:

7:20 AM

a 1 hour ....

b 25 minutes ....

c 50 minutes ....

d 3 hours and 15 minutes ....

4 The time now is 5 minutes past 6 in the evening.
Write down what the time was:

a 1 hour earlier ....

b 20 minutes earlier ....

c 2 hours and 45 minutes earlier ....

5 Look at the time on the clock.

The time is ....

Write down what the time will be after:

a 50 minutes ....

b 1 hour and 10 minutes ....

am

6 8:15 AM

Lucy looked at the clock.
She had missed her school bus by 10 minutes.

What time did her school bus leave? ....

7 4:40 PM

The clock shows the time Amelie finished swimming.
She swam for 35 minutes.

Amelie started to swim at ....

8 Find the time 2 hours and 40 minutes after 10:50 am. ....

The time changed from "am" to "pm" at ....

9 Find the time 10 hours before 7:15 am. ....

Was this time on the same day? Explain your answer.

....

....

The **units** of time we use are **days**, **hours**, **minutes**, and **seconds**.

Exercise 14

Write which unit of time you would use to measure the time:

a to blow your nose ______________

b for each period of a basketball game ______________

c you spend at school on Tuesday ______________

d to run a lap around a tennis court ______________

e to write a message in a birthday card ______________

Exercise 15

Sometimes we use more than one unit of time.

Write *two* units of time you would use together to measure:

a the time to run 1500 metres ______________ ______________

b the length of a movie ______________ ______________

c the length of a song ______________ ______________

d the length of a flight from Doha to Singapore

______________ ______________

We can **convert** from one unit of time to another.

1 day = 24 hours
1 hour = 60 minutes
1 minute = 60 seconds

Exercise 16

Complete:

$1 \times 60 =$ ______ 1 hour = 60 minutes

$2 \times 60 =$ ______ 2 hours = ______ minutes

$3 \times 60 =$ ______ 3 hours = ______ minutes

$4 \times 60 =$ ______ 4 hours = ______ minutes

$5 \times 60 =$ ______ 5 hours = ______ minutes

$6 \times 60 =$ ______ 6 hours = ______ minutes

Exercise 17

Complete:

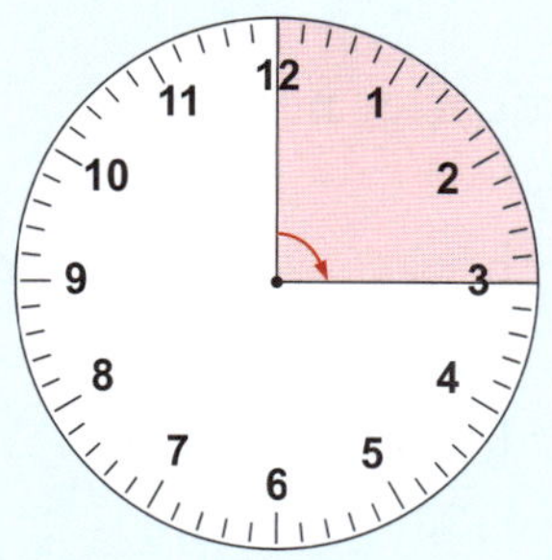

a quarter of an hour = ______ minutes

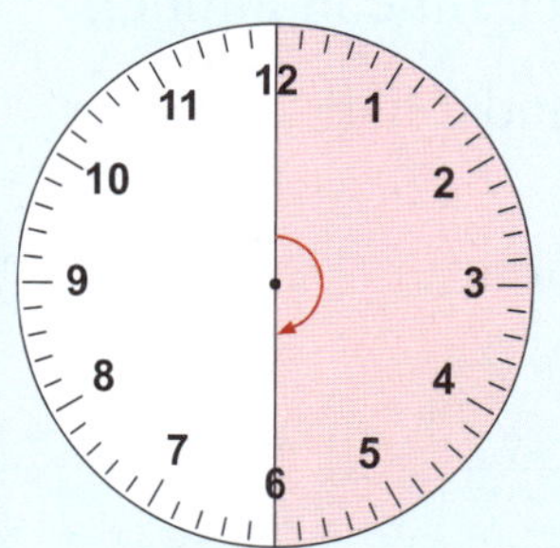

a half of an hour = ______ minutes

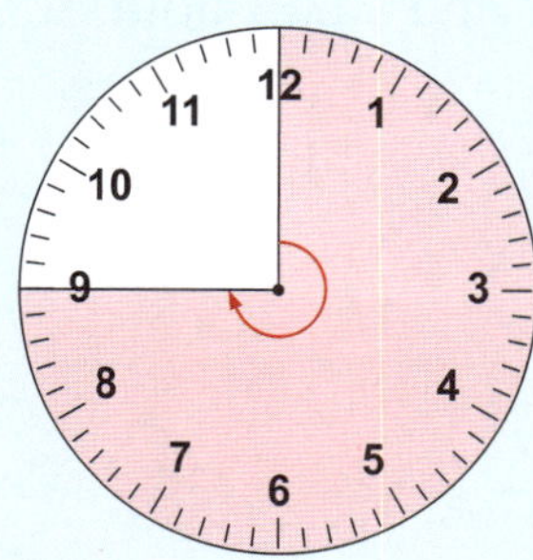

three quarters of an hour = ______ minutes

1 hour and 25 minutes
= 60 minutes + 25 minutes
= 85 minutes

Exercise 18

Write each time in minutes:

a 1 hour and 15 minutes

= ______ minutes + ______ minutes

= ____________________

b 1 hour and 35 minutes

= ______ minutes + ______ minutes

= ____________________

c 1 hour and 48 minutes

= ____________________

\+ ____________________

= ____________________

d 2 hours and 10 minutes

= ____________________

\+ ____________________

= ____________________

Exercise 19

Write each time in minutes:

a 1 and a half hours

= 1 hour + ______ minutes

= ______ minutes + ______ minutes

= ____________________

b 2 and a quarter hours

= ______ hours + ______ minutes

= ______ minutes + ______ minutes

= ____________________

c 1 and three quarter hours

= ______ hour + ______ minutes

= ____________________

\+ ____________________

= ____________________

d 2 and a half hours

= ______ hours + ______ minutes

= ____________________

\+ ____________________

= ____________________

130 minutes − 60 minutes = 70 minutes
70 minutes − 60 minutes = 10 minutes
So, 130 minutes = 2 hours and 10 minutes.

Exercise 20

Complete:

a 85 minutes − 60 minutes = ______ minutes

So, 85 minutes = ______ hour and ______ minutes.

b 95 minutes − 60 minutes = ______ minutes

So, 95 minutes = ________________________________.

c 125 minutes − 60 minutes = ______ minutes

______ minutes − 60 minutes = ______ minutes

So, 125 minutes = ________________________________.

Exercise 21

Write each time using hours and minutes:

a 75 minutes ________________________________

____

____

b 145 minutes ________________________________

____

____

____

Exercise 22

Complete:

$1 \times 60 =$ ______	1 minute = 60 seconds
$2 \times 60 =$ ______	2 minutes = ______ seconds
$3 \times 60 =$ ______	3 minutes = ______ seconds
$4 \times 60 =$ ______	4 minutes = ______ seconds
$5 \times 60 =$ ______	5 minutes = ______ seconds
$6 \times 60 =$ ______	6 minutes = ______ seconds

Exercise 23

Complete:

a quarter of a minute = ______ seconds

a half of a minute = ______ seconds

three quarters of a minute = ______ seconds

Exercise 24

Write each time in seconds:

a 1 minute and 40 seconds

= ______ seconds + ______ seconds

= ______ seconds

b 2 minutes and 19 seconds

= ____________

+ ____________

= ____________

c 3 and a half minutes

= ______ minutes + ______ seconds

= ______ seconds + ______ seconds

= ______ seconds

d 1 and a quarter minutes

= ______ minute + ______ seconds

= ______ seconds + ______ seconds

= ______ seconds

Exercise 25

Complete:

a $1 \times 24 =$ ______

$2 \times 24 =$ ______

$3 \times 24 =$ ______

b 1 day = 24 hours

2 days = ______ hours

3 days = ______ hours

Exercise 26

This is Sarah's school **timetable** for Monday.

Sarah's English lesson starts at 10:15 am and finishes at 11:00 am.

The lesson lasts for 45 minutes.

Lesson	*Start time*
Roll call	9:00 am
Mathematics	9:15 am
English	10:15 am
Break	11:00 am
Art	11:15 am
Lunch	12:15 pm
Sport	1:15 pm
History	2:20 pm
Home time	3:00 pm

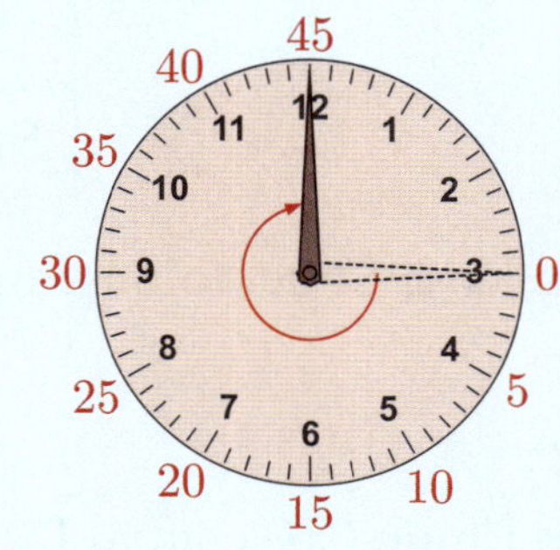

a Sarah's Mathematics lesson begins at ______________.

b Sarah's Art lesson ends at ______________.

c Sarah's school day starts at ______________ and ends at ______________.

d Sarah's lunch starts at ______________ and ends at ______________.

Sarah's lunch lasts for ______________________________.

e Sarah's History lesson starts at ______________ and ends at ______________.

Sarah's History lesson lasts for ______________________________.

f Sarah's Sport lesson starts at ______________ and ends at ______________.

Sarah's Sport lesson lasts for ________ hour and ________ minutes.

Exercise 27

This is the timetable for Raoul's school bus.

Stop	*Departs*
Rose Street	7:45 am
Station Road	7:56 am
Emile Crescent	8:02 am
Brussels Gardens	8:10 am
Dethy Road	8:15 am
Victory Avenue	8:21 am
Berkmans Terrace	8:27 am

a What does the word "depart" mean?

....

....

b At what time does the bus depart from Emile Crescent? ....

c The bus *arrives* at each stop 2 minutes *before* its departure time.

Complete this table with arrival times.

Stop	*Arrives*
Rose Street	
Station Road	
Emile Crescent	
Brussels Gardens	
Dethy Road	
Victory Avenue	
Berkmans Terrace	

d How long is the journey from Rose Street to Station Road?

....

e How long is the journey from Emile Crescent to Dethy Road?

....

f If the bus departs from Dethy Road 5 minutes late, at what time will the bus arrive at Victory Avenue?

....

g If the bus arrives at school 6 minutes after departing Berkmans Terrace, at what time will Raoul arrive at school?

....

Discussion

Why do you have a timetable at school?

What would happen if your class did not have a timetable?

Activity

Think about your own class timetable.

a What time do you start school in the morning? ______

b What time is your lunch break? ______

How long does it last? ______

c How many lessons do you have in your day? ______

d What time does the school day finish? ______

e When is your school assembly? ______

Revision

1 Complete:

a Another name for midday is ______.

b Night-time is between ______ and ______ the next day.

c The present is between the ______ and the ______.

2 What time is shown on each clock? Write your answer in words.

a 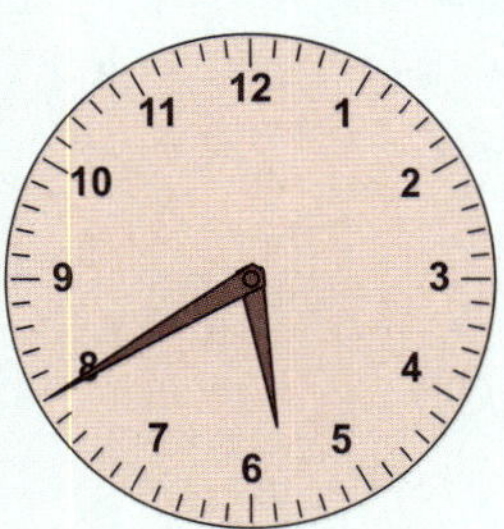

______

b 

______

______

c 

______

3 Draw hands on each clock face to show the time:

a 10 minutes to 9

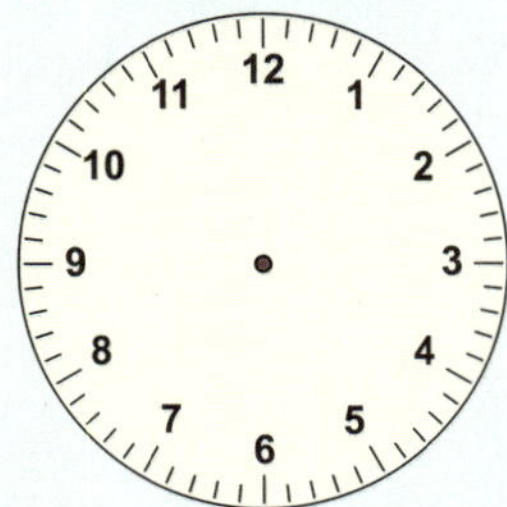

b half past 3

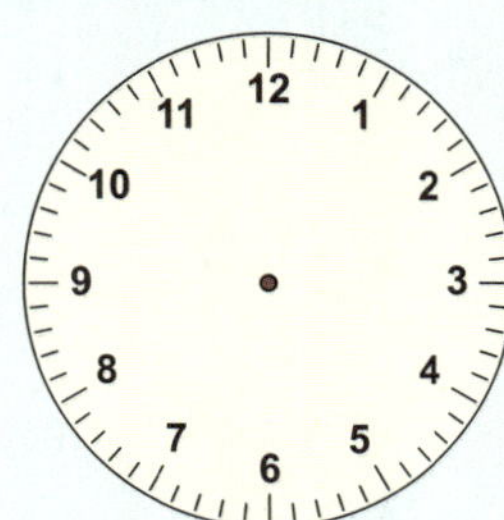

c 25 minutes past 11

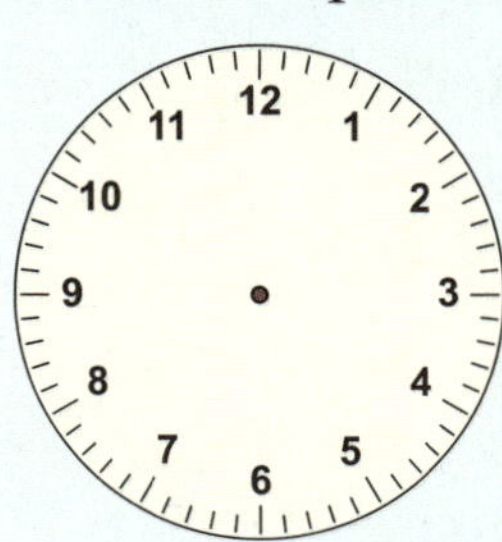

4 Write as a digital time:

a

b

c

d

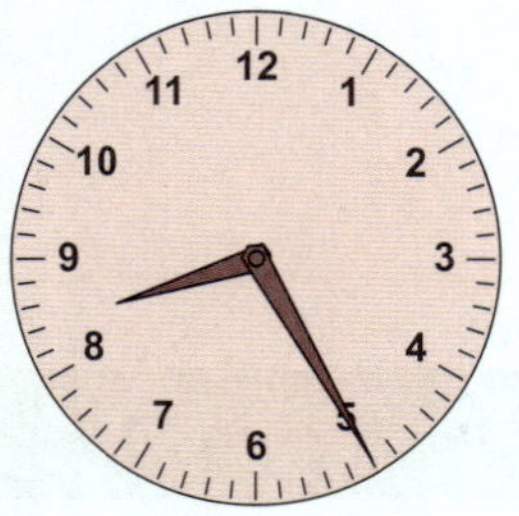

e

f

5 Draw each time on the clock.

a 2:00

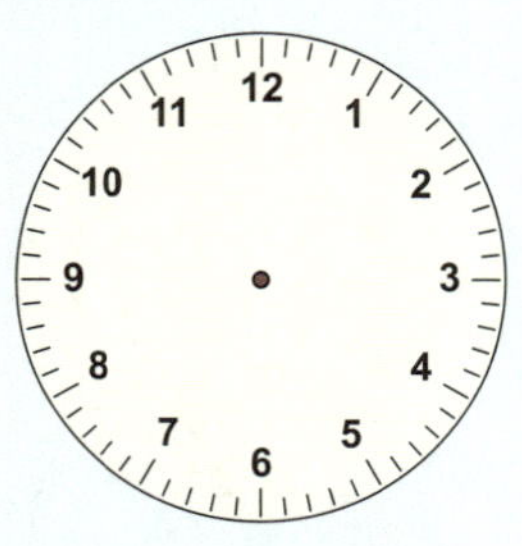

b 7:35

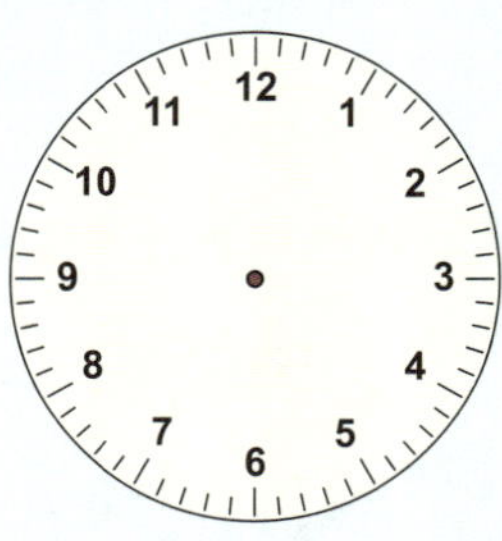

c 4:10

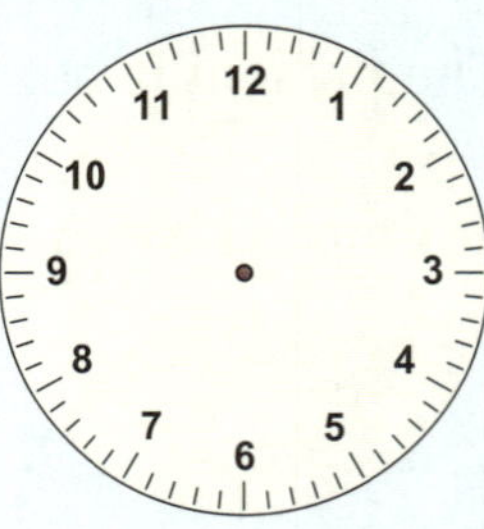

d 9:50

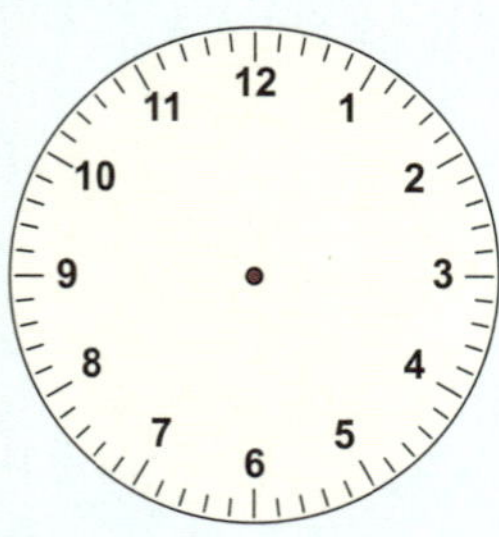

e 10:08

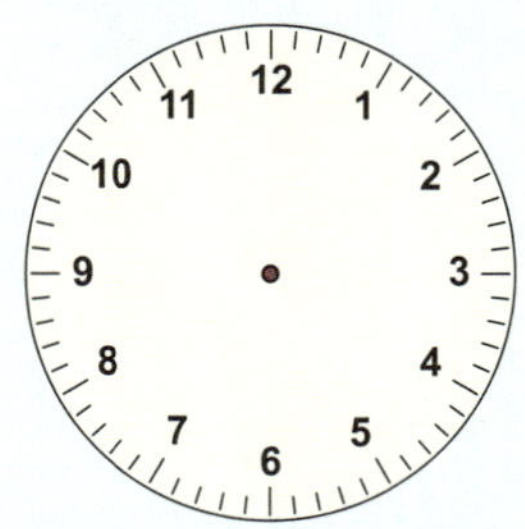

f 3:48

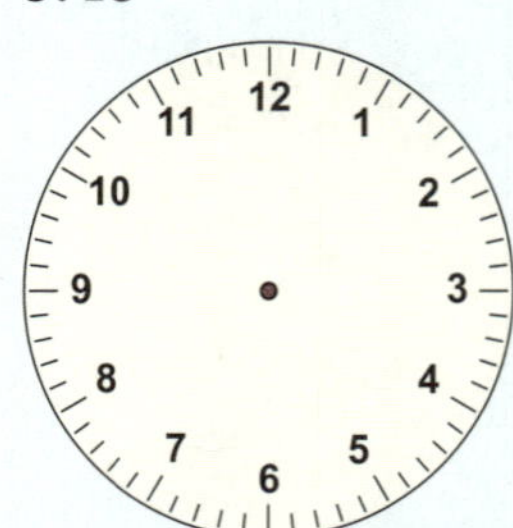

6 Write each time as you would say it:

a 8:45 ....

b 12:27 ....

c 2:08 ....

d 7:52 ....

7 Write each time using "to" or "past". Write whether it is morning, afternoon, or evening.

a 4:40 PM ______________ in the ______________.

b 9:20 AM ______________ in the ______________.

c 10:55 PM ______________ in the ______________.

d 1:05 AM ______________ in the ______________.

8 Write as a digital time:

a 25 minutes past 3 in the afternoon ______________

b 10 minutes to 8 in the evening ______________

c 25 minutes to 1 in the morning ______________

9 Write *two* units of time you would use together to measure:

a the length of a theatre show ______________ ______________

b the time to swim 400 metres ______________ ______________

10 Complete:

1 day = ______ hours

1 hour = ______ minutes

1 minute = ______ seconds

11 Write each time in minutes:

a 1 hour and 25 minutes

= ______ minutes

\+ ______ minutes

= ______________

b 2 and three quarter hours

= ______ hours + ______ minutes

= ______ minutes + ______ minutes

= ______________

12 Complete:

140 minutes − 60 minutes = ______ minutes

______ minutes − 60 minutes = ______ minutes

So, 140 minutes = ______ hours and ______ minutes.

13 Write 115 minutes using hours and minutes.

....

....

14 Write each time in seconds:

a 1 minute and 16 seconds

= ______ seconds + ______ seconds

= ______________

b 2 and a half minutes

= ______ minutes + ______ seconds

= ______ seconds + ______ seconds

= ______________

15 This is Eryn's school timetable for Wednesday.

Lesson	*Start time*
Science	8:40 am
Music	9:35 am
Recess	10:30 am
English	11:00 am
Art	11:40 am
Lunch	12:20 pm
Sport	1:20 pm
Mathematics	2:10 pm
Home time	3:00 pm

a Eryn's Music lesson begins at ______________.

b Eryn's lunch ends at ______________.

c Eryn's Science lesson starts at ______________ and ends at ______________.

Eryn's Science lesson lasts for ______________.

d Eryn's Mathematics lesson starts at ______________ and ends at ______________.

Eryn's Mathematics lesson lasts for ______________.

e Eryn's recess lasts for ______________.

CHAPTER 9: TURNS AND ANGLES

- In one hour, the minute hand makes a **full turn** around the clock.

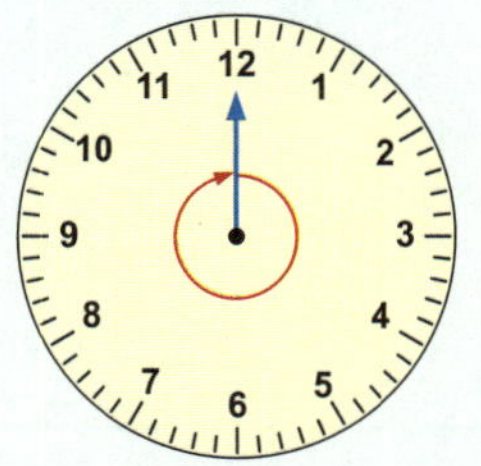

- In a half hour, the minute hand makes a **half turn**.

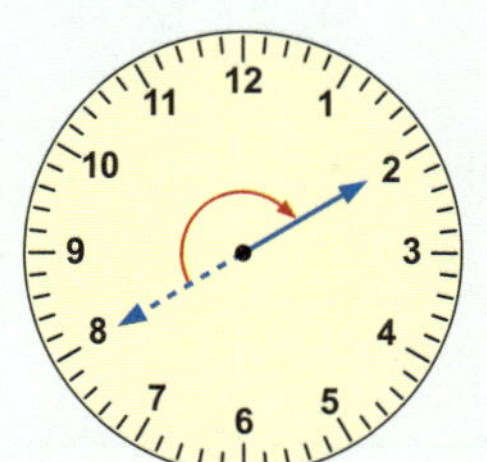

- In a quarter of an hour, the minute hand makes a **quarter turn**.

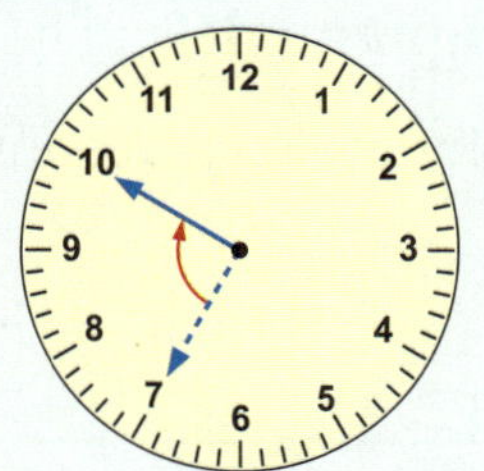

Exercise 1

Describe each turn:

a

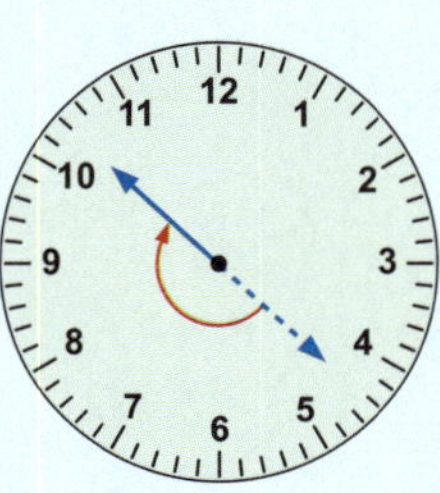

____

b

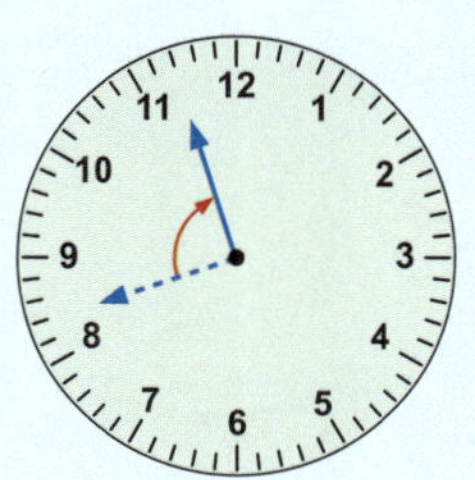

____

c

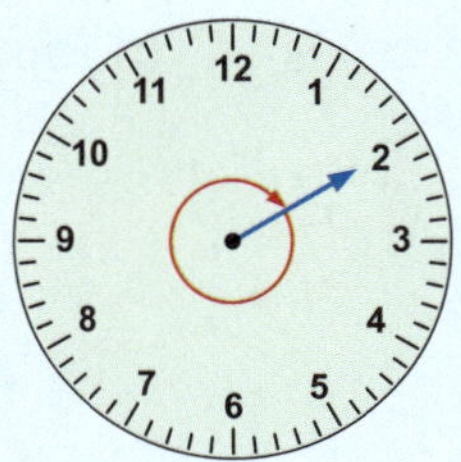

____

d

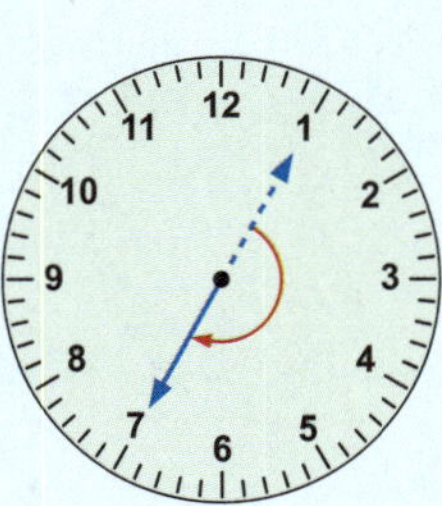

____

e

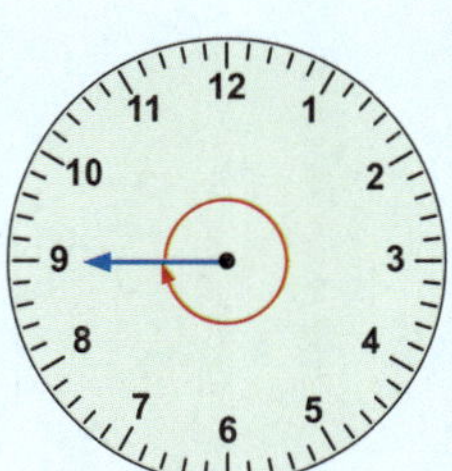

____

f

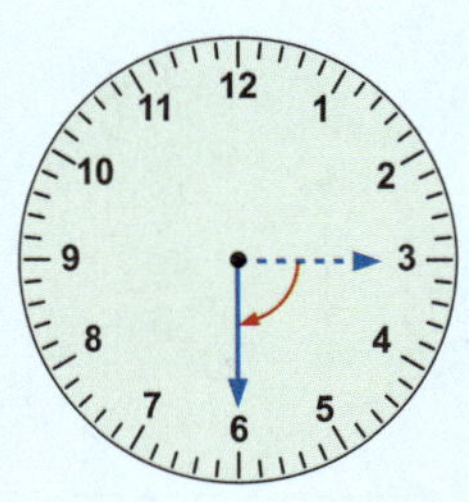

____

Exercise 2

How many quarter turns are there in a full turn? ______

How many half turns are there in a full turn? ______

How many quarter turns are there in a half turn? ______

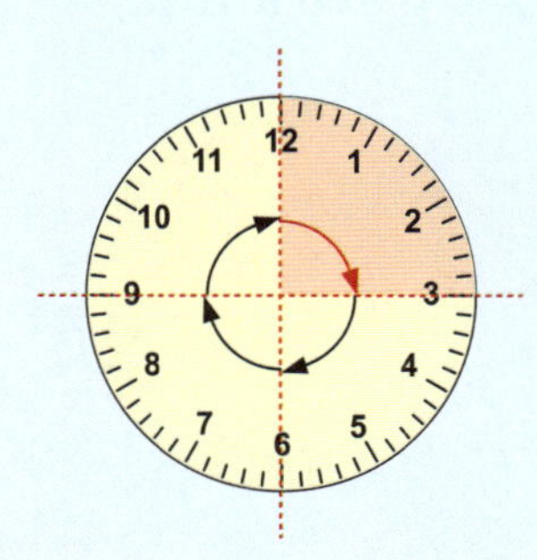

An **angle** is formed when two straight lines meet.

The **size** of an angle is the amount one line must *turn* so it overlaps the other.

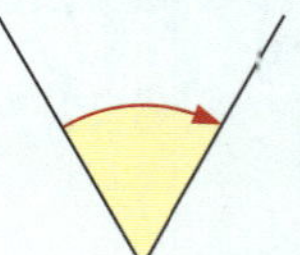

At 3 o'clock, the angle between the hour and minute hands is a **right angle**.

A right angle corresponds to a quarter turn.

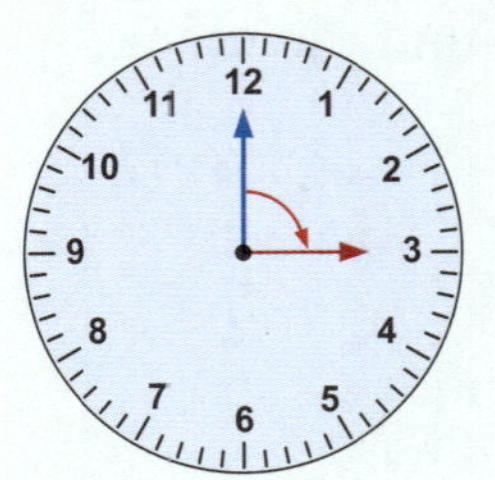

Exercise 3

Look at each clock.

Place a tick ✓ in the box if the angle between the hands is a right angle.

Place a cross ✗ in the box if it is not.

a 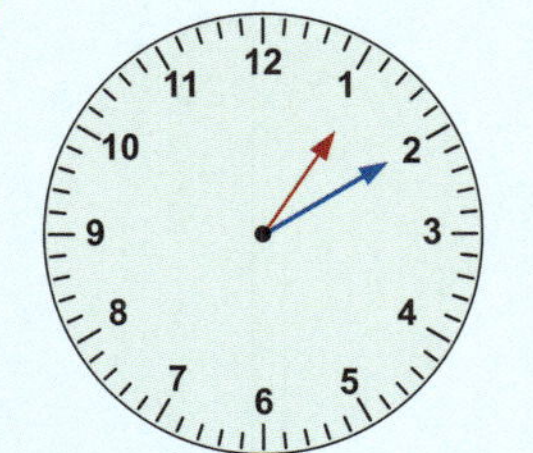

b 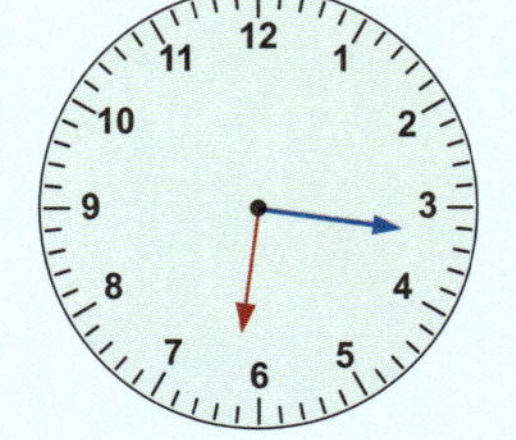

c 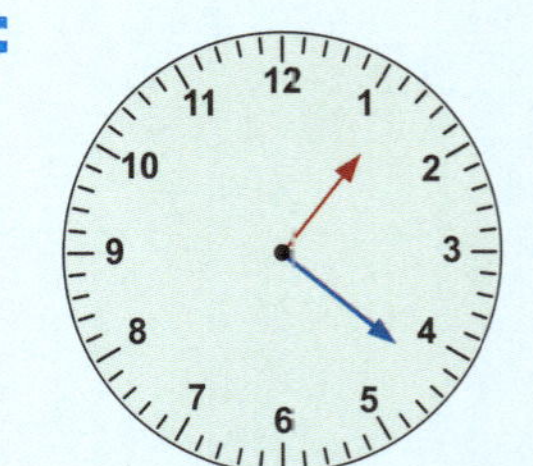

d 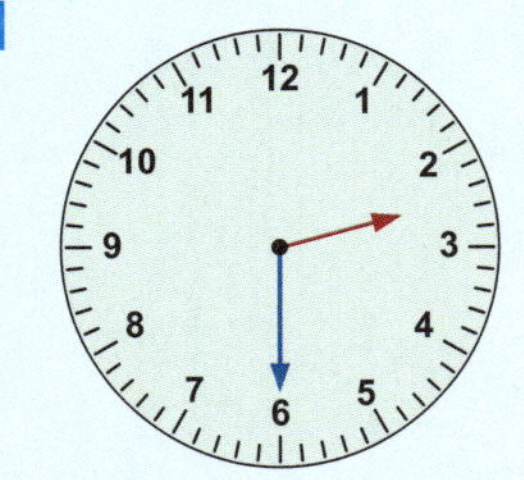

e 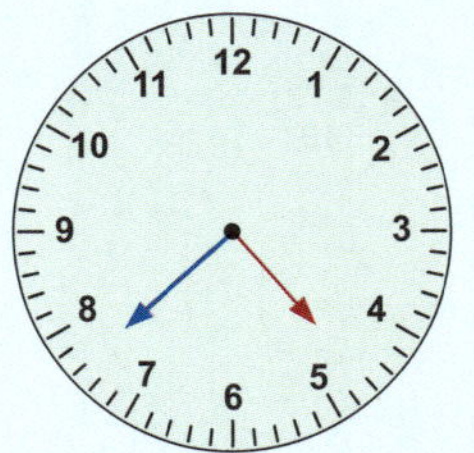

f 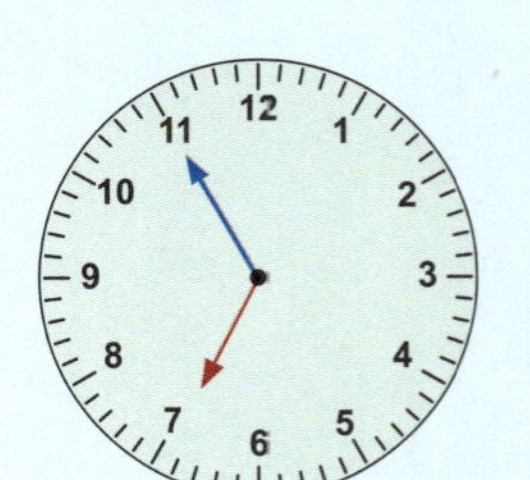

Exercise 4

Draw *two* arrows which make a right angle with the given arrow.

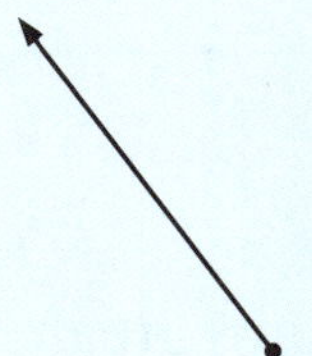

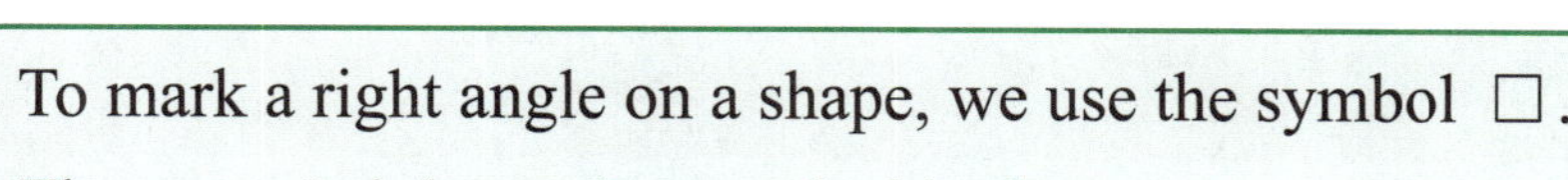

To mark a right angle on a shape, we use the symbol □.

There are 3 right angles *inside* this shape.

There are 2 right angles on the *outside* of this shape.

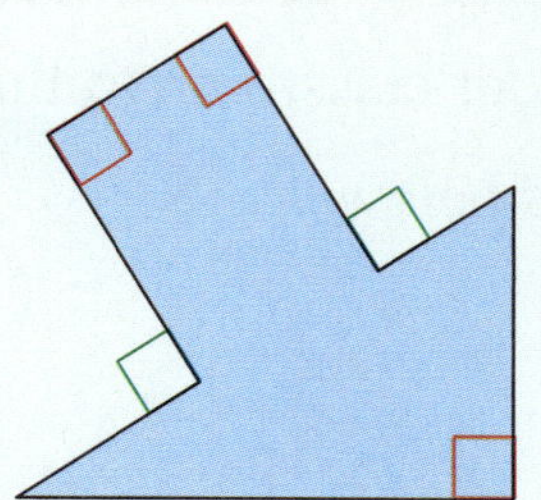

Exercise 5

Mark the right angles which are *inside* each shape:

a

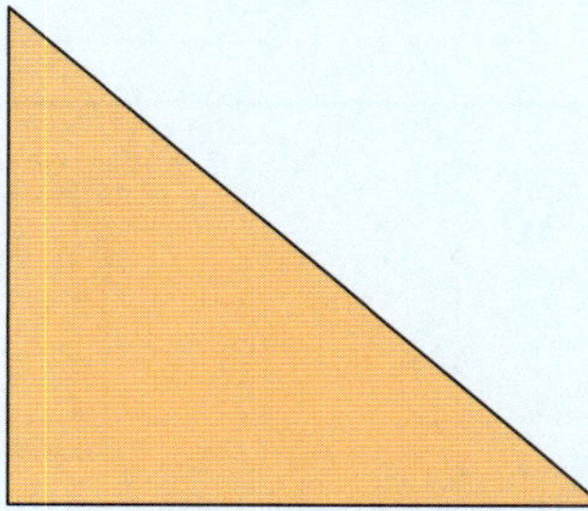

b

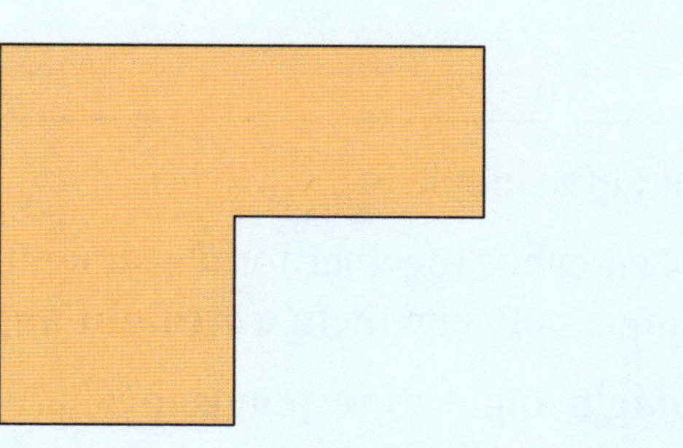

Exercise 6

Mark *all* of the right angles on the inside and outside of each shape:

a

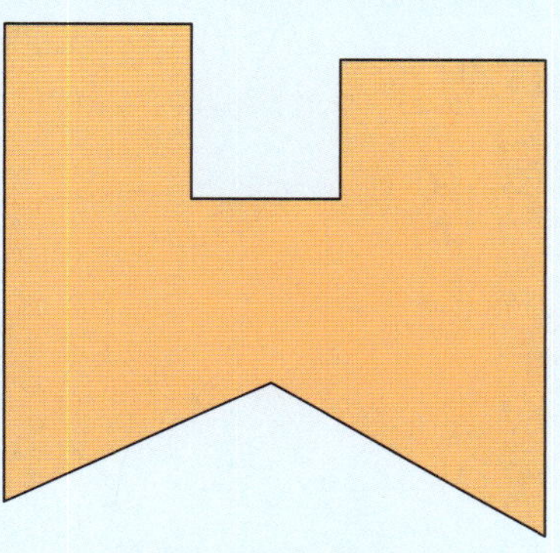

b

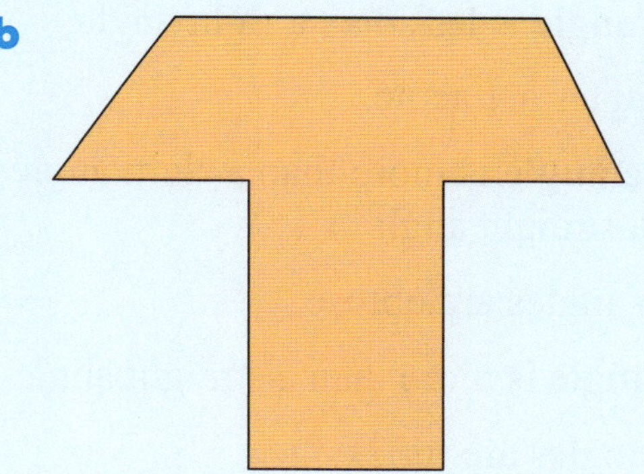

Activity

Look carefully at the picture alongside.

Mark all right angles you see.

Discussion

Look around your classroom to find items which have right angles.

List eight items below.

1 ______________ **2** ______________

3 ______________ **4** ______________

5 ______________ **6** ______________

7 ______________ **8** ______________

Look at the clock.

The two hands together form a straight line, so we call the angle between them a **straight angle**.

A straight angle corresponds to a half turn.

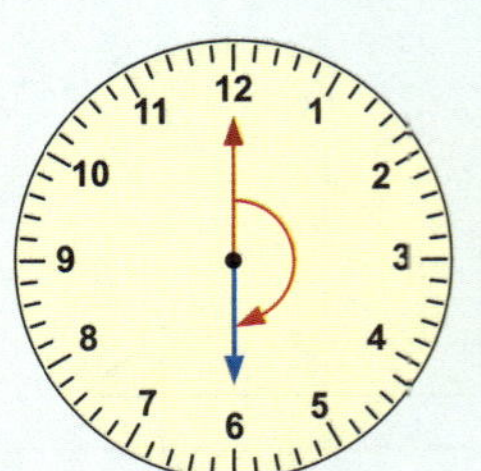

Look at the star.

An **acute angle** is less than a right angle.

The red angles are acute.

An **obtuse angle** is more than a right angle but less than a straight angle.

The green angles are obtuse.

A **reflex angle** is more than a straight angle.

The blue angles are reflex.

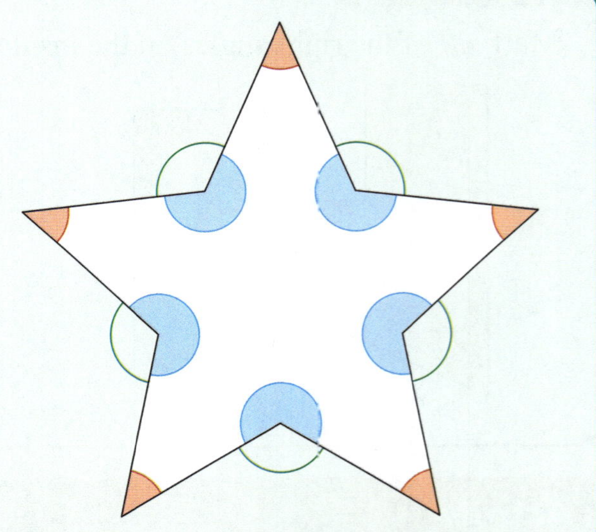

Exercise 7

Label each angle as an *acute angle*, a *right angle*, an *obtuse angle*, a *straight angle*, or a *reflex angle*:

a

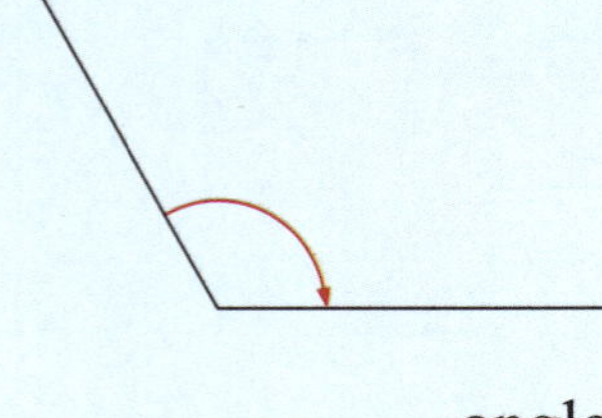

______________ angle

b

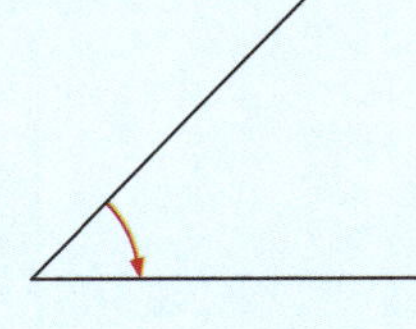

______________ angle

c

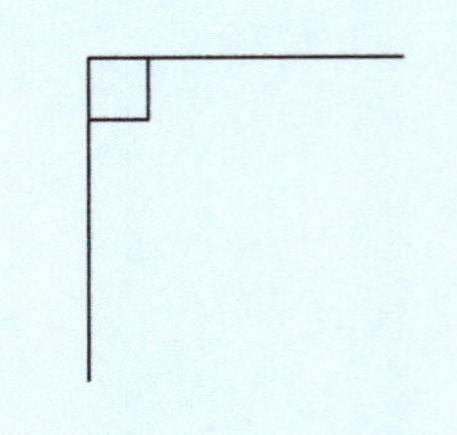

______________ angle

d

________________ angle

e

________________ angle

f

________________ angle

g

________________ angle

h

________________ angle

i

________________ angle

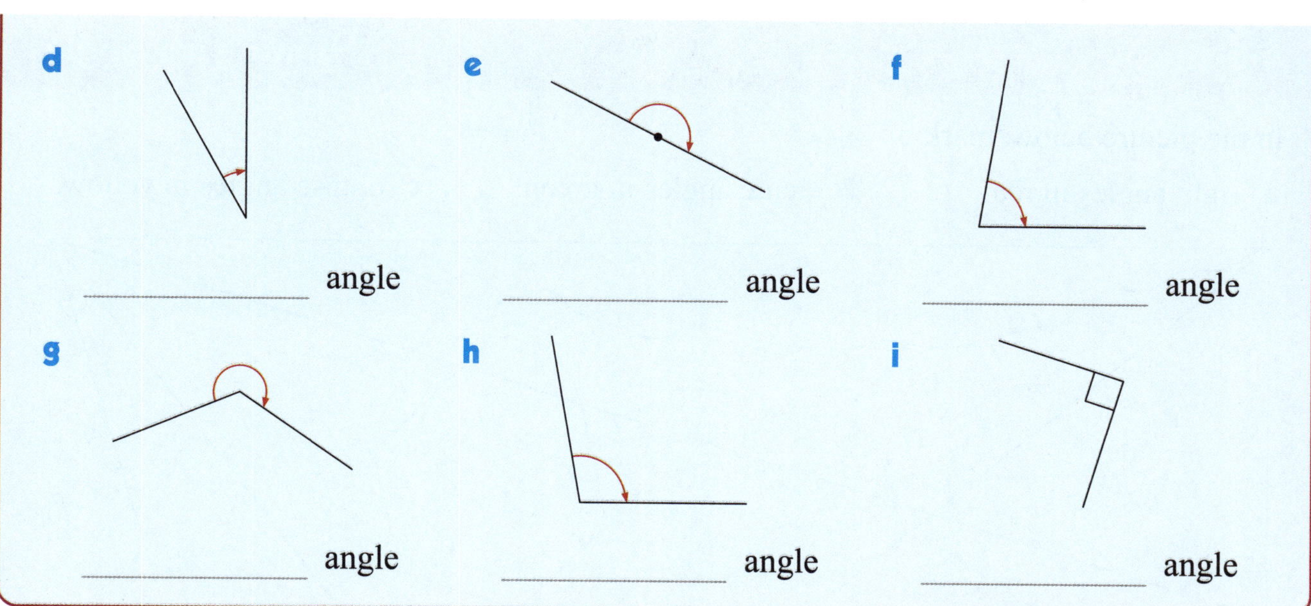

Exercise 8

Describe each angle as an acute angle, a right angle, an obtuse angle, or a reflex angle.

a

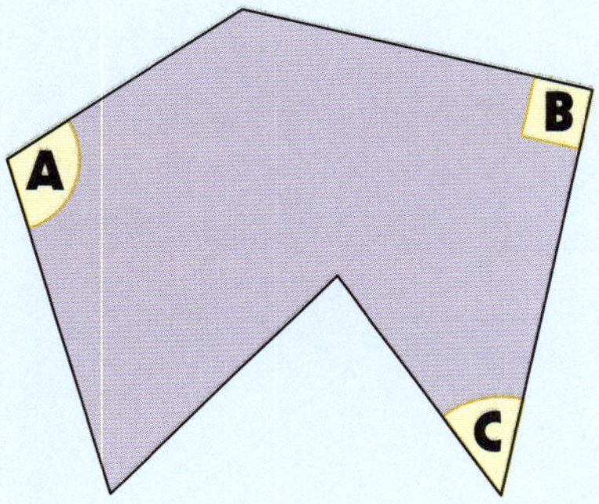

A is ________________________________

B is ________________________________

C is ________________________________

b

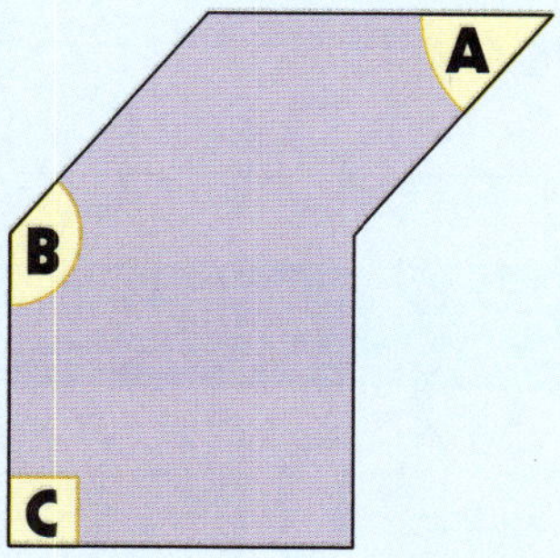

A is ________________________________

B is ________________________________

C is ________________________________

c

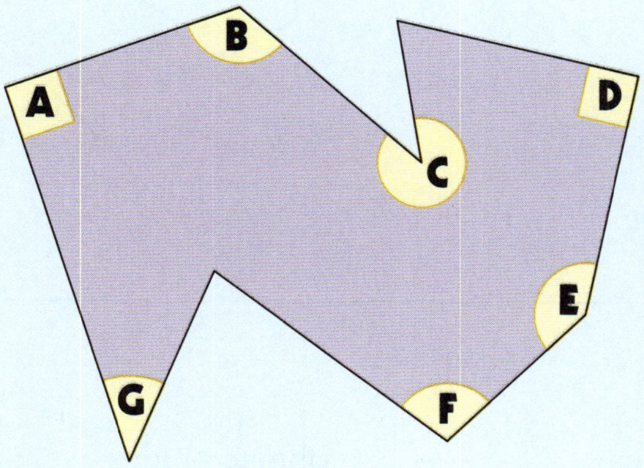

A is ________________________

B is ________________________

C is ________________________

D is ________________________

E is ________________________

F is ________________________

G is ________________________

Activity — Hunting for angles

In the picture below, mark:

a right angles in red **b** acute angles in green **c** obtuse angles in yellow.

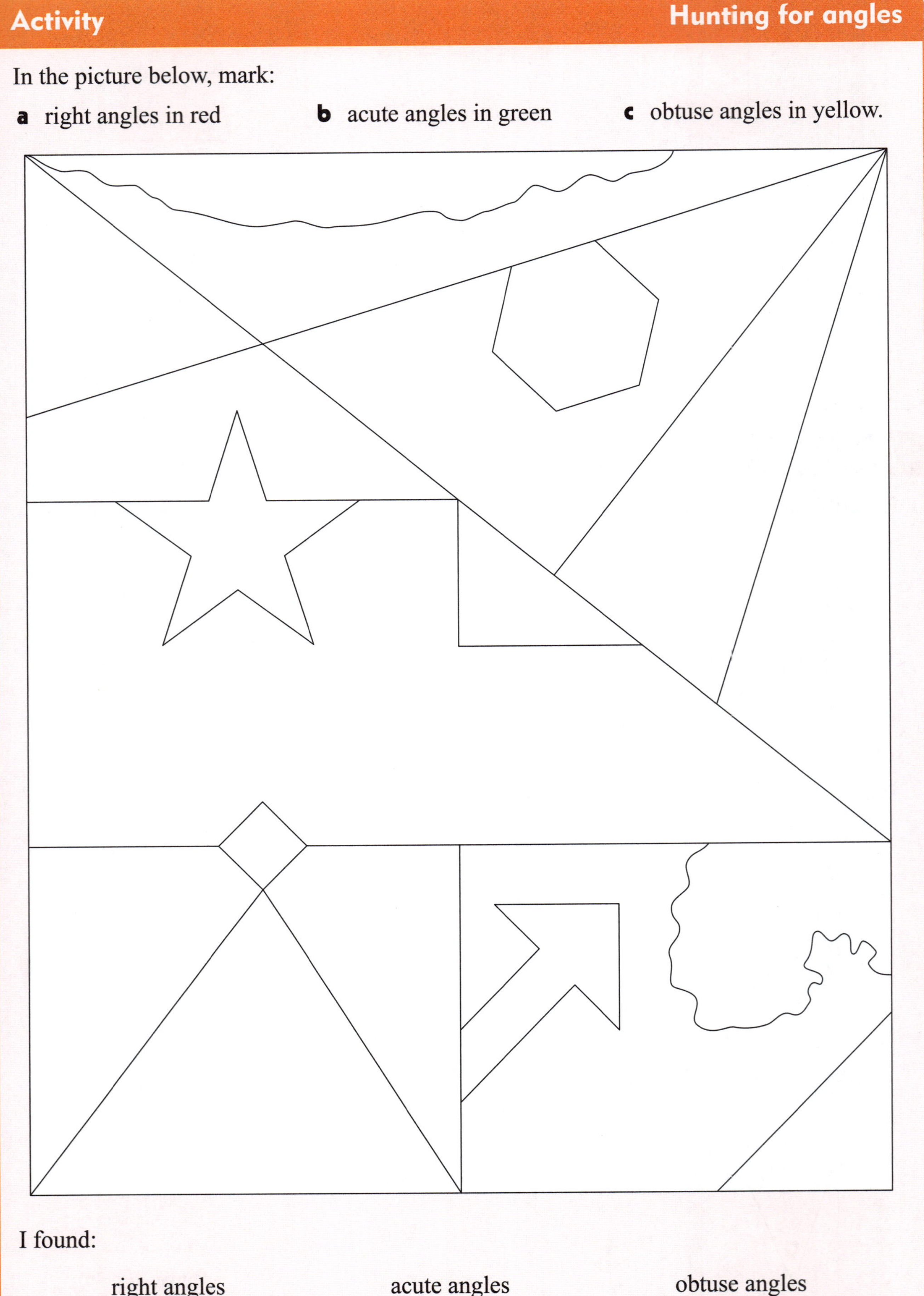

I found:

_______ right angles _______ acute angles _______ obtuse angles

Puzzle

Draw a shape which has 4 angles inside it.

There should be 2 acute angles, a reflex angle, and a right angle.

Label each angle clearly.

Revision

1 Match each turn with its description:

full turn ● ●

half turn ● ●

quarter turn ● ●

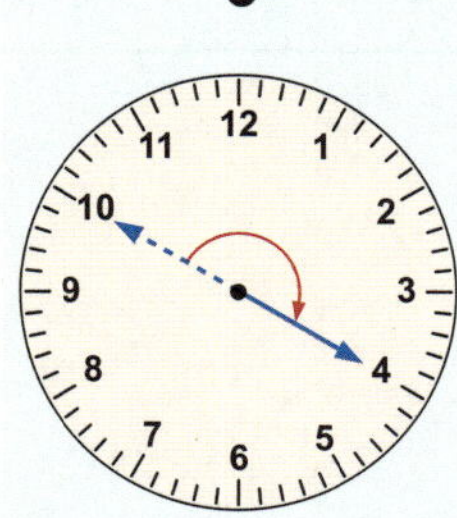

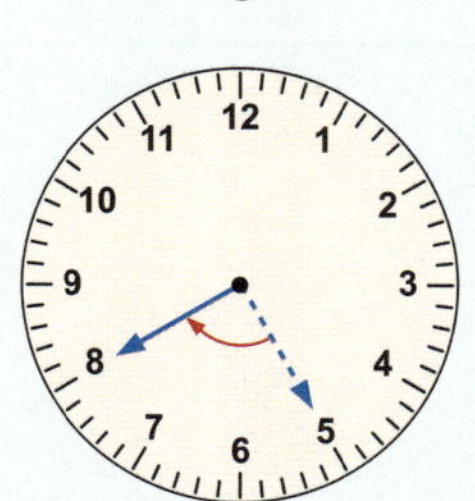

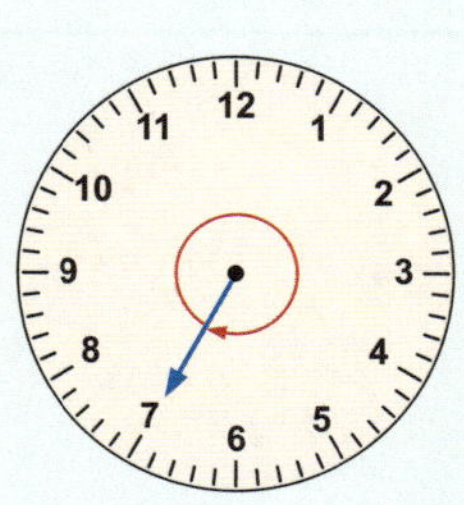

2 Complete:

A right angle has the same size as a .... turn.

3 Draw an arrow which makes a straight angle with the given arrow.

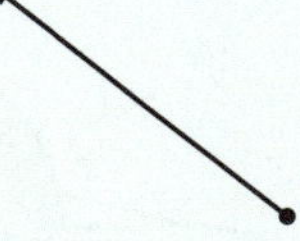

4 Mark *all* of the right angles on the inside and outside of this shape.

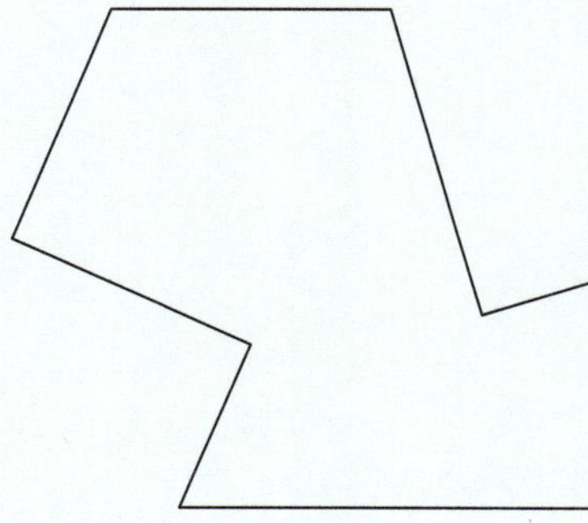

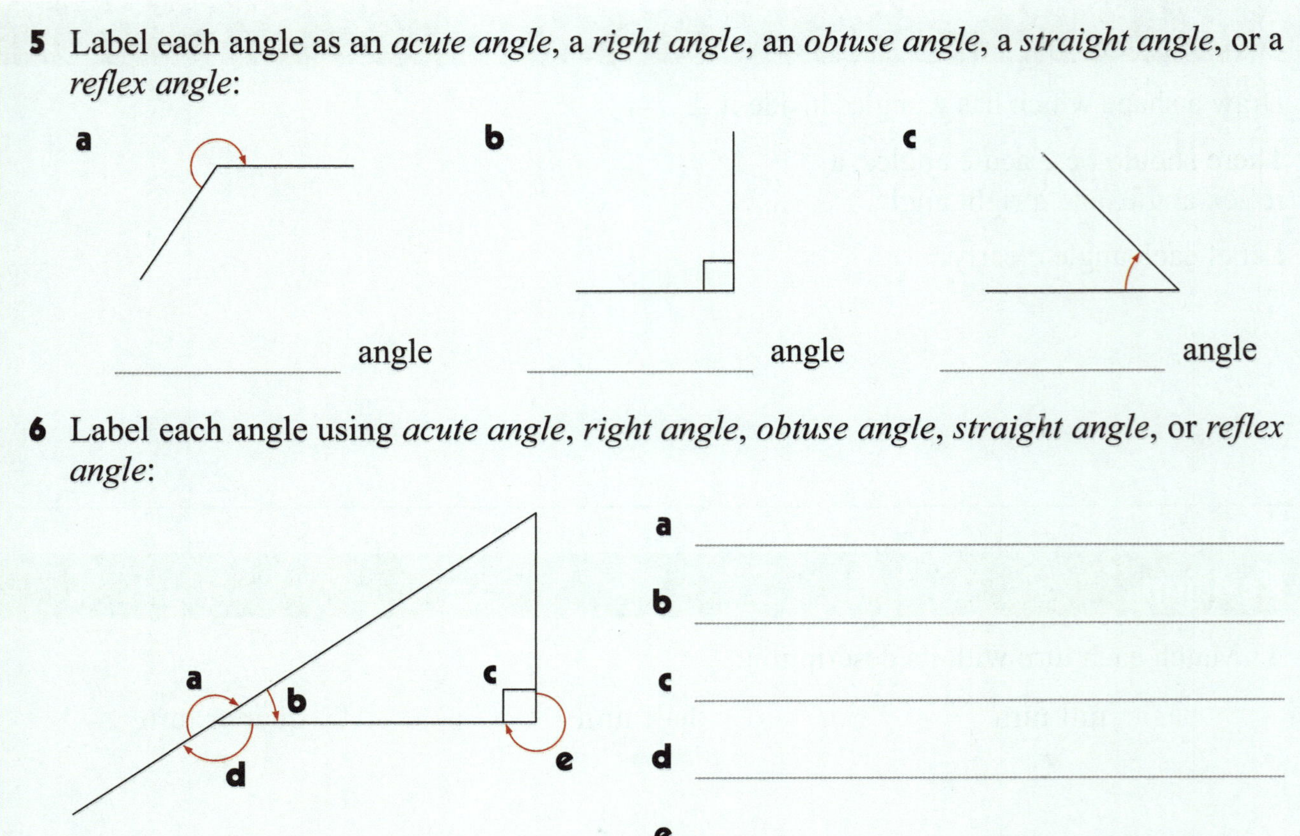

5 Label each angle as an *acute angle*, a *right angle*, an *obtuse angle*, a *straight angle*, or a *reflex angle*:

a ______________ angle

b ______________ angle

c ______________ angle

6 Label each angle using *acute angle*, *right angle*, *obtuse angle*, *straight angle*, or *reflex angle*:

a ______________________________

b ______________________________

c ______________________________

d ______________________________

e ______________________________

CHAPTER 10: DECIMAL NUMBERS

We have seen how whole numbers are written using digits in different place value columns.

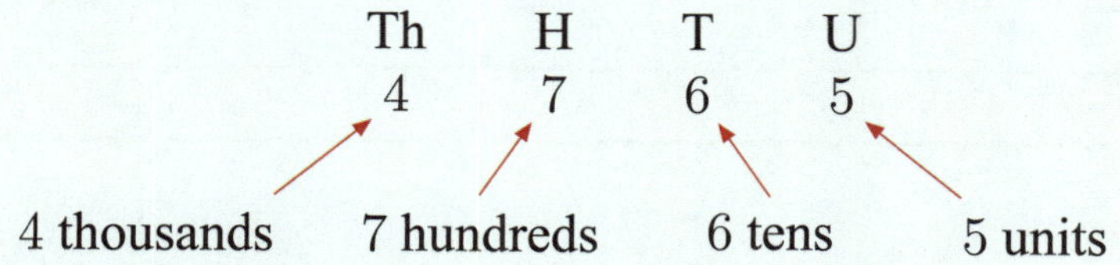

Fractions of a whole can also be written using digits in different place value columns.

To the *right* of the units place, we write a **decimal point**. This separates the *wholes* on its left from the *fractions* on its right.

The first place after the decimal point is for **tenths**.

One tenth is written as $\frac{1}{10}$ or 0.1

no units — decimal point — one tenth

In some countries, the decimal point is written as a comma. 0,1

We read 0.1 as "zero point one" or "zero decimal one".

A number written with a decimal point is called a **decimal number**.

Exercise 1

Complete the table.

Decimal	*Words*	*Meaning*	*Fraction*
0.4	zero point four	zero wholes and four tenths	$\frac{4}{10}$
0.3			
0.6			
0.9			

Exercise 2

Write these decimal numbers using numerals.

a zero point one ______

b zero point three ______

c zero decimal eight ______

d zero decimal five ______

Exercise 3

Write as a decimal number:

a two tenths = ______

b seven tenths = ______

c $\frac{5}{10}$ = ______

d $\frac{8}{10}$ = ______

Discussion

Why is the decimal point so important?

Without the decimal point, we would not know the ______ ______ of any digit.

Exercise 4

Match the decimal number to the correct fraction.

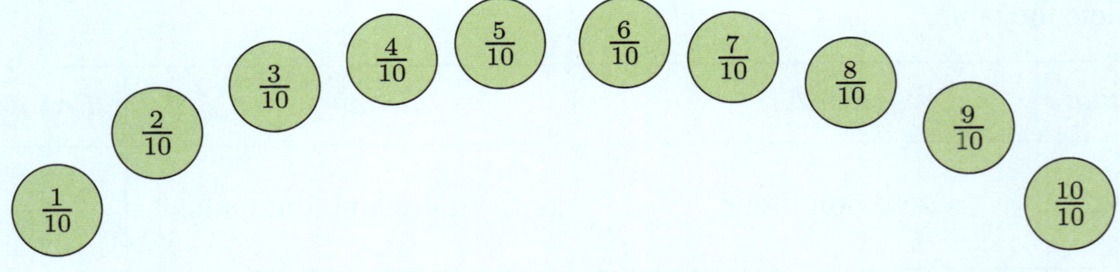

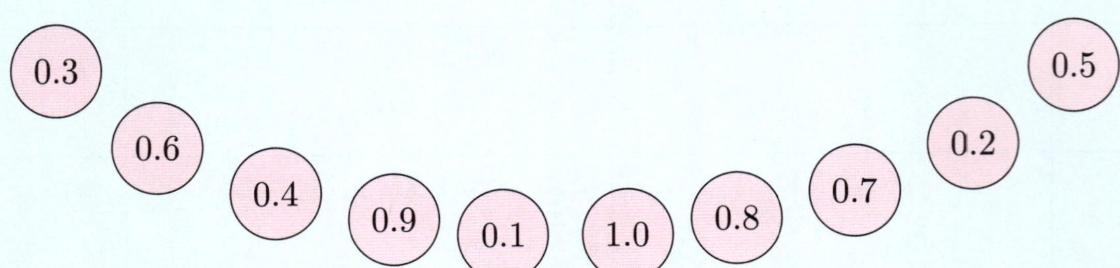

Exercise 5

Write as a fraction and as a decimal:

a

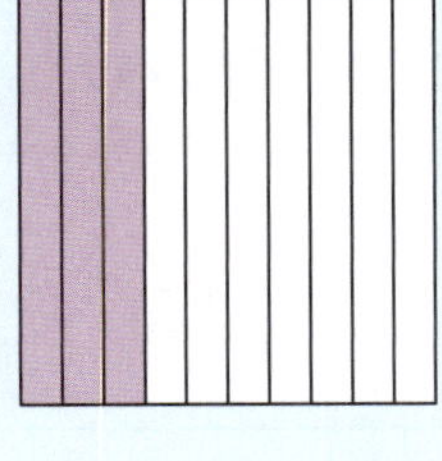

______ or ______

b

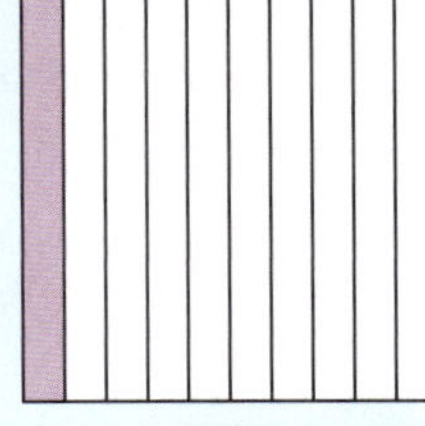

______ or ______

c

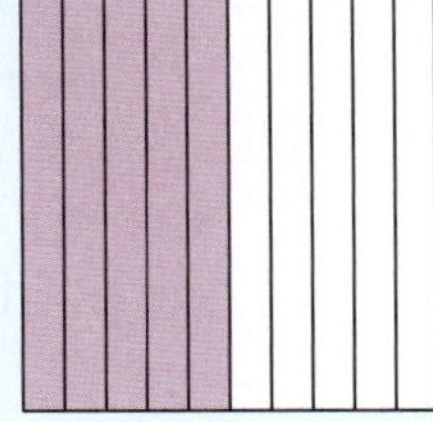

______ or ______

d

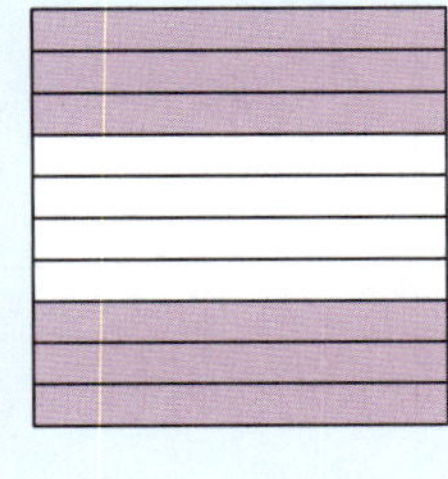

______ or ______

e

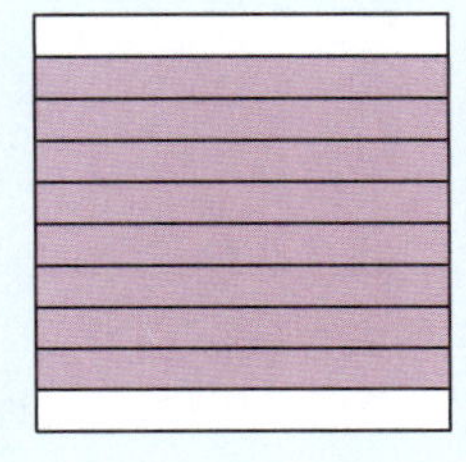

______ or ______

f

______ or ______

Exercise 6

Shade the amount given.

a 0.4

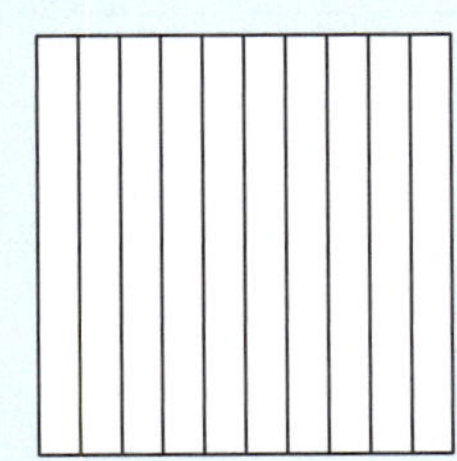

b 0.7

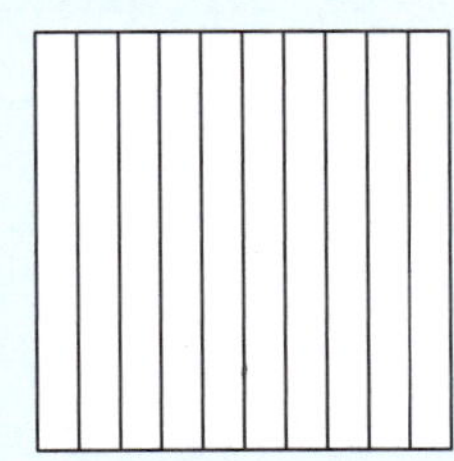

c 1.0

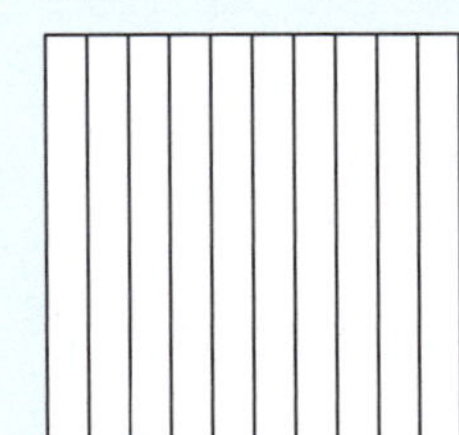

d 0.3

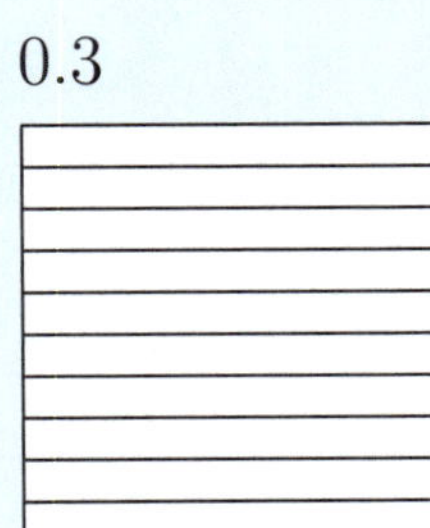

e 0.8

f 0.6

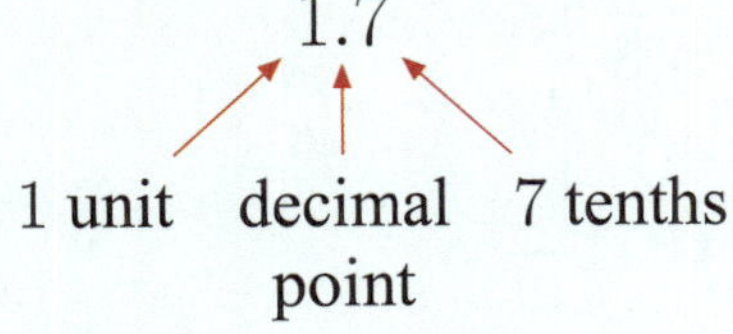

"one point seven"

$$1.7 = 1\frac{7}{10}$$

Exercise 7

Complete the table.

Decimal	*Words*	*Meaning*	*Mixed number*
1.7	one point seven	one whole and seven tenths	$1\frac{7}{10}$
1.4			
2.5			
		two wholes and two tenths	
	four point seven		
			$5\frac{3}{10}$

Exercise 8

Write as a mixed number:

a 2.6 ______ **b** 2.9 ______ **c** 3.8 ______

d 7.4 ______ **e** 8.3 ______ **f** 10.9 ______

Exercise 9

Write as a decimal number:

a $2\frac{7}{10}$ ______ **b** $3\frac{4}{10}$ ______ **c** $8\frac{9}{10}$ ______

d $12\frac{3}{10}$ ______ **e** $6\frac{6}{10}$ ______ **f** $11\frac{5}{10}$ ______

Exercise 10

Write as a mixed number and as a decimal:

a 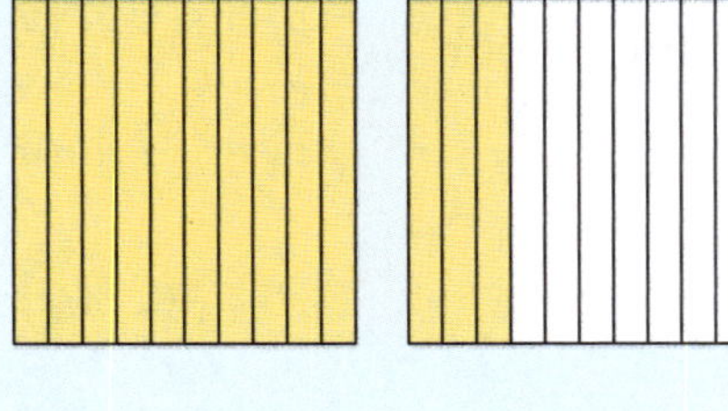

b 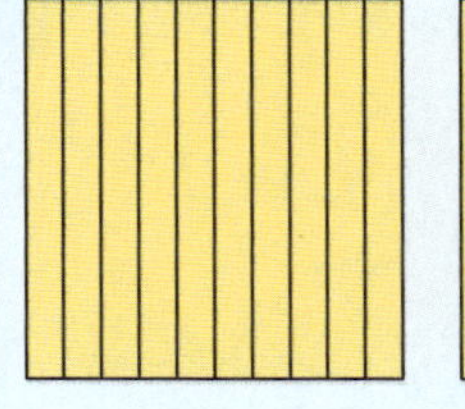

______ or ______   ______ or ______

c 

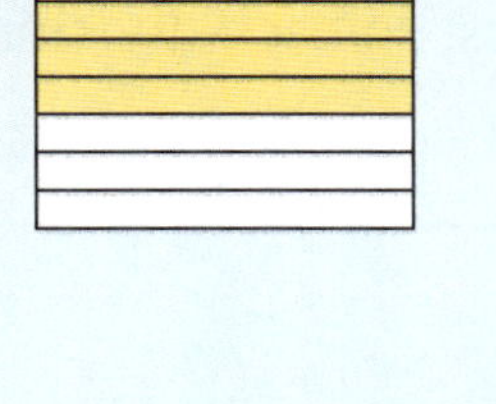

______ or ______

d 

______ or ______

The decimal point tells us which place a digit is in.

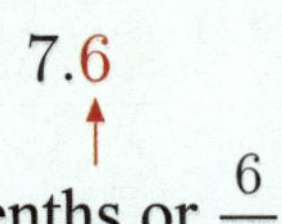

7.6 — 6 tenths or $\frac{6}{10}$

16.9 — 6 units or 6

67.8 — 6 tens or 60

Exercise 11

Give the place value of the digit 8.

a 8.4 ______   **b** 7.8 ______

c 18.9 ______   **d** 19.8 ______

e 84.1 ______   **f** 801.2 ______

Exercise 12

Give the place value of the digit 4.

a four point nine ____________

b seventeen point four ____________

c twenty four point eight ____________

d four hundred point three ____________

This number line begins at zero and ends at 2.

Each whole is divided in 10 equal spaces, so each division represents 0.1 or $\frac{1}{10}$.

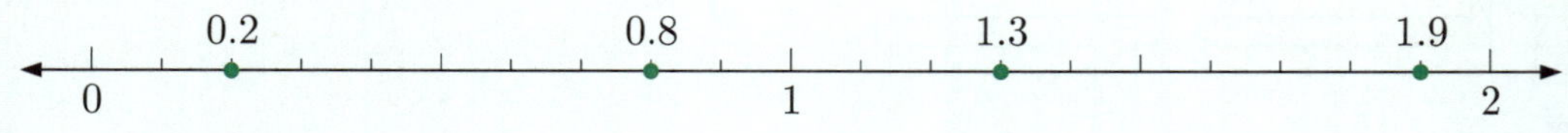

To reach 0.8 or $\frac{8}{10}$, we count eight divisions from zero.

Exercise 13

Label the decimal numbers.

a

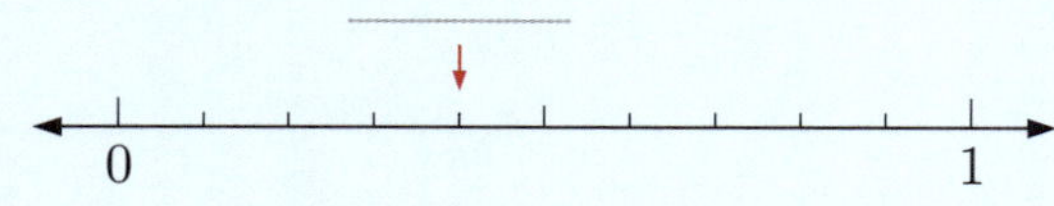

b

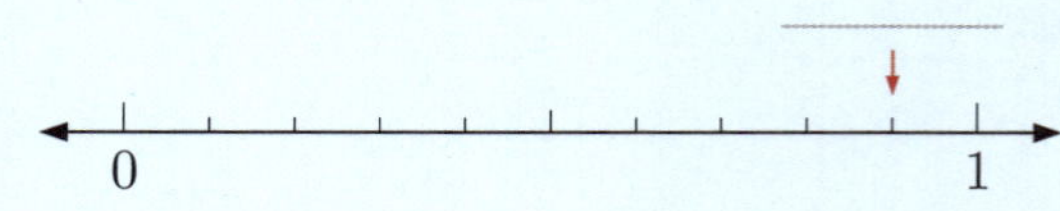

c

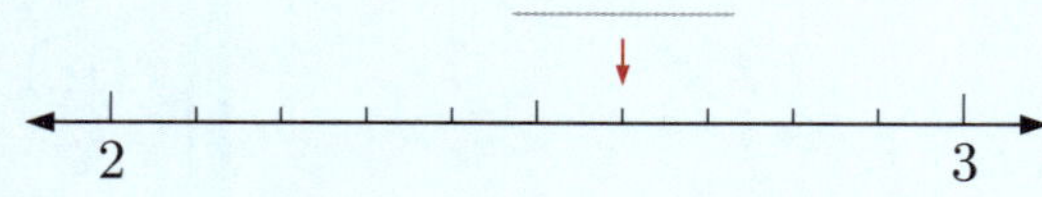

d

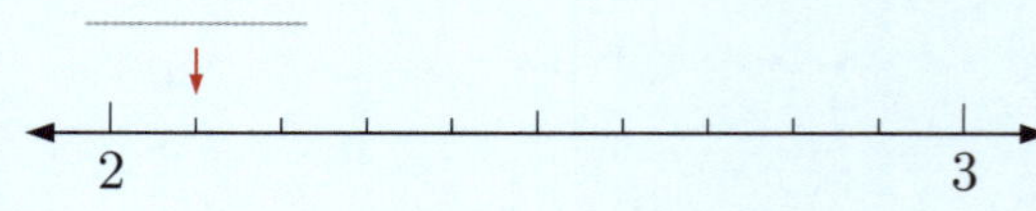

e

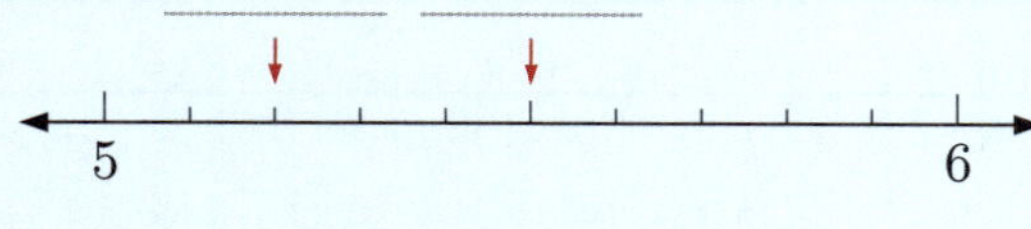

f

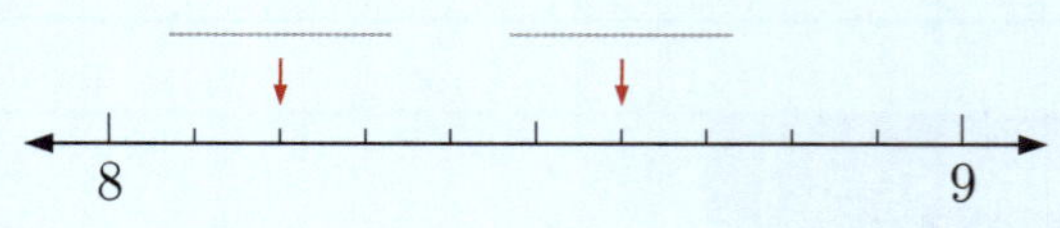

g

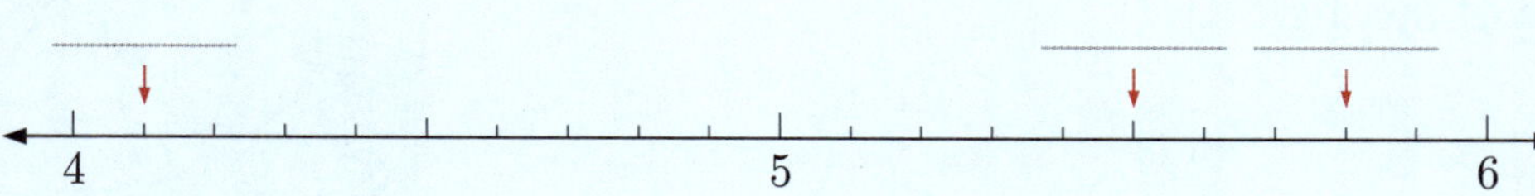

h

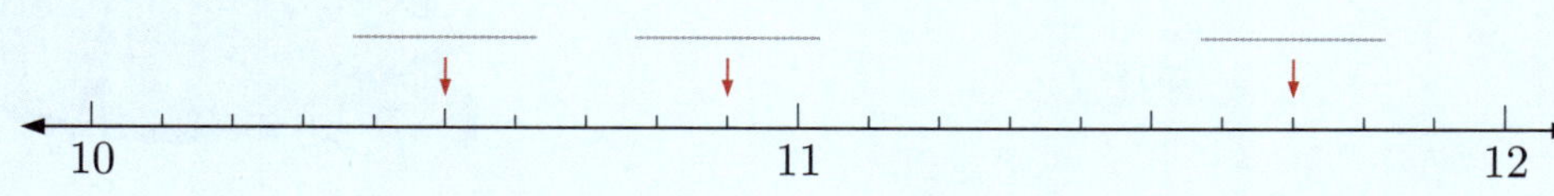

Exercise 14

Mark each decimal number on the number line:

a 1.3, 1.7, 2.1, 2.9

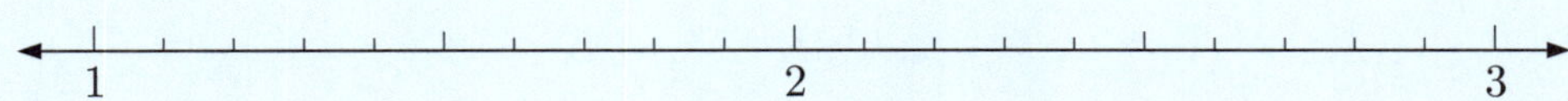

b 3.7, 3.8, 4.4, 4.8

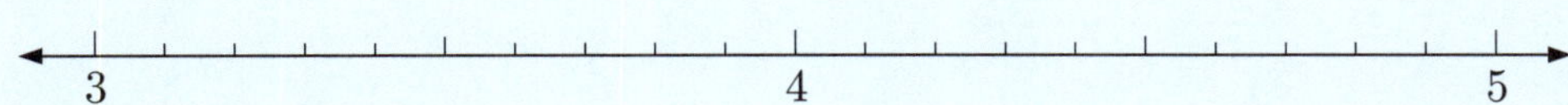

c 5.6, 6.2, 6.9, 7.3, 7.7

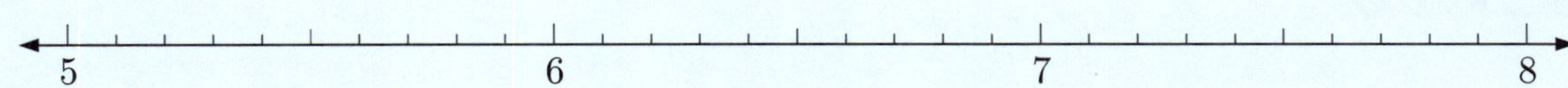

Exercise 15

Complete each number line:

a

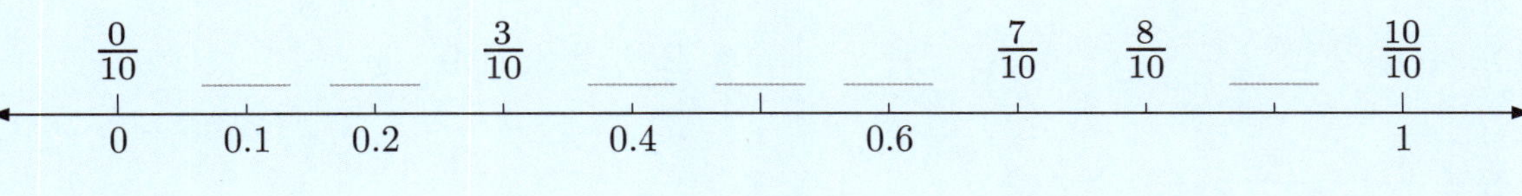

b

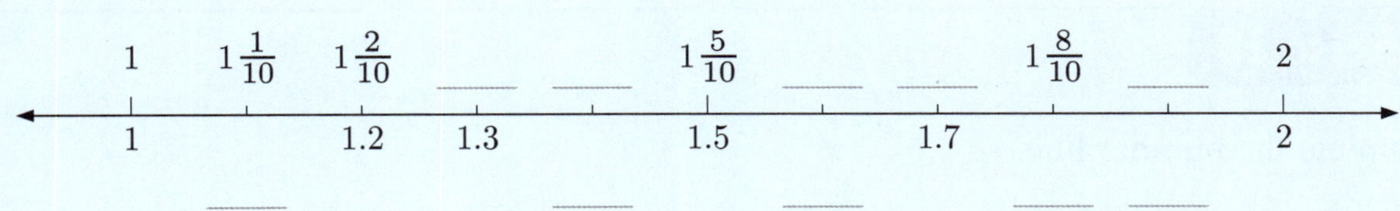

On a number line, the *least* number is to the *left*,
and the *greatest* number is to the *right*.

We can write 2.4, 2.7, 2.1 in order.

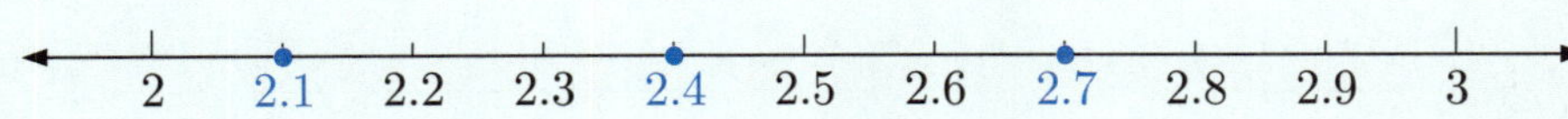

From least to greatest, we write 2.1, 2.4, 2.7.

From greatest to least, we write 2.7, 2.4, 2.1.

Exercise 16

0 0.1 0.2 0.3 0.4 0.5 0.6 0.7 0.8 0.9 1 1.1 1.2 1.3 1.4 1.5 1.6 1.7 1.8 1.9 2

Use the number line to order the decimal numbers from least to greatest.

a 0.3, 0.8, 0.6

....

b 1.2, 1.1, 1.5

....

c 0.9, 1.4, 0.7

....

d 1.6, 1.2, 0.9

....

Exercise 17

1 1.1 1.2 1.3 1.4 1.5 1.6 1.7 1.8 1.9 2 2.1 2.2 2.3 2.4 2.5 2.6 2.7 2.8 2.9 3

Use the number line to order the decimal numbers from greatest to least.

a 1.2, 1.6, 1.9

....

b 2.5, 2.1, 2.7

....

c 1.8, 1.1, 2.2

....

d 2.6, 2.8, 1.9

....

Exercise 18

Complete the number line.

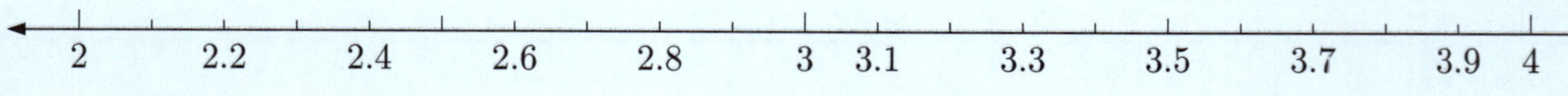

Arrange from least to greatest:

a 2.8, 2.4, 2.6

....

b 3.1, 2.6, 2.9

....

c 2.6, 3.4, 2.2, 3.8

....

d 4.0, 3.3, 3.1, 2.9

....

To order decimals *without* using a number line, look at the largest place value first.

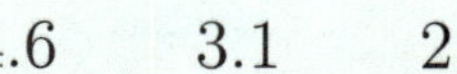

4.6 3.1 2.9 4.2

The largest place value is units.

2.9 has the least units.

3.1 has the next least units.

4.6 and 4.2 both have 4 units.

4.2 has fewer tenths than 4.6 .

From least to greatest, the numbers are: 2.9, 3.1, 4.2, 4.6 .

Exercise 19

Order each group of decimal numbers from least to greatest:

a 1.3, 1.1, 1.5 ....

b 2.4, 2.6, 1.9 ....

c 3.2, 3.1, 2.9, 2.8 ....

d 4.2, 1.2, 2.7, 1.5 ....

Exercise 20

Order each group of decimal numbers from greatest to least:

a 2.5, 1.8, 2.8 ....

b 3.1, 3.6, 2.7 ....

c 2.7, 1.8, 3.7, 2.9 ....

d 2.6, 3.2, 2.4, 2.9 ....

Exercise 21

Order from least to greatest: $3\frac{2}{10}$, 2.8, 3.4, $\frac{29}{10}$, $2\frac{7}{10}$

....

To add tenths, we can count up on a number line.

$0.2 + 0.4 = 0.6$

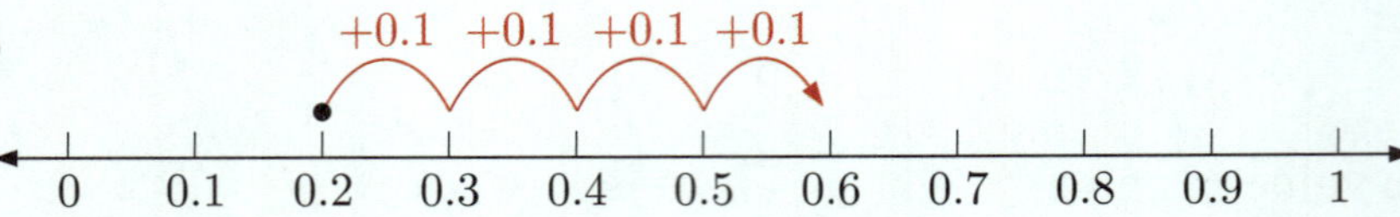

Exercise 22

Use the number line to complete each addition:

a $0.1 + 0.3 =$ ....

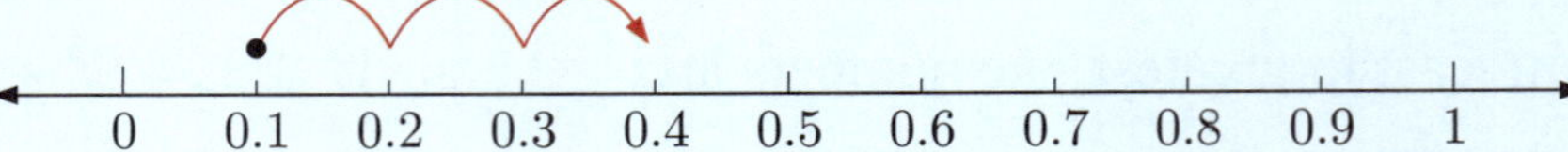

b $0.3 + 0.2 =$ ....

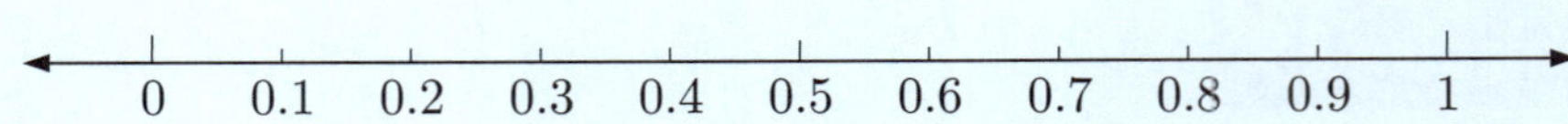

c $0.4 + 0.5 =$ ....

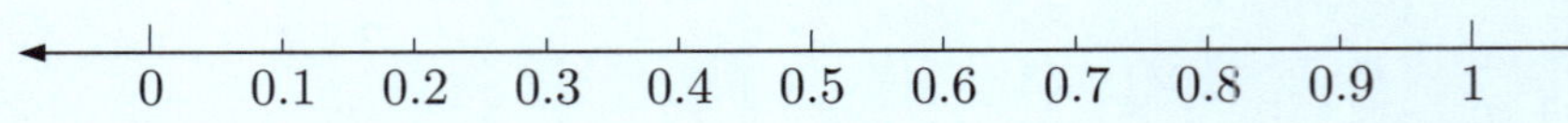

d $0.6 + 0.2 =$ ....

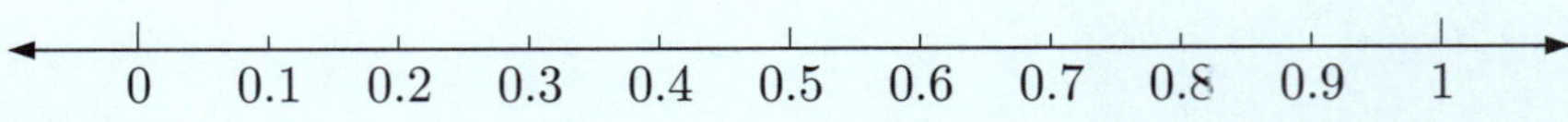

Exercise 23

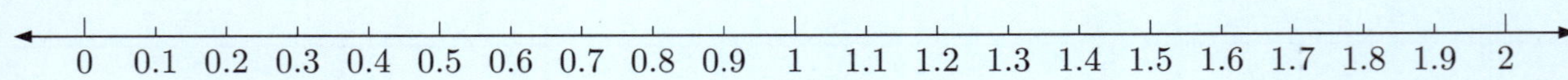

Use the number line to complete each addition:

a $0.6 + 0.5 =$ .... **b** $0.7 + 0.4 =$ .... **c** $1.1 + 0.3 =$ ....

d $1.2 + 0.4 =$ .... **e** $1.3 + 0.5 =$ .... **f** $1.4 + 0.5 =$ ....

Exercise 24

Complete these additions.

a $1 + 4 =$ .... **b** $5 + 2 =$ .... **c** $3 + 8 =$ ....

$0.1 + 0.4 =$ .... $0.5 + 0.2 =$ .... $0.3 + 0.8 =$ ....

d $7 + 5 =$ .... **e** $11 + 4 =$ .... **f** $6 + 12 =$ ....

$0.7 + 0.5 =$ .... $1.1 + 0.4 =$ .... $0.6 + 1.2 =$ ....

Exercise 25

Complete each addition mentally:

a $0.3 + 0.3 =$ ______ **b** $0.5 + 0.3 =$ ______ **c** $0.4 + 0.9 =$ ______

d $0.8 + 0.2 =$ ______ **e** $0.7 + 1.1 =$ ______ **f** $0.9 + 0.8 =$ ______

g $1.2 + 0.6 =$ ______ **h** $0.2 + 1.7 =$ ______ **i** $1.3 + 0.9 =$ ______

Exercise 26

0 0.1 0.2 0.3 0.4 0.5 0.6 0.7 0.8 0.9 1 1.1 1.2 1.3 1.4 1.5 1.6 1.7 1.8 1.9 2

Complete each addition, using the number line if necessary:

a $0.2 + 0.2 + 0.2 =$ ______ **b** $0.2 + 0.3 + 0.3 =$ ______

c $0.1 + 0.2 + 0.4 =$ ______ **d** $0.4 + 0.4 + 0.4 =$ ______

e $0.4 + 0.3 + 0.4 =$ ______ **f** $0.3 + 0.5 + 0.5 =$ ______

g $0.5 + 0.6 + 0.4 =$ ______ **h** $0.1 + 0.6 + 0.6 =$ ______

$0.9 - 0.5 = 0.4$

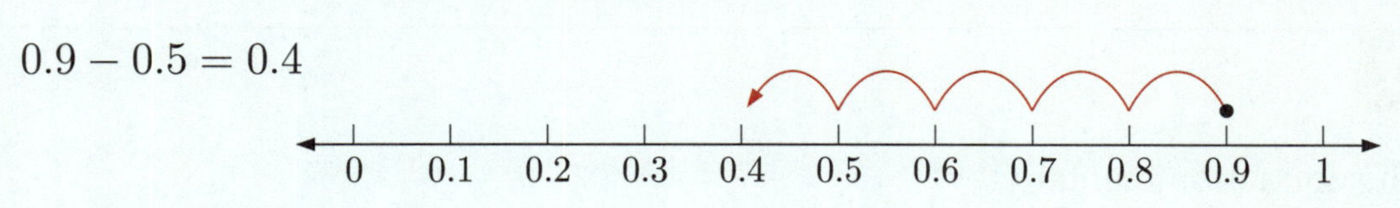

Exercise 27

Use the number line to complete each subtraction:

a $0.8 - 0.5 =$ ______

0 0.1 0.2 0.3 0.4 0.5 0.6 0.7 0.8 0.9 1

b $0.9 - 0.3 =$ ______

0 0.1 0.2 0.3 0.4 0.5 0.6 0.7 0.8 0.9 1

c $1 - 0.7 =$ ______

0 0.1 0.2 0.3 0.4 0.5 0.6 0.7 0.8 0.9 1

Exercise 28

0 0.1 0.2 0.3 0.4 0.5 0.6 0.7 0.8 0.9 1 1.1 1.2 1.3 1.4 1.5 1.6 1.7 1.8 1.9 2

Use the number line to complete each subtraction:

a $0.5 - 0.2 =$ ______ **b** $0.6 - 0.3 =$ ______ **c** $0.4 - 0.4 =$ ______

d $0.7 - 0.5 =$ ______ **e** $1.2 - 0.2 =$ ______ **f** $1.3 - 0.4 =$ ______

g $1.5 - 0.6 =$ ______ **h** $1.8 - 0.9 =$ ______ **i** $1.9 - 1.2 =$ ______

Exercise 29

Complete these subtractions.

a $8 - 4 =$ ______
$0.8 - 0.4 =$ ______

b $6 - 5 =$ ______
$0.6 - 0.5 =$ ______

c $10 - 3 =$ ______
$1.0 - 0.3 =$ ______

d $12 - 7 =$ ______
$1.2 - 0.7 =$ ______

e $15 - 8 =$ ______
$1.5 - 0.8 =$ ______

f $16 - 14 =$ ______
$1.6 - 1.4 =$ ______

Exercise 30

Complete each subtraction mentally:

a $0.8 - 0.3 =$ ______ **b** $0.6 - 0.4 =$ ______ **c** $1.0 - 0.5 =$ ______

d $1.5 - 1.3 =$ ______ **e** $1.3 - 0.7 =$ ______ **f** $1.5 - 0.2 =$ ______

g $1.1 - 0.8 =$ ______ **h** $1.4 - 0.9 =$ ______ **i** $1.7 - 1.1 =$ ______

The second place after the decimal point is for **hundredths**.

0.07

units tenths hundredths

"zero point zero seven"

$$0.07 = \frac{7}{100}$$

Exercise 31

Write as a decimal number:

a 1 hundredth ______

b 5 hundredths ______

c $\frac{3}{100} =$ ______

d $\frac{9}{100} =$ ______

2.64

units — decimal point — tenths — hundredths

"two point six four"

Exercise 32

Give the place value of the digit 7.

a 3.74 ______

b 7.36 ______

c 4.07 ______

d 17.93 ______

e 0.67 ______

f 72.13 ______

Exercise 33

Give the place value of the digit 1.

a zero point two one ______

b three point one five ______

c eighteen point zero three ______

d twenty one point four nine ______

Exercise 34

Write as a decimal number:

a 1 whole, 3 tenths, and 5 hundredths ______

b 2 wholes and 7 hundredths ______

c 4 wholes, 9 tenths, and 3 hundredths ______

d 3 wholes, 2 tenths, and 9 hundredths ______

The diagram shows 1 whole and 35 hundredths.

$$1\frac{35}{100}$$

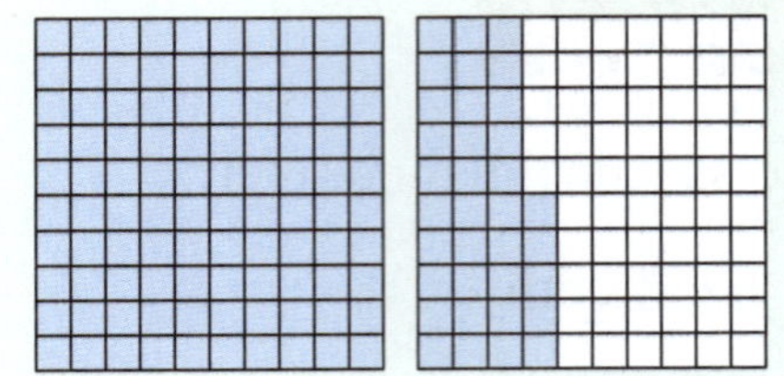

How can we write this mixed number as a decimal number?

30 hundredths is the same as 3 tenths.

As a decimal number, we write

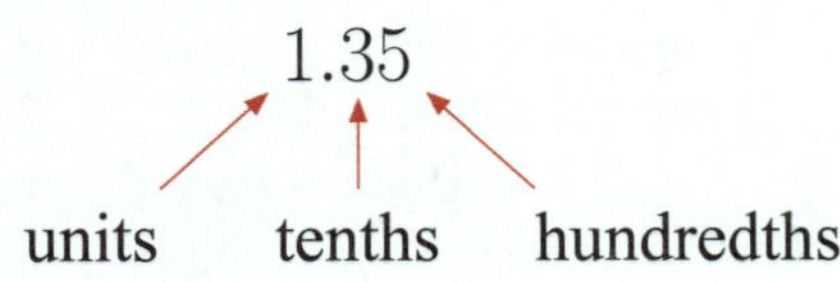

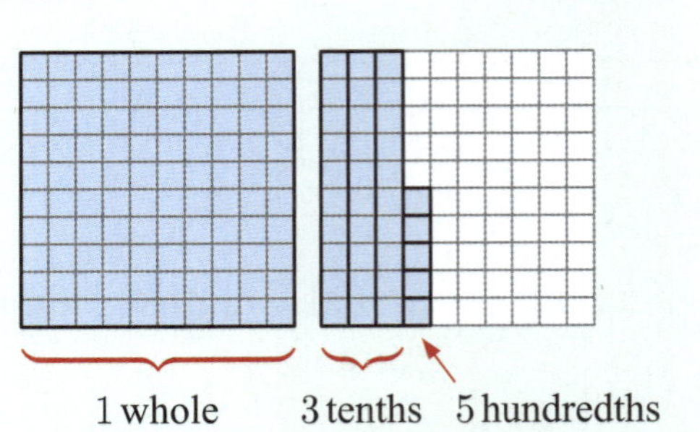

Discussion

To show 1.35 using blocks, we use:

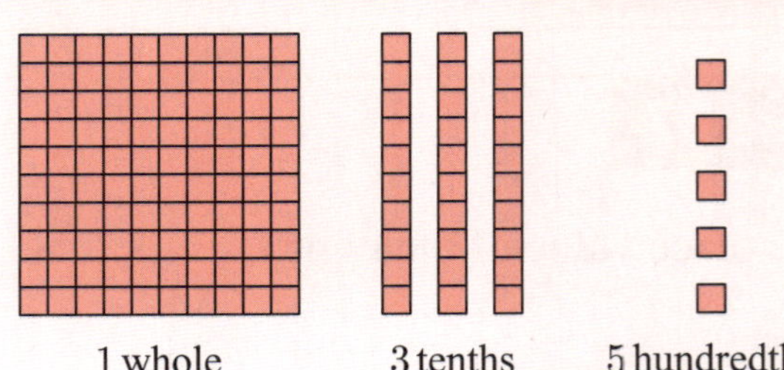

What decimal numbers are shown in each diagram?

a 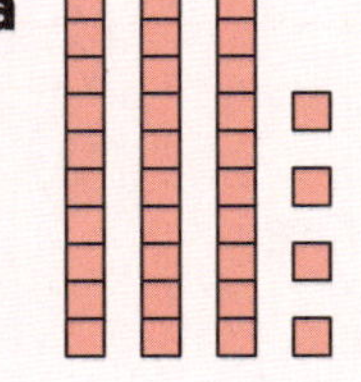

no wholes

____ tenths

____ hundredths

0.34

b 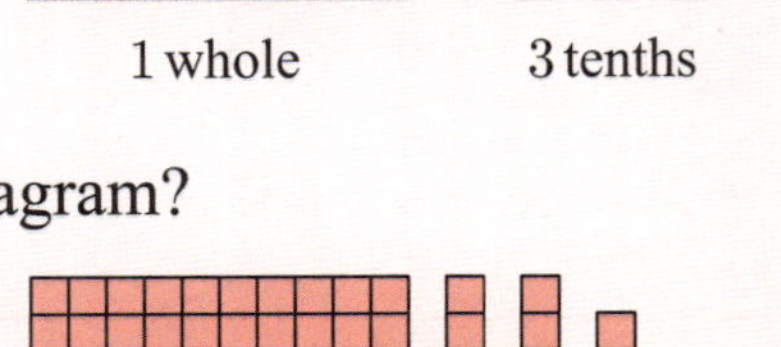

____ whole

____ tenths

____ hundredths

____

c 

____ whole

____ tenths

____ hundredths

____

d 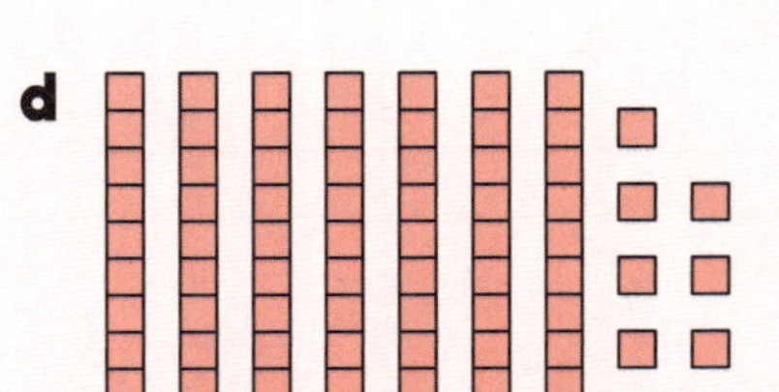

____ whole

____ tenths

____ hundredths

____

Exercise 35

What decimal number is shown?

a 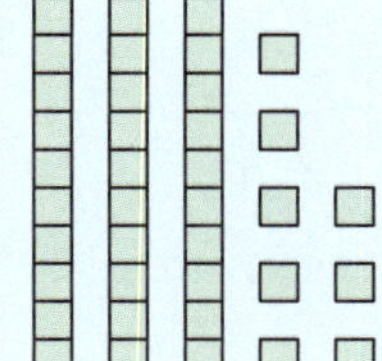

b 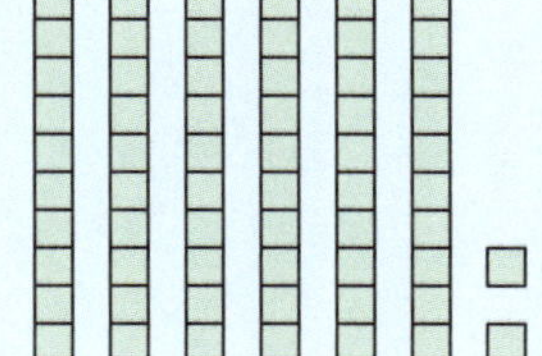

c 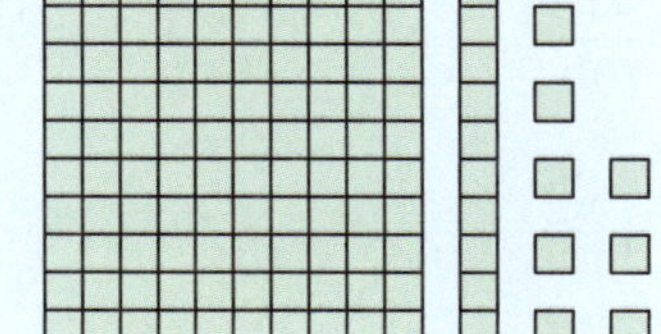

d 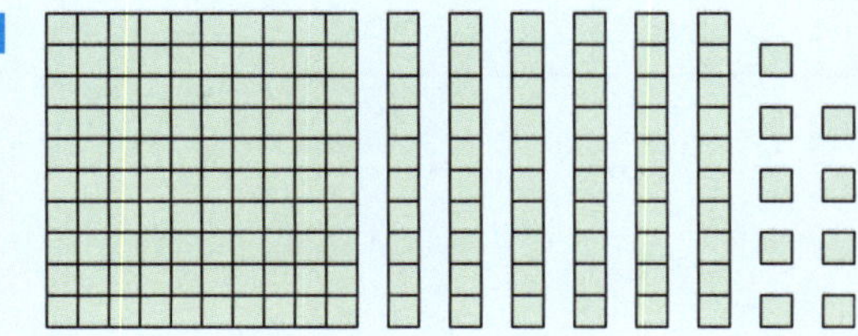

e 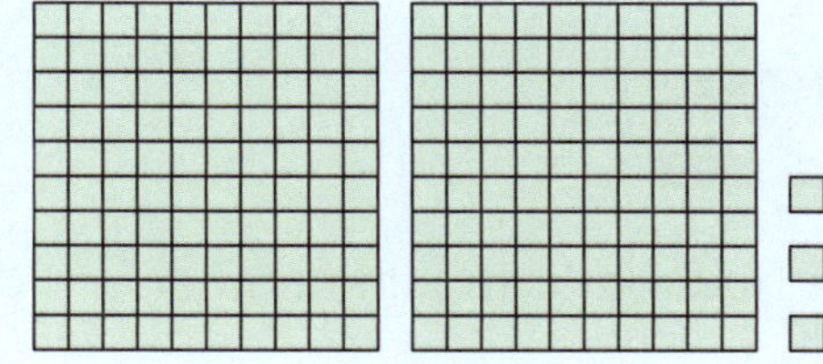

Exercise 36

Complete the table.

Fraction	*Diagram*	*Decimal*
$\frac{43}{100}$		
$\frac{27}{100}$		

Discussion

How can we quickly convert between a fraction with denominator 100, and a decimal number?

Exercise 37

Write as a decimal number:

a $\frac{17}{100}$ ______ **b** $\frac{84}{100}$ ______ **c** $\frac{9}{100}$ ______

d $2\frac{35}{100}$ ______ **e** $1\frac{7}{100}$ ______ **f** $5\frac{43}{100}$ ______

g $10\frac{28}{100}$ ______ **h** $13\frac{29}{100}$ ______ **i** $15\frac{2}{100}$ ______

Exercise 38

Write as a fraction or mixed number:

a 0.25 ______ **b** 0.68 ______ **c** 0.91 ______

d 0.02 ______ **e** 1.28 ______ **f** 1.93 ______

g 2.49 ______ **h** 7.63 ______ **i** 7.04 ______

Exercise 39

Write as a decimal number:

a $1\frac{6}{10}$ ______ **b** $10\frac{8}{10}$ ______ **c** $5\frac{3}{10}$ ______

$1\frac{6}{100}$ ______ $10\frac{8}{100}$ ______ $5\frac{3}{100}$ ______

Exercise 40

Complete these additions.

a $3+6=$ ______ **b** $7+5=$ ______ **c** $4+15=$ ______

$0.3+0.6=$ ______ $0.7+0.5=$ ______ $0.4+1.5=$ ______

$0.03+0.06=$ ______ $0.07+0.05=$ ______ $0.04+0.15=$ ______

Exercise 41

Complete these subtractions.

a $9 - 4 =$ ______

$0.9 - 0.4 =$ ______

$0.09 - 0.04 =$ ______

b $13 - 8 =$ ______

$1.3 - 0.8 =$ ______

$0.13 - 0.08 =$ ______

c $17 - 2 =$ ______

$1.7 - 0.2 =$ ______

$0.17 - 0.02 =$ ______

Revision

1 Complete the table.

Decimal	*Words*	*Meaning*	*Fraction*
	zero point seven		
		zero wholes and two tenths	

2 Write as a fraction and as a decimal:

a 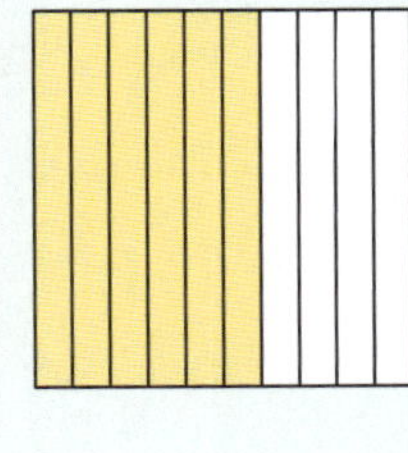

______ or ______

b 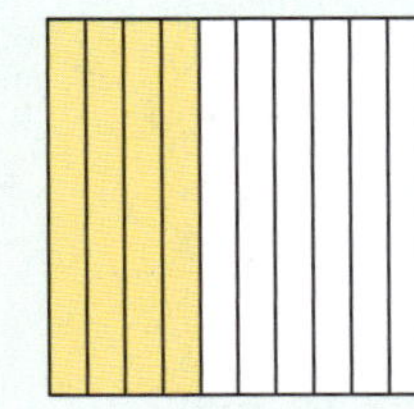

______ or ______

3 Complete the table.

Decimal	*Words*	*Meaning*	*Mixed number*
		two wholes and eight tenths	
4.3			
	five point six		

4 Write as a mixed number and as a decimal number:

__________ or __________

5 Give the place value of the digit 3.

a 3.2 __________

b 6.3 __________

c thirty point seven __________

d seventeen point three __________

6 Label the decimal numbers.

a

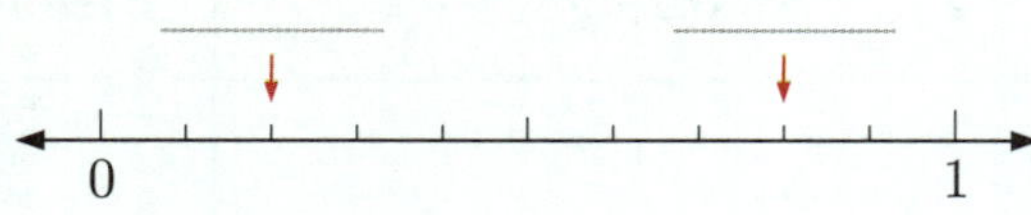

b

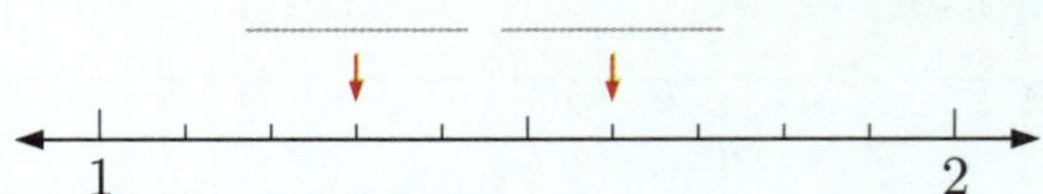

7 **a** Complete the number line.

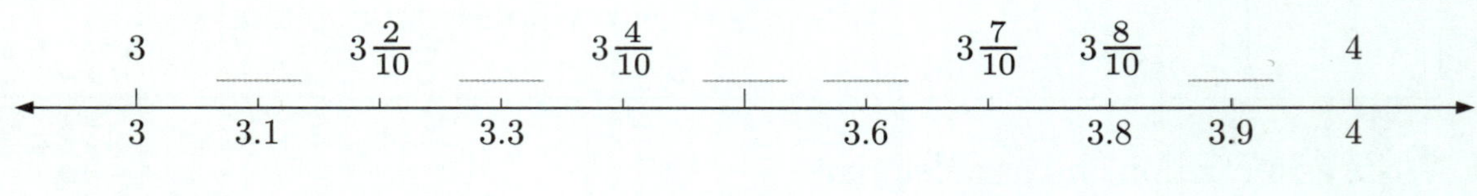

b Arrange from least to greatest: 3.3, 3.8, 3.1

____

c Arrange from greatest to least: 3.6, 3.9, 3.4

____

8 Order each group of decimal numbers from least to greatest:

a 2.9, 2.5, 3.1 __________

b 4.8, 3.9, 4.2, 3.5 __________

9 Use the number line to find each answer:

a $0.2 + 0.8 =$ __________

0 0.1 0.2 0.3 0.4 0.5 0.6 0.7 0.8 0.9 1

b $0.7 - 0.3 =$ __________

0 0.1 0.2 0.3 0.4 0.5 0.6 0.7 0.8 0.9 1

10 Number line: 0 0.1 0.2 0.3 0.4 0.5 0.6 0.7 0.8 0.9 1 1.1 1.2 1.3 1.4 1.5 1.6 1.7 1.8 1.9 2

Use the number line to find each answer:

a $0.5 + 0.9 =$ ______ **b** $0.6 + 1.2 =$ ______

c $1.1 + 0.6 =$ ______ **d** $0.5 + 0.3 + 0.7 =$ ______

e $0.9 - 0.8 =$ ______ **f** $1.4 - 0.8 =$ ______

g $1.8 - 0.9 =$ ______ **h** $1.6 - 1.3 =$ ______

11 Complete:

a $5 + 4 =$ ______ **b** $12 + 5 =$ ______

$0.5 + 0.4 =$ ______ $1.2 + 0.5 =$ ______

c $6 + 2 + 7 =$ ______ **d** $7 - 2 =$ ______

$0.6 + 0.2 + 0.7 =$ ______ $0.7 - 0.2 =$ ______

e $14 - 9 =$ ______ **f** $19 - 15 =$ ______

$1.4 - 0.9 =$ ______ $1.9 - 1.5 =$ ______

12 Complete mentally:

a $0.7 + 0.2 =$ ______ **b** $0.3 + 0.8 =$ ______ **c** $1.3 + 0.4 =$ ______

d $0.8 - 0.4 =$ ______ **e** $1.2 - 0.7 =$ ______ **f** $1.9 - 0.5 =$ ______

g $0.6 + 0.9 =$ ______ **h** $1.8 - 1.2 =$ ______ **i** $1.4 - 0.8 =$ ______

13 Give the place value of the digit 6.

a 2.65 ______ **b** 3.16 ______

c 62.05 ______ **d** twelve point zero six ______

e forty six point seven five ______

14 Write as a decimal number:

a 2 wholes, 6 tenths, and 8 hundredths **b** 9 wholes and 5 hundredths

______ ______

15 Complete the table.

Fraction	*Diagram*	*Decimal*
$\frac{6}{100}$		

16 Write as a decimal number:

a $\frac{65}{100}$ ________ **b** $2\frac{3}{100}$ ________ **c** $12\frac{74}{100}$ ________

17 Write as a fraction or mixed number:

a 3.14 ________ **b** 0.06 ________ **c** 6.95 ________

18 Complete:

a

$15 - 6 =$ ________

$1.5 - 0.6 =$ ________

$0.15 - 0.06 =$ ________

b

$12 + 7 =$ ________

$1.2 + 0.7 =$ ________

$0.12 + 0.07 =$ ________

CHAPTER 11: LENGTH

A **length** or **distance** is a measure of how far it is between two points.

A **desk ruler** is used to measure short lengths.

0cm 1 2 3 4 5 6 7 8 9 10 11 12 13 14 15

The ruler is a *number line*. The distance from 0 to 1 on the number line is the **unit** of measurement.

For a desk ruler, the unit is a special length called a **centimetre**.

We often write centimetre as **cm**.

Exercise 1

How long is this computer mouse? ______ cm

0cm 1 2 3 4 5 6 7 8 9 10 11 12 13 14 15

Exercise 2

Measure each line carefully. Write your answer in the space provided.

a ________________________________ ______ cm

b ________ ______ cm

c ____ ______ cm

d ____ ______ cm

Exercise 3

Use your ruler and pencil to draw a line with length:

a 5 cm

b 10 cm

c 8 cm

d 1 cm

e 13 cm

Exercise 4

Measure each length carefully.

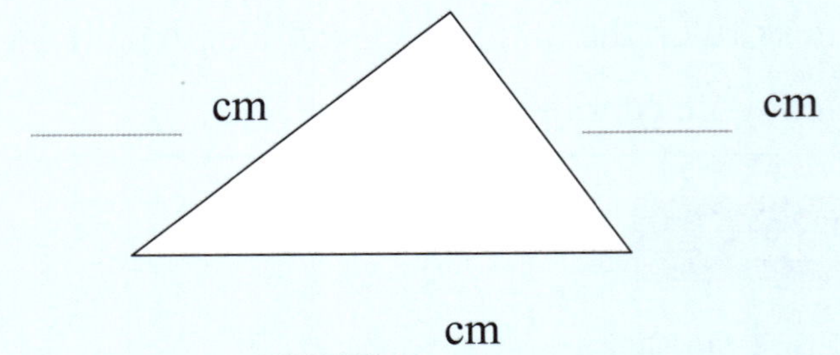

Each centimetre on the ruler is divided into 10 minor divisions.

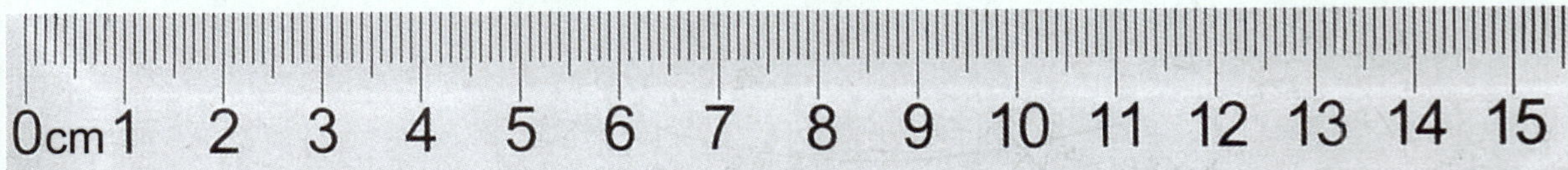

Each minor division is $\frac{1}{10}$ or 0.1 cm, which is 1 **millimetre** or 1 **mm**.

The millimetre is another **unit** of length.

$$\mathbf{1\ cm = 10\ mm}$$

$$\mathbf{\frac{1}{10}\ cm = 1\ mm}$$

Exercise 5

Complete the numbers on the ruler.

cm	0	1	____	3	4	____	____	7	____	____	____	11	12
mm	____	10	20	____	____	50	____	____	80	____	100	____	120

Discussion

Why is it important to write the units with every measurement?

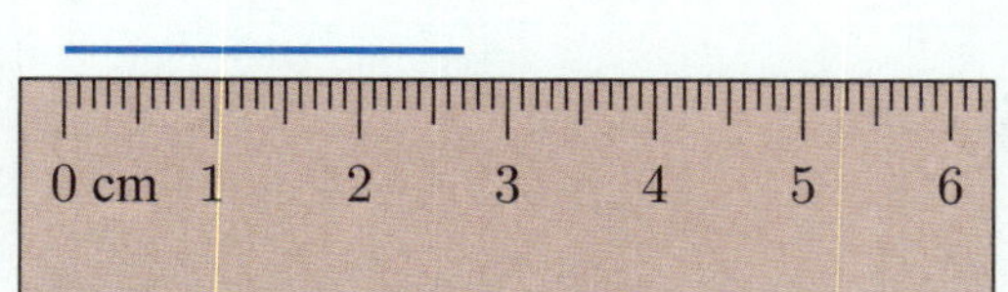

- The line is $2\frac{7}{10}$ or 2.7 cm long.
- Since 2 cm = 20 mm, the line is also 20 mm + 7 mm = 27 mm long.

Exercise 6

Read each measurement in centimetres and in millimetres:

a

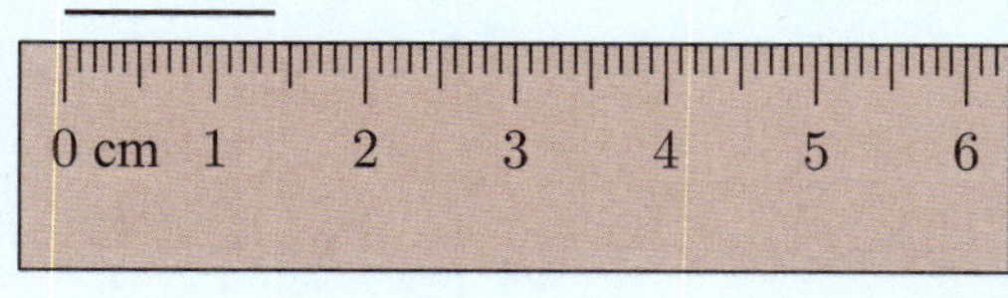

_______ cm or _______ mm

b

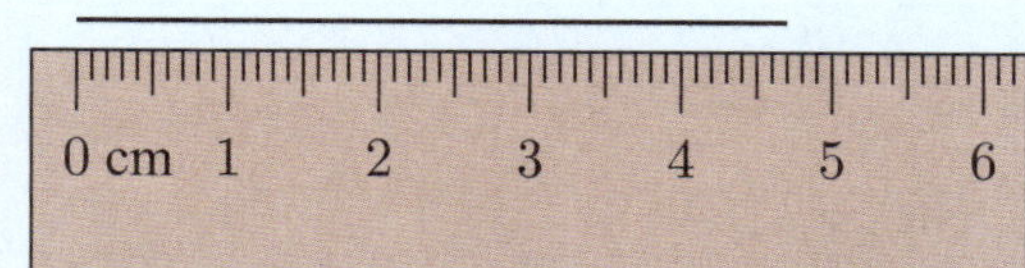

_______ cm or _______ mm

Exercise 7

Measure each line. Write your answer in centimetres.

a ______________________________

_______ cm

b ____________________

_______ cm

c ____________________________________

_______ cm

d ___________

_______ cm

Exercise 8

Measure each line. Write your answer in millimetres.

a ______________________

_______ mm

b ______________

_______ mm

c _____________________________

_______ mm

d _____________

_______ mm

Exercise 9

Complete the table.

	6 cm 2 mm	$6\frac{2}{10}$ cm	6.2 cm	62 mm
a	3 cm 5 mm		3.5 cm	
b		$1\frac{9}{10}$ cm		19 mm
c	4 cm 1 mm			41 mm
d		$8\frac{7}{10}$ cm		
e			11.4 cm	114 mm
f	12 cm 8 mm			

Exercise 10

Measure the length of each object. Write your answer in centimetres.

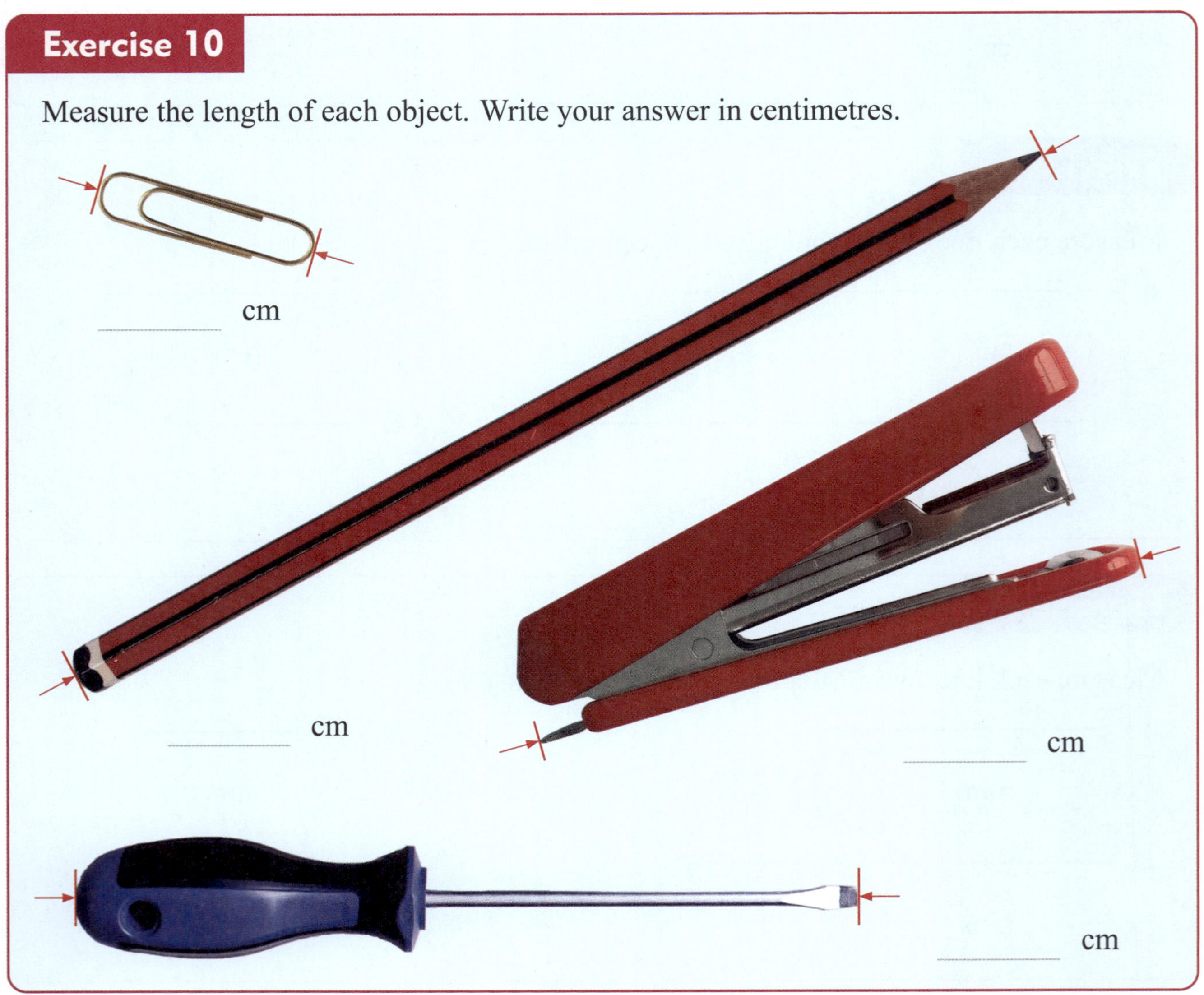

Exercise 11

Measure the length of each object. Write your answer in millimetres.

Activity

Your teacher will provide 4 books for you to measure.

What to do:

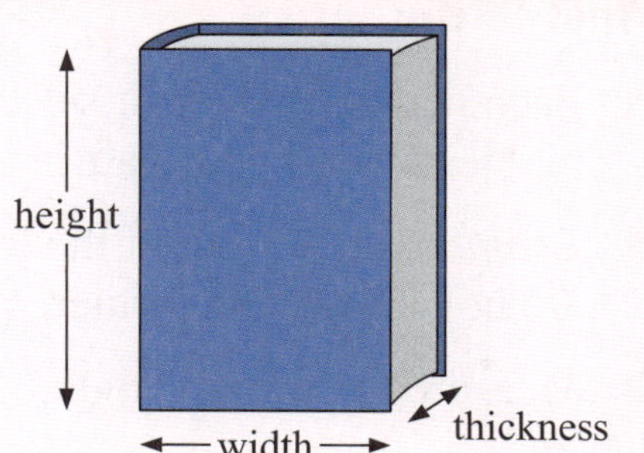

1 Measure the height, width, and thickness of each book in millimetres.

Book	*Height*	*Width*	*Thickness*
A			
B			
C			
D			

2 Complete: The tallest book is ....

The narrowest book is ....

The thickest book is ....

Exercise 12

Estimate the length of each object in mm. Then check your estimate by measuring.

a

Estimate = ____________

Actual length = ____________

b

Estimate = ____________

Actual length = ____________

Activity — Estimating and measuring

You will need:

a partner, 3 small classroom objects each, a desk ruler.

What to do:

1. Each person in the pair selects 3 small objects from the classroom. Write the names of all 6 objects from your pair in the table.
2. *Estimate* the length of the longest part of each object. Record your estimate in the table. Do not show your partner.
3. Measure the actual length of each object. Record the length in the table.
4. Compare your estimate with the actual length.
5. Discuss with your class how to determine who has the most accurate estimate.

Object	*Estimate* (cm)	*Actual length* (cm)

To measure larger distances, we use another different unit called a **metre**.

We often write metre as **m**.

$$1\text{ m} = 100\text{ cm}$$

$$\frac{1}{100}\text{ m} = 1\text{ cm}$$

Your teacher might have a **metre ruler** which is 100 cm or 1 m long.

To measure distances in metres, we can use:

- a tape measure
- a trundle wheel

Exercise 13

Write whether you would measure each length in millimetres, centimetres, or metres.

a

b

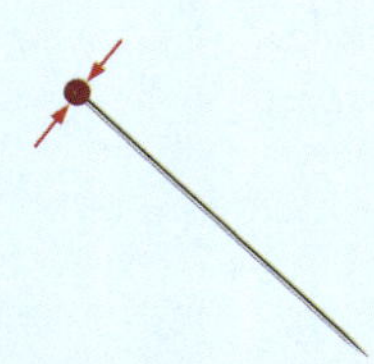

c

d

e

f

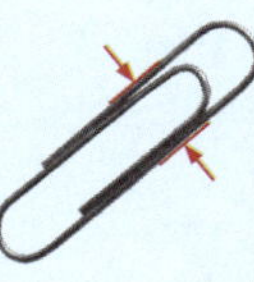

Exercise 14

Cynthia threw a ball 22 m. Her friend Sylvia threw the ball 27 m.

a Who threw the ball farther? ____________

b How much farther did this person throw the ball? ______

c Lucy threw a ball 7 m farther than Sylvia. How far did Lucy throw her ball?

______

Exercise 15

The distance around the athletics track is 400 m.

a Joe ran three laps of the track. How far did he run? ________

b How many laps around the track do you need in order to run 800 m? ______

Exercise 16

Complete:

1 m = ______ cm	______ m = 600 cm
2 m = ______ cm	7 m = ______ cm
______ m = 300 cm	______ m = 800 cm
4 m = ______ cm	______ m = 900 cm
______ m = 500 cm	10 m = ______ cm

Exercise 17

1 m 30 cm = 100 cm + 30 cm = 130 cm

a 1 m 95 cm = 100 cm + ______ cm = ______ cm

b 2 m 54 cm = ______ cm + 54 cm = ______ cm

c 4 m 8 cm = ______ cm + ______ cm = ______ cm

d 7 m 63 cm = ______ cm + ______ cm = ______ cm

Discussion

Your teacher will show you a measuring tape and a trundle wheel.

Discuss how:

- a tape measure is read in metres and centimetres
- a trundle wheel is read as a decimal or mixed number of metres.

The grizzly bear is 2 m 16 cm tall.

To measure in m, the bear is

$$2\frac{16}{100} \text{ or } 2.16 \text{ m tall.}$$

To measure in cm, the bear is

$$\begin{aligned} 2 \text{ m } 16 \text{ cm} &= 200 \text{ cm} + 16 \text{ cm} \\ &= 216 \text{ cm tall.} \end{aligned}$$

Exercise 18

Complete this table.

	1 m 10 cm	$1\frac{10}{100}$ m	1.10 m	110 cm
a	2 m 30 cm		2.30 m	
b	1 m 65 cm	$1\frac{65}{100}$ m		
c		$3\frac{20}{100}$ m	3.20 m	
d	2 m 18 cm			218 cm
e	4 m 15 cm			
f		$2\frac{49}{100}$ m		
g			6.75 m	
h				302 cm

Exercise 19

Write down the distance measured by the tape measure.

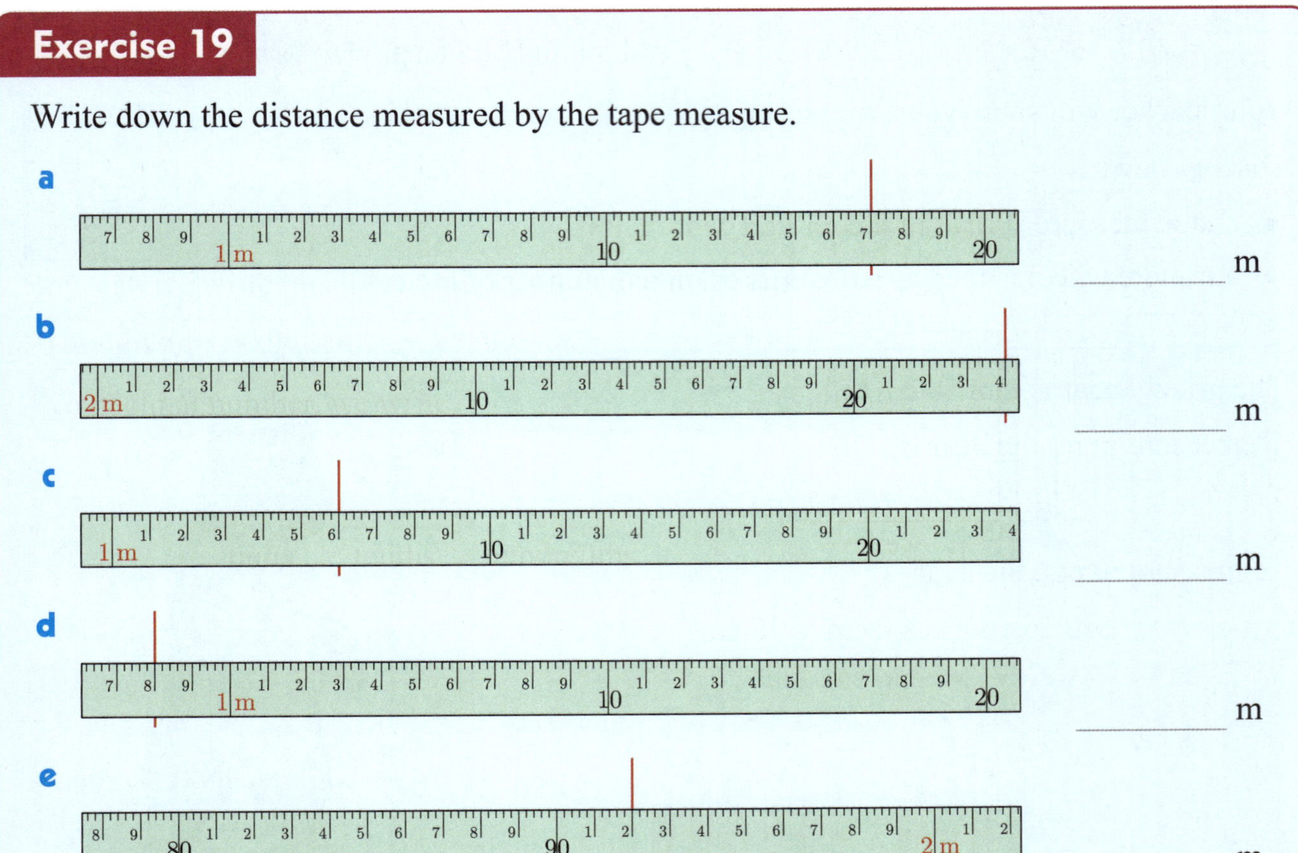

a .... m

b .... m

c .... m

d .... m

e .... m

Activity

Estimate the length and width of your classroom.

Measure the distances as a class to check your estimates.

	Estimate (m)	*Actual distance* (m)
length		
width		

Activity — A question of sport

What Olympic sports might you be watching, if the winning score is a *distance*:

a less than 10 m

b between 10 m and 80 m

c more than 80 m?

Activity — Step and jump

You will need:

a metre ruler or tape measure, chalk or coloured tape.

What to do:

1 Divide the class into groups with 5 or 6 in each.

2 Each group should mark a line on the ground outside.

3 In turn, each person places their feet on the ground with their toes just touching the line.

The person then steps as far forward as possible with one foot, leaving the other in position.

Another member of the group marks the spot on the ground where the person's toes finish.

Measure the distance from the start line to the marked line and record the result in the table below.

Name	*Step length* (cm)	*Jump length* (cm)

4 Measure how far each person can jump, starting with both feet *together* behind the line.

5 Compare your group's results with the rest of the class.

Who has the longest step and longest jump in the class?

longest step ....

longest jump ....

Discussion

What *units* would you use to measure the length of a highway between two cities? Is a metre large enough?

For very long distances, we can use a unit called a **kilometre**.

We often write kilometre as **km**.

$$\mathbf{1\ km = 1000\ m}$$

$$\frac{1}{1000}\ \mathbf{km = 1\ m}$$

Discussion

How can you *measure* a distance in kilometres?

Exercise 20

Complete:

1 km = ______ m

______ km = 5000 m

______ km = 2000 m

6 km = ______ m

3 km = ______ m

7 km = ______ m

______ km = 4000 m

______ km = 8000 m

Exercise 21

Jamal ran 12 km last week and 17 km this week.

a How far has Jamal run in total? ______ km

b How much farther did Jamal run this week than last week? ______ km

Exercise 22

Joohee travels 13 km each way between home and work each day.

a How many 13 km trips does Joohee make in 4 days? ______

b How far does Joohee travel in 4 days? ______

Activity

You will need: a trundle wheel

What to do:

1 Use the trundle wheel to measure the distance around a large sporting field such as a football field.

The distance around ______ is about ______ m.

2 Complete:

1 lap = ________ m

2 laps = ________ m

3 laps = ________ m

4 laps = ________ m

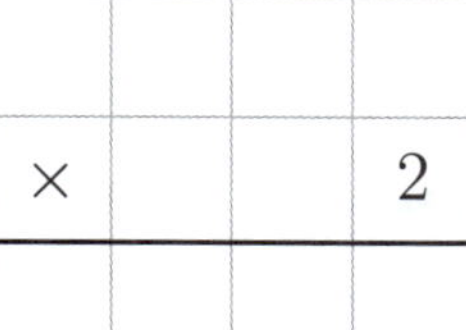

× 3

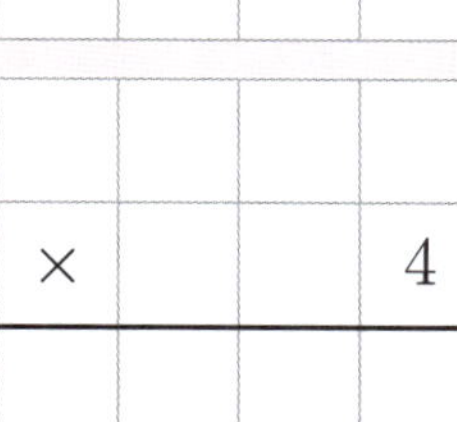

3 Complete:

1 km is about ______ laps and ______ more metres around the ______________________.

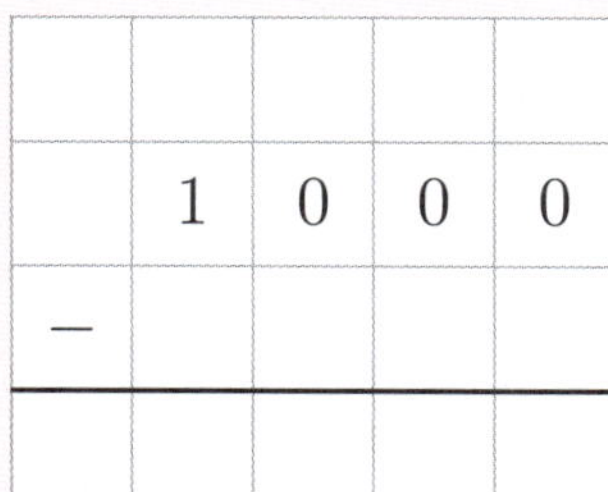

Revision

1 Measure each line carefully. Write your answer in centimetres and millimetres.

a ________ cm or ________ mm

b ________ cm or ________ mm

c ________ cm or ________ mm

2 Measure the length of each object. Write your answer in the correct units.

________ mm

________ cm

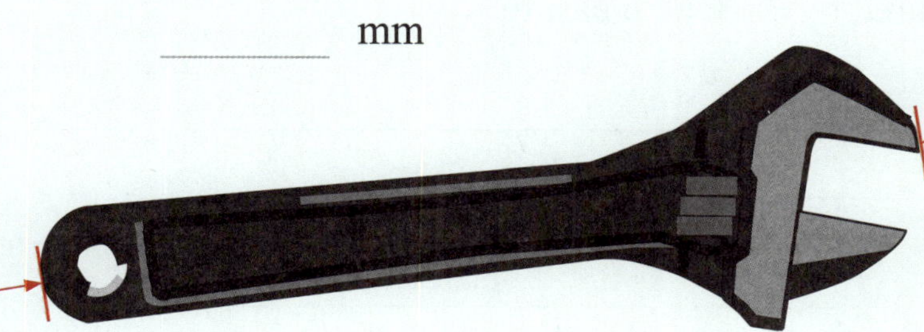

________ cm

3 Aaron takes 6 steps which are 90 cm each.

In total, Aaron has moved ________ cm or ________ m.

4 Complete the table.

	6 cm 2 mm	$6\frac{2}{10}$ cm	6.2 cm	62 mm
a	4 cm 3 mm			
b			10.6 cm	
c				124 mm

5 Complete:

3 m 67 cm = ... cm + 67 cm = ... cm

8 m 12 cm = 800 cm + ... cm = ... cm

6 Carla kicked a ball 38 m. Sammy kicked a ball 24 m.

a Who kicked the ball farther? ....

b How much farther did this person kick the ball? ....

7 Complete the table.

	1 m 10 cm	$1\frac{10}{100}$ m	1.10 m	110 cm
a		$2\frac{35}{100}$ m		
b				380 cm
c			5.65 m	

8 Write down the distance measured by the tape measure.

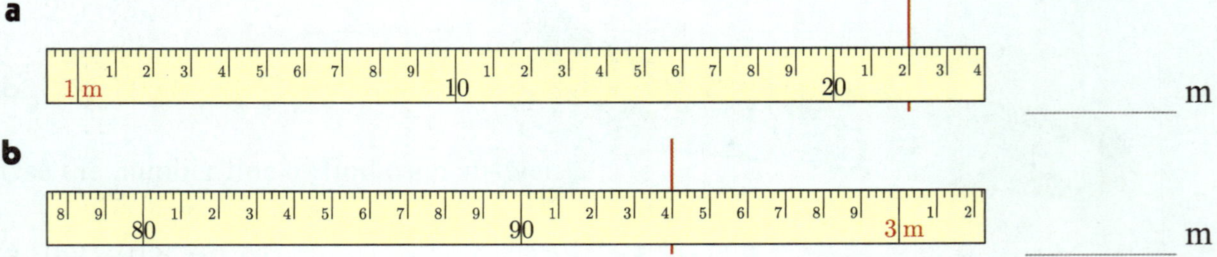

a .... m

b .... m

9 Felicity rides her bicycle 24 km to her grandparents, and 27 km home.

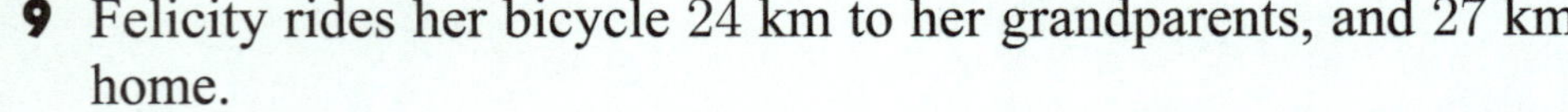

Felicity rode .... km in total.

CHAPTER 12: SHAPE

A shape is **2-dimensional** if it can be drawn on a **surface**.

To correctly identify shapes, we look at their sides, corners, and angles. These are *properties* of the shape.

Exercise 1

These shapes are all made with straight sides.

Complete the table.

Name	*Diagram*	*Sides*	*Corners*
triangle			3
quadrilateral			
pentagon			
hexagon		6	
octagon			

Exercise 2

In this Exercise we look at **triangles**.

a A triangle has ______ sides and ______ corners.

b Measure each side length.

Are the side lengths equal?

______

Are there any right angles?

______

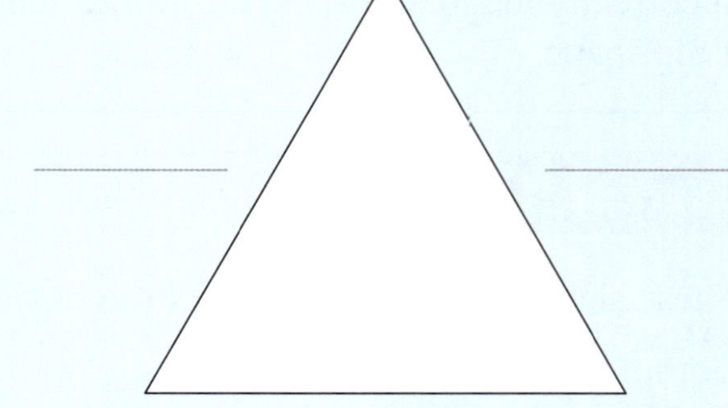

c Measure each side length.

Are the side lengths equal?

______

Label the right angle on the triangle.

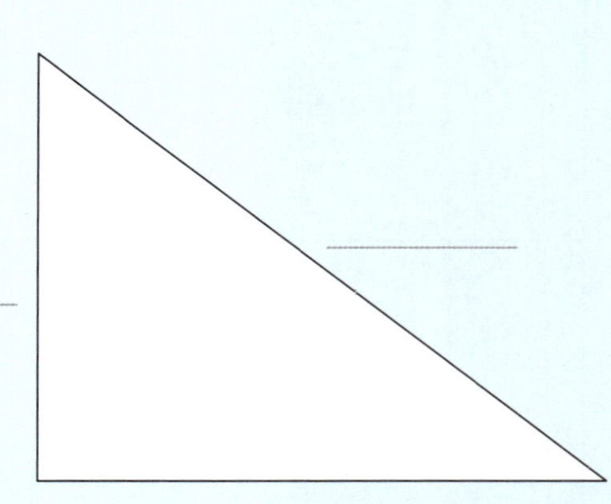

d Draw a triangle which has *two* equal sides.

e Draw a triangle which contains an obtuse angle.

Discussion

Is it possible to draw a triangle with two right angles? ________

Exercise 3

a This is a **circle**.

A circle is drawn with a single curve.

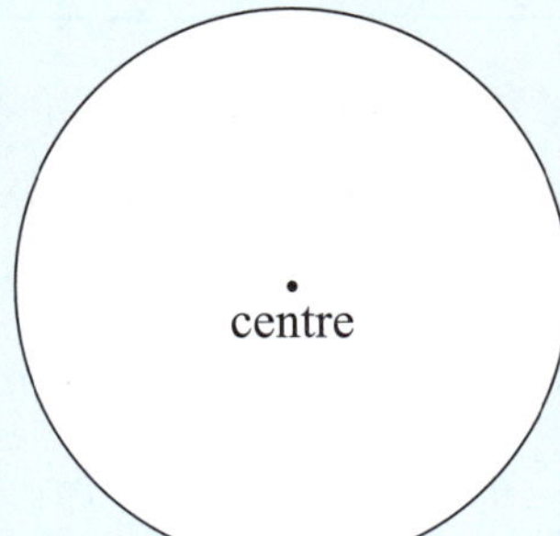

Does a circle have any corners? ________

Measure the distance from the centre to the circle.

____

Is it always the same distance from the centre to *any point* on the circle? ________

b This is an **ellipse**.

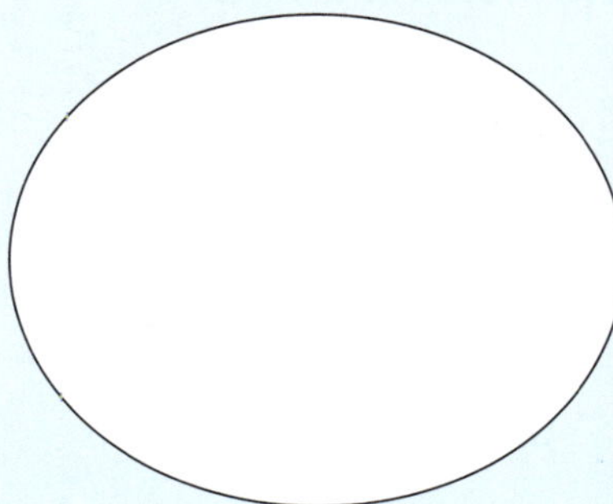

An ellipse is drawn with a single ____________.

Does an ellipse have any corners? ________

Discussion

How is an ellipse different from a circle?

Exercise 4

This is a **semi-circle**.

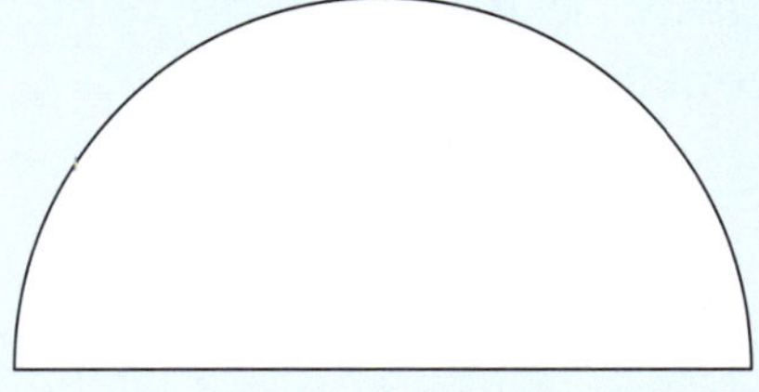

A semi-circle is made with a ____________ and a straight ____________.

How many corners does a semi-circle have? ______

A semi-circle is half of a ________________.

This is a **square**.

It has:

- 4 sides which all have the same length
- 4 corners which are all right angles.

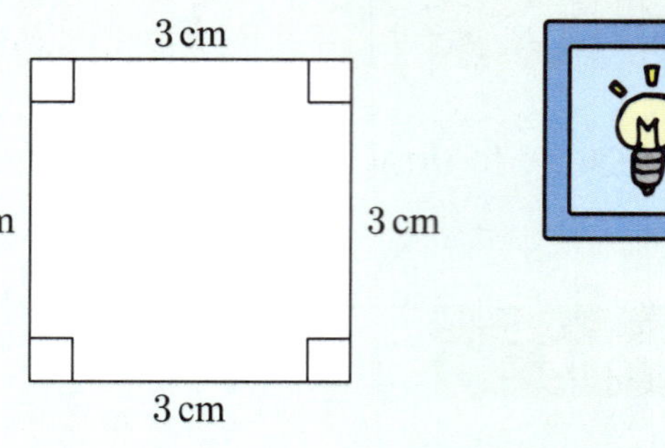

Discussion

Do *all* squares have these properties?

- 4 sides which all have the same length
- 4 corners which are all right angles.

Exercise 5

This is a **rectangle**.

a A rectangle has ______ sides.

b Measure each side length.

The opposite sides of the rectangle have

the ____________ ____________ .

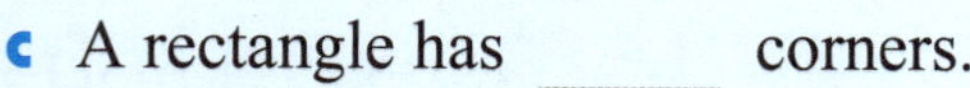

c A rectangle has ______ corners.

d Describe the angles at the corners of the rectangle.

____

e Does *every* rectangle have these properties? ________

Activity

Draw a quadrilateral which contains a reflex angle.

Exercise 6

This is a **rhombus**.

A rhombus has ______ sides and ______ corners.

Measure each side length.

Are the side lengths equal?

______

These properties are true for *all* rhombuses.

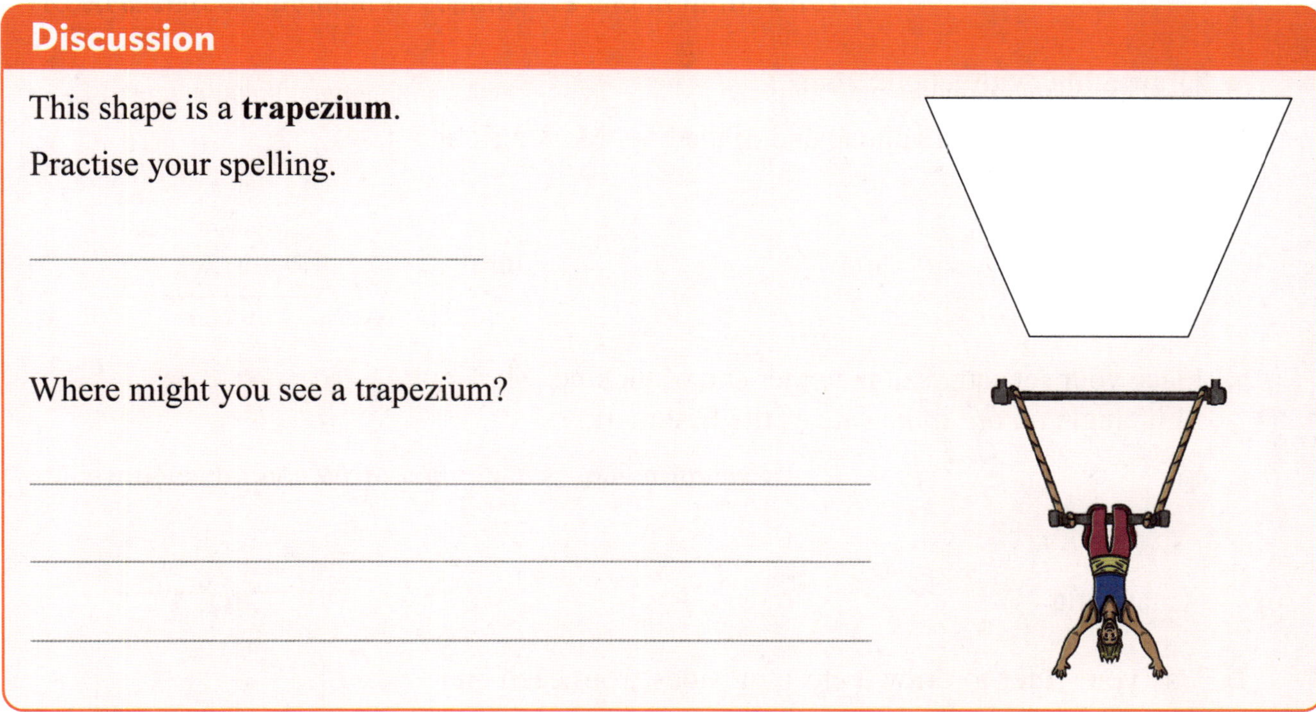

Discussion

This shape is a **trapezium**.

Practise your spelling.

____

Where might you see a trapezium?

____

____

____

Activity

Draw a quadrilateral which contains a right angle and *two* obtuse angles.

Activity — Drawing shapes

You will need: a set square, a ruler, paper

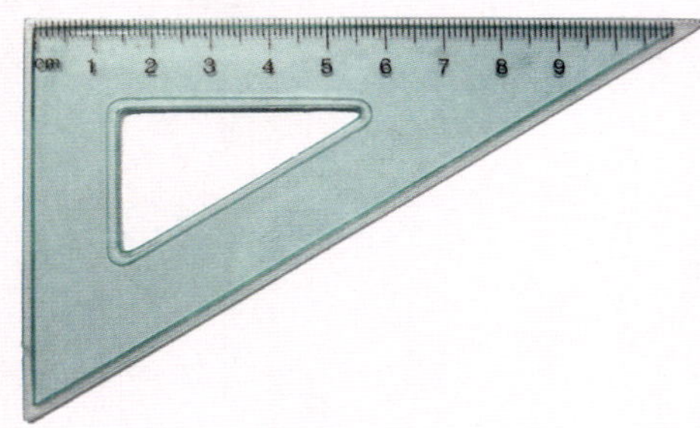

A *set square* is a triangle which has a right angle in one corner. We use it to help draw right angles.

What to do:

1 Follow these steps to draw a square with side length 3 cm on a piece of paper.

a Draw a line with length 3 cm.

b Place your set square at one end of the line. Mark a right angle.

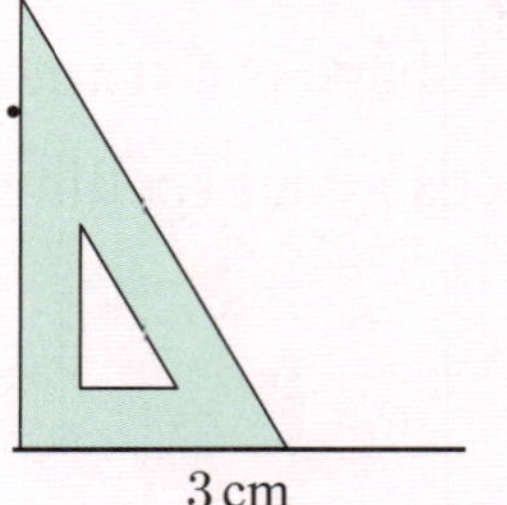

c Place your set square at the other end of the line. Mark a right angle *on the same side* as the first mark.

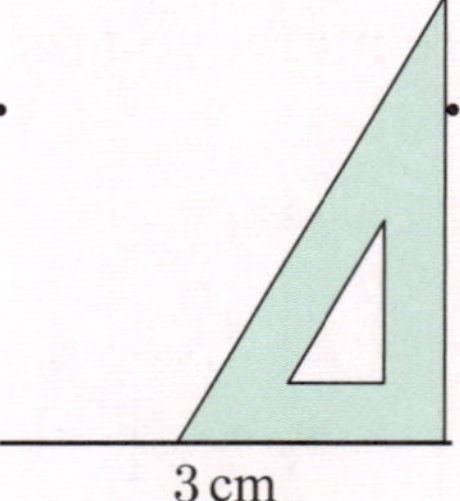

d Use your ruler to draw 3 cm long sides from each end through its corresponding mark.

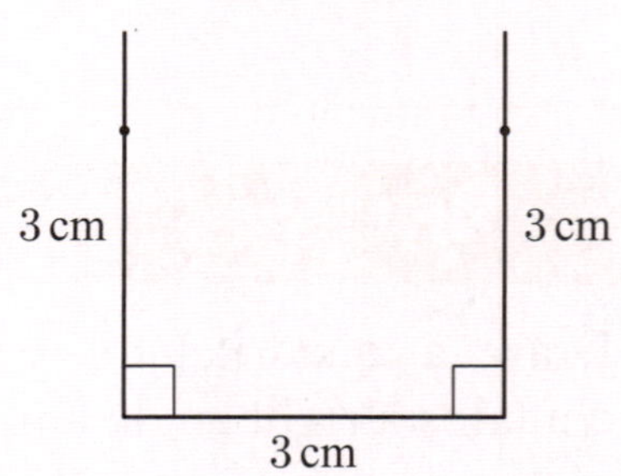

e Draw in the fourth side.

f *Check* that the top is also 3 cm long, and that the top two angles are right angles.

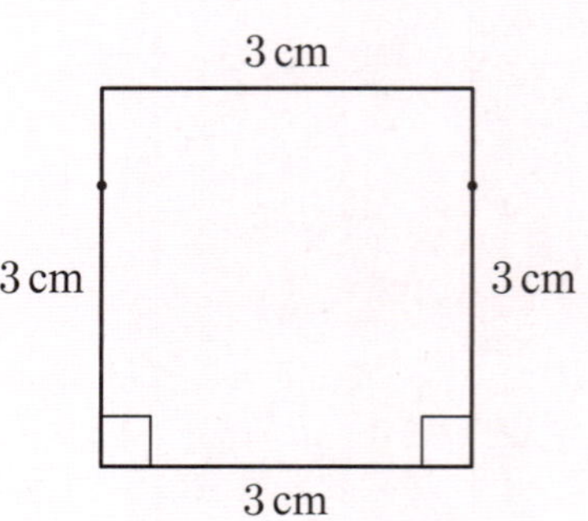

2 Draw using your set square:

i a square with side length 5 cm **ii** a rectangle with side lengths 2.5 cm and 6 cm.

Exercise 7

A
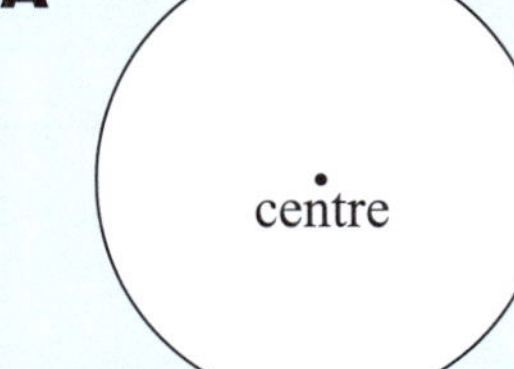

B
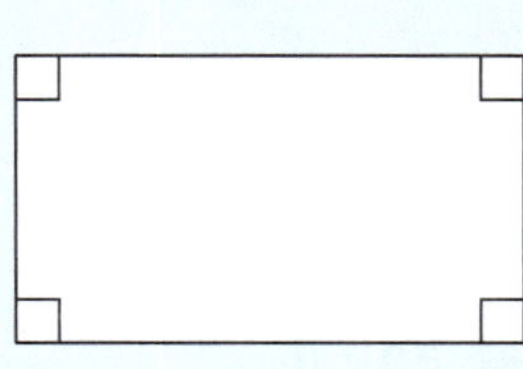

C
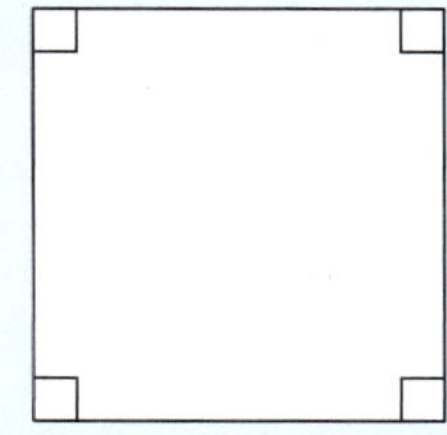

D
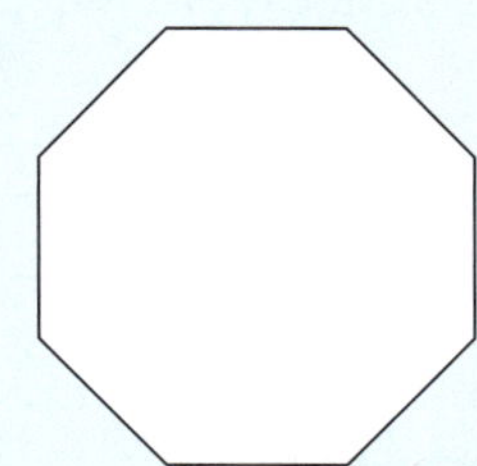

Read each description. Match it to the correct diagram and name the shape.

a This shape has 8 side and 8 corners. All 8 angles inside the shape are obtuse.

Diagram .... Name ....

b This shape has 4 sides and 4 corners. All 4 angles are right angles. The sides are all the same length.

Diagram .... Name ....

c This shape has a curve which is always the same distance from its centre.

Diagram .... Name ....

d This shape has 4 sides and 4 corners. All 4 angles are right angles. The sides are *not* all the same length.

Diagram .... Name ....

Exercise 8

Complete:

a A rectangle is a special quadrilateral.

It is special because all of its angles are ______________________.

b A square is a special rectangle.

It is special because all of its sides have the same ______________________.

c A rhombus is a special quadrilateral.

It is special because all of its sides have the same ______________________.

d A square is a special rhombus.

It is special because all of its angles are ______________________.

Exercise 9

Another word for a shape is a **figure**.

Practise your writing. ______________________

We can join shapes together to make other figures.

This figure is made up of a rectangle, a triangle, and a semi-circle.

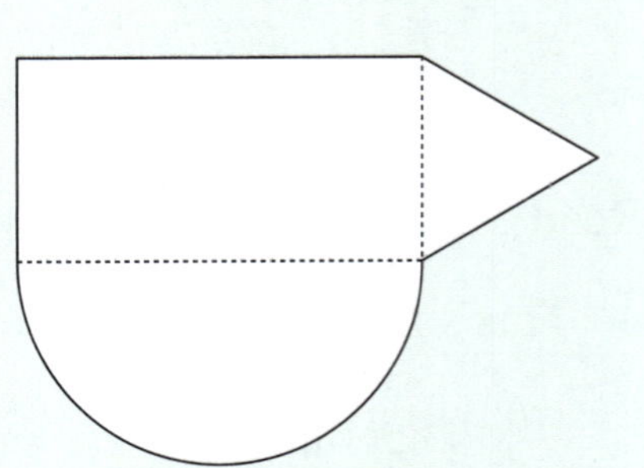

Exercise 10

List the shapes which make up this figure.

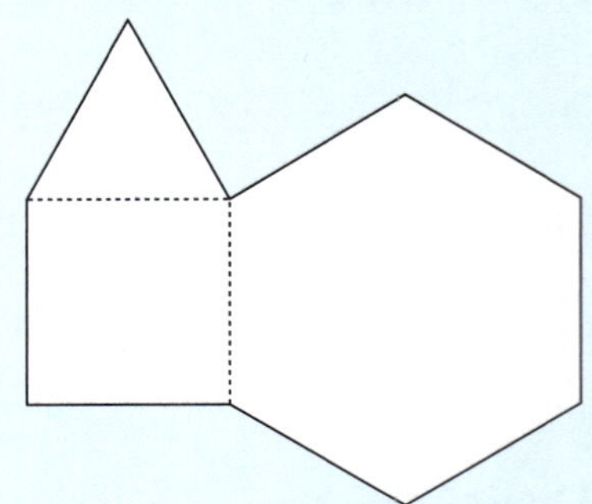

Activity

You will need: 1 hexagon, 2 triangles, 1 square from a pattern block set

If you do not have pattern blocks, click the icon and print the template onto card.

What to do:

1. Arrange the four blocks as the figure shown in the photograph.

   We can draw this figure as:

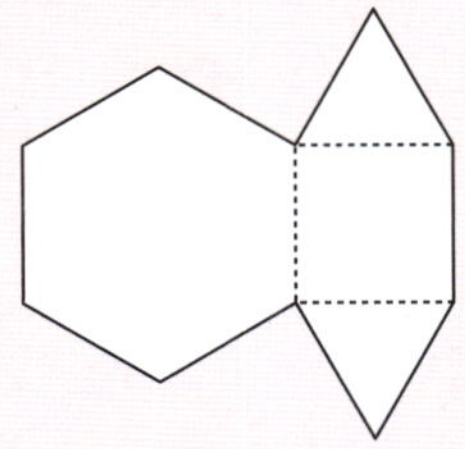

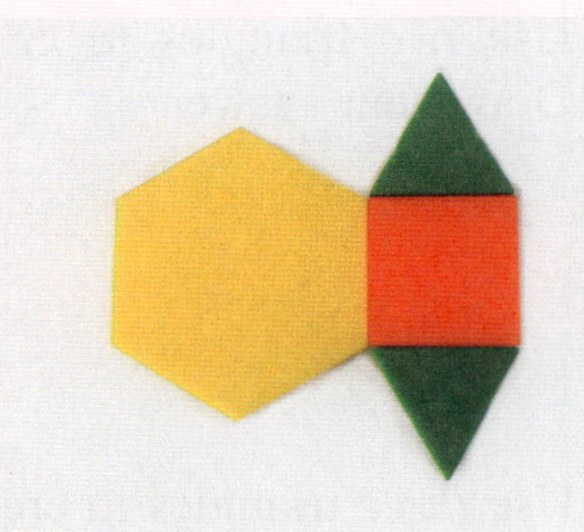

2. Draw *two* more different figures that you can make with these four blocks.

Exercise 11

Draw a line to show how the shape can be made using:

a a rectangle and a triangle

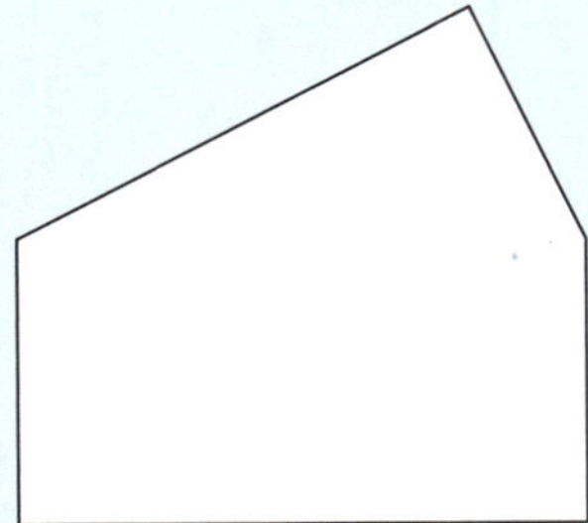

b a square and a hexagon.

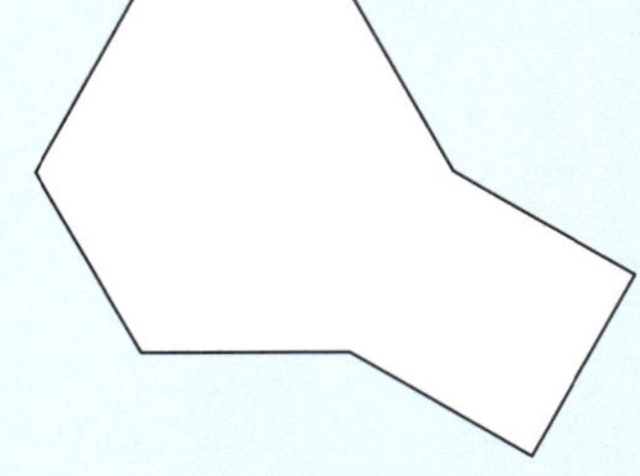

Activity

You will need: 6 triangles from a pattern block set

If you do not have pattern blocks, click the icon and print the template onto card.

What to do:

1. Use *two* triangles to create a rhombus. Draw your answer.
2. Use *three* triangles to create a trapezium. Draw your answer.
3. Use *four* triangles to create a larger triangle. Draw your answer.
4. Use all *six* triangles to create a hexagon. Draw your answer.

Exercise 12

Draw lines to show how the figure can be made using the given shapes.

a rectangle, triangle, semi-circle

b square, semi-circle, triangle

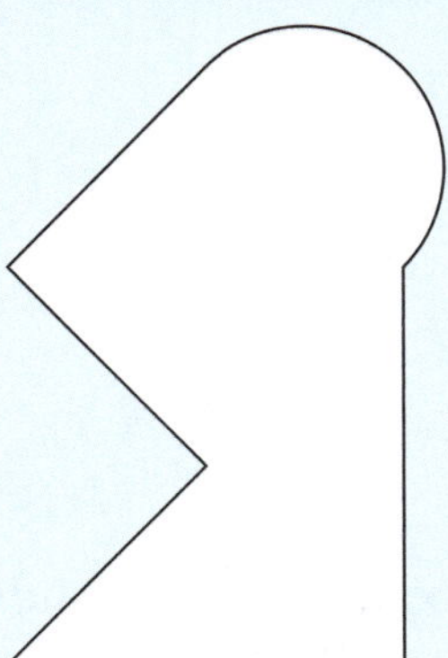

Exercise 13

Draw lines to show how this figure can be made using *three* common shapes. List the shapes below.

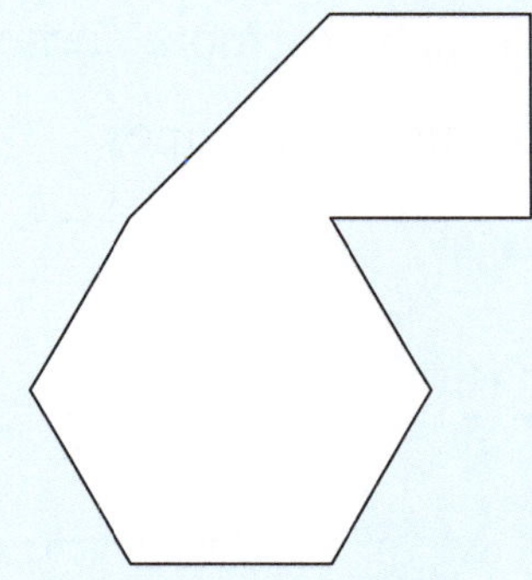

Revision

1 **a** Measure the sides of this shape.

Do the sides have the same length?

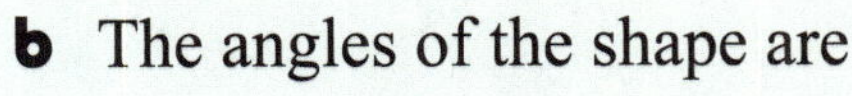

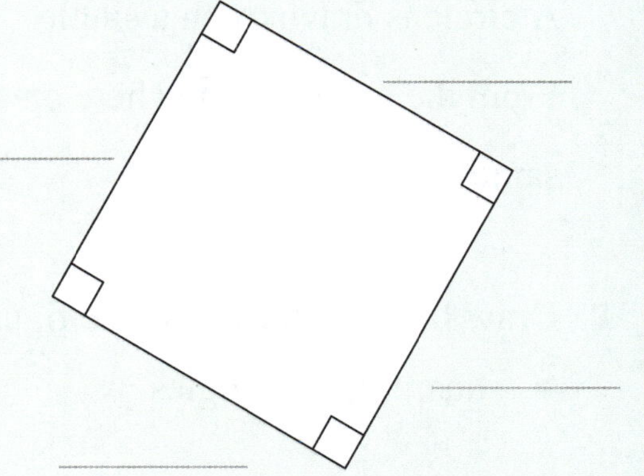

b The angles of the shape are ....

c The shape is a ....

2 Complete:

a A rhombus is a quadrilateral which has all .... side lengths.

b A rectangle is a quadrilateral whose angles are all ....

3 Draw:

a a triangle whose side lengths are all different **b** a semi-circle.

4 This shape has ______ sides and ______ corners.

This shape is a ____________________.

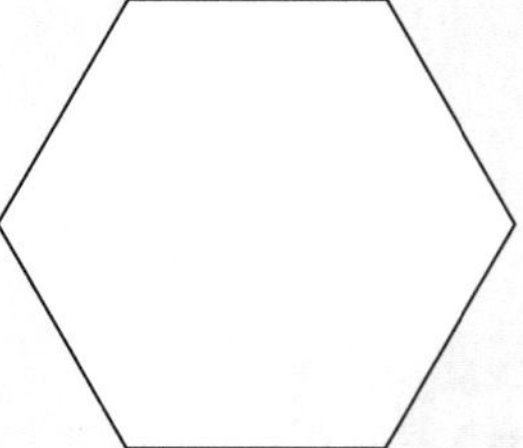

5 How many more corners does an octagon have than a pentagon? ______

6 Name each shape:

a

b

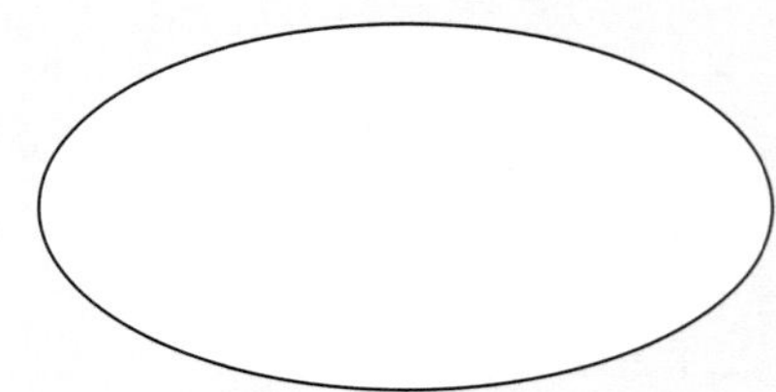

____________________ ____________________

7 Complete:

A circle is drawn with a single ____________________.

From the centre to anywhere on the circle, it is always the same ____________________.

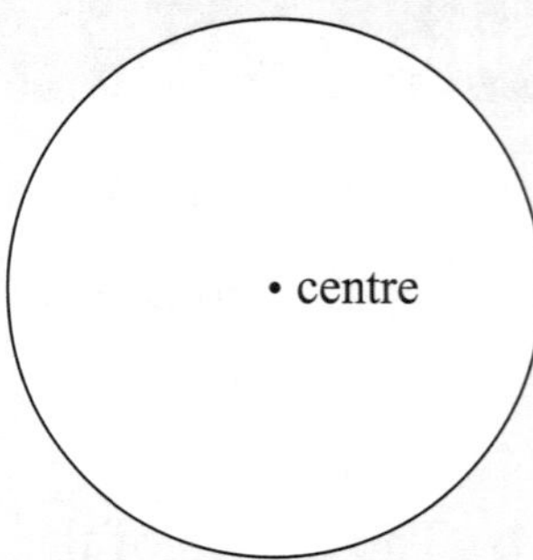

8 Draw lines to show how the figure can be made using the given shapes.

a square, two triangles

b rectangle, semi-circle, triangle

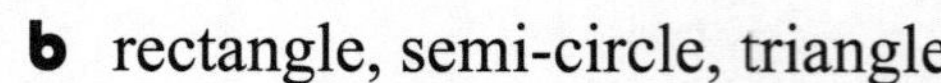

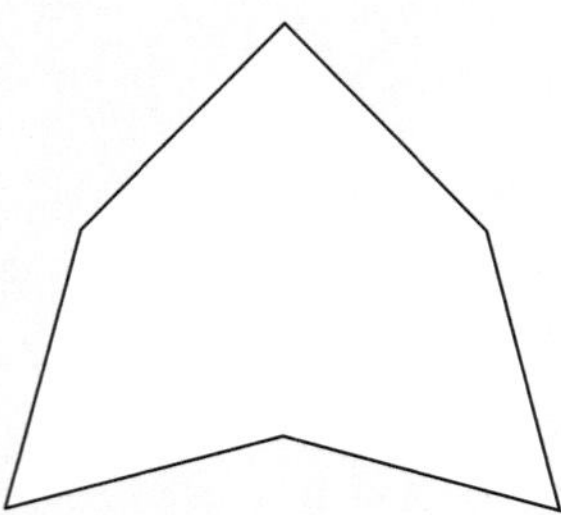

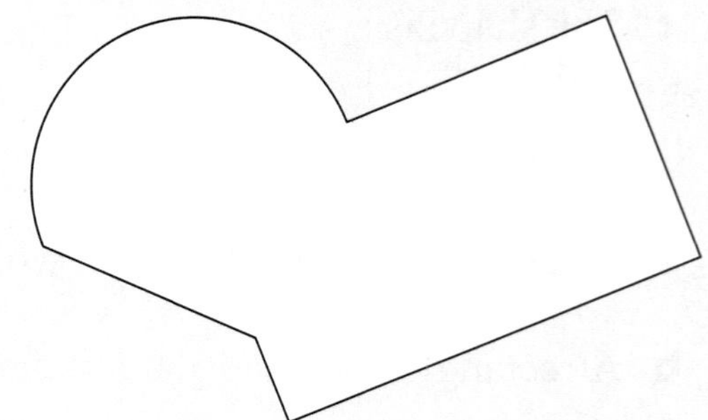

CHAPTER 13: AREA

Discussion

Look at these shapes.

Which shape is *larger*?

What is it about the shape that makes it larger?

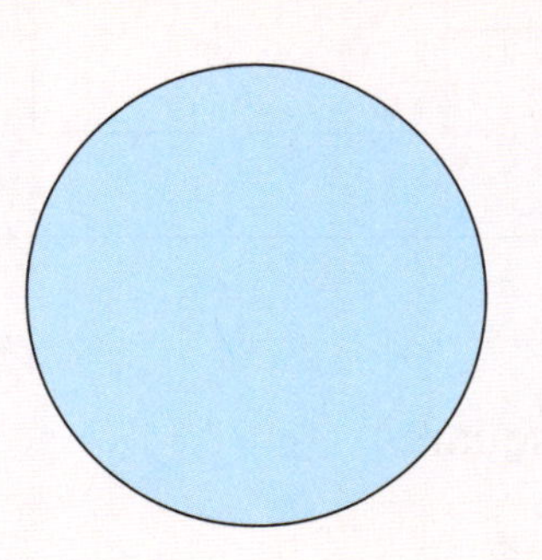

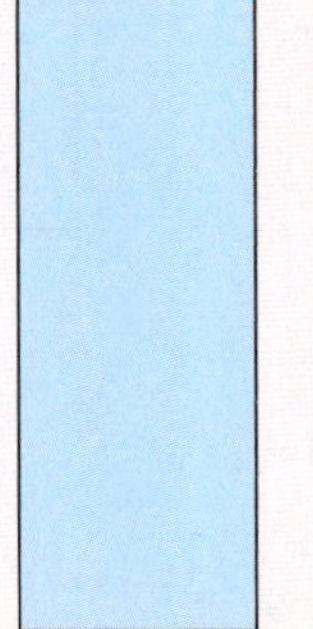

When we talk about the size of a 2-dimensional shape, we are usually interested in how much *surface* it covers.

Area is a measurement of the size of a surface.

To measure the area of a figure, we can draw a grid of squares on top of it.

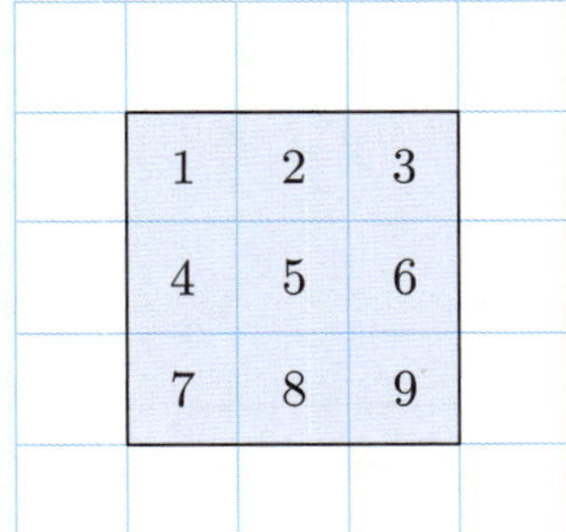

9 squares cover this figure.

The area of this figure is 9 squares.

Exercise 1

Count the squares to find the area of each figure:

a 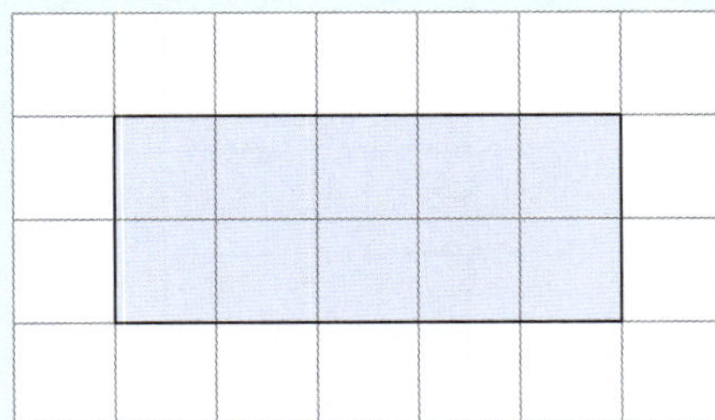

_______ squares

b 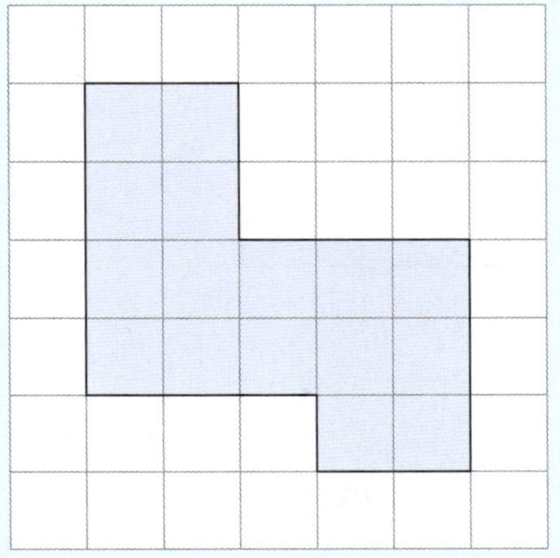

_______ squares

Exercise 2

Count the squares to find the area of each figure:

a

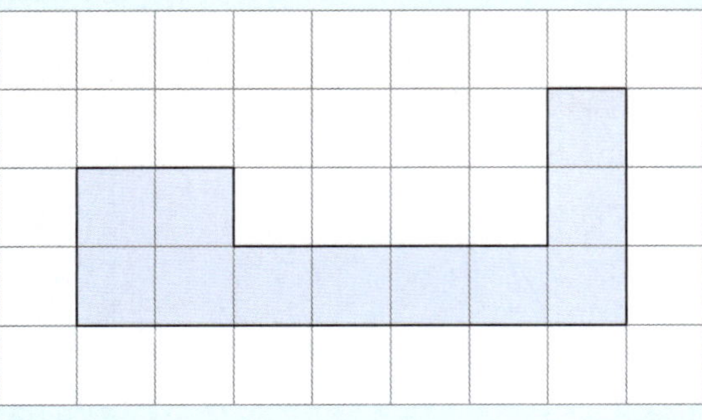

_____ squares

b

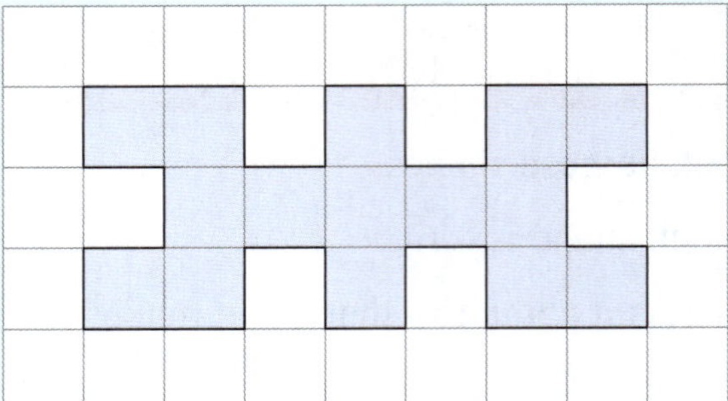

_____ squares

c

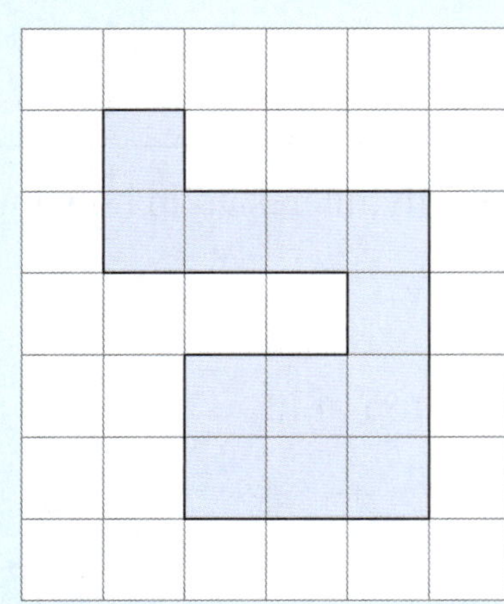

_____ squares

d

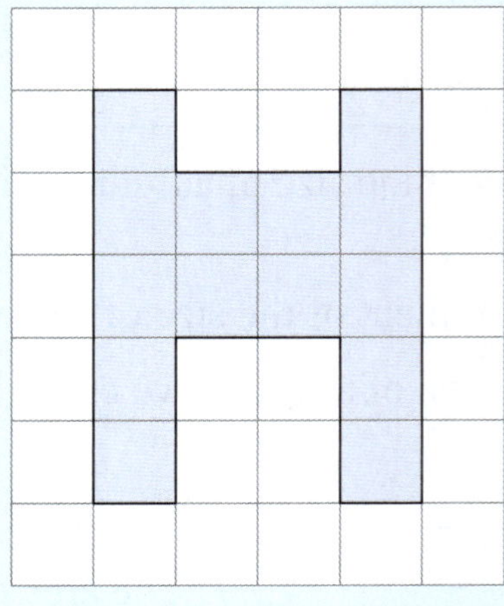

_____ squares

Exercise 3

Count the squares to find the area of each figure:

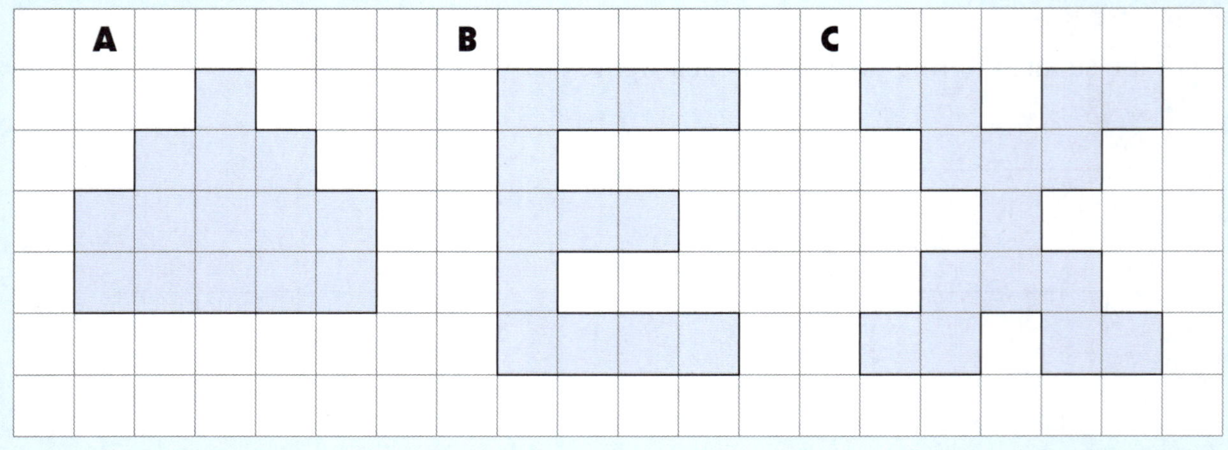

_____ squares _____ squares _____ squares

The largest figure is _____. The smallest figure is _____.

Exercise 4

Draw a *rectangle* on the grid with area:

a 12 squares

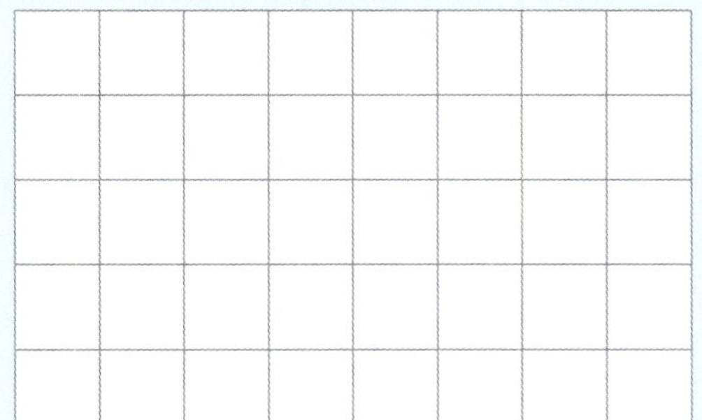

b 15 squares.

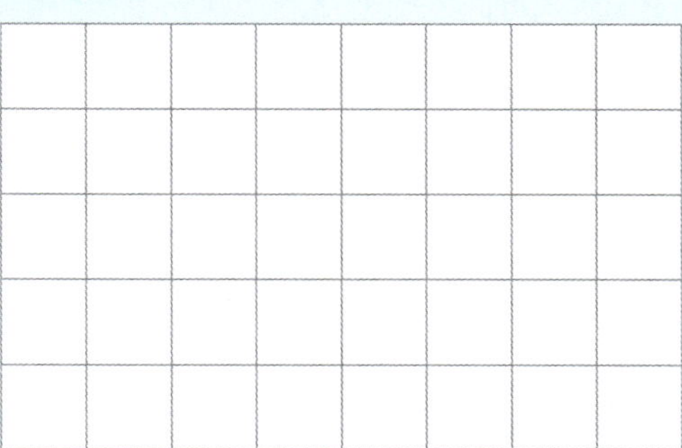

Activity

In Japan, the area of a room is sometimes measured using *tatami* mats.

Each mat is twice as long as it is wide.

You will need: a set of printable sheets, scissors

What to do:

1. Print out your sheets, then carefully cut out the tatami mats.
2. Arrange the mats onto each room template. The mats must not overlap the border of the room or each other.
3. Record the area of each room.

Room 1	*Room 2*	*Room 3*
______ mats	______ mats	______ mats

4. Which room has the largest area? ______________
5. Are there any rooms which have the same *area*? Does this mean that these rooms have the same *shape*?

   ....

   ....

6. Do you think this is a good way to measure the area of a room? Can we measure the area of *any* room using this method?

   ....

   ....

Exercise 5

Draw a *square* on the grid with area:

a 9 squares

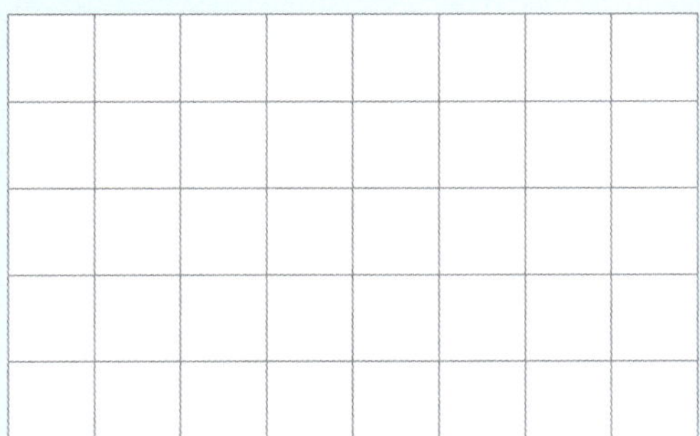

b 16 squares.

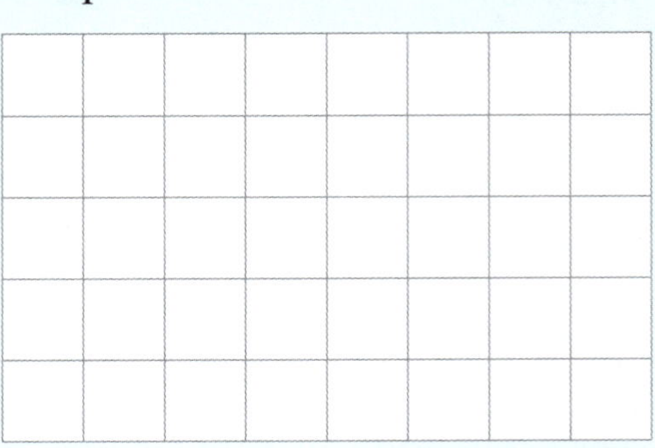

Exercise 6

Draw a figure on the grid with area:

a 11 squares

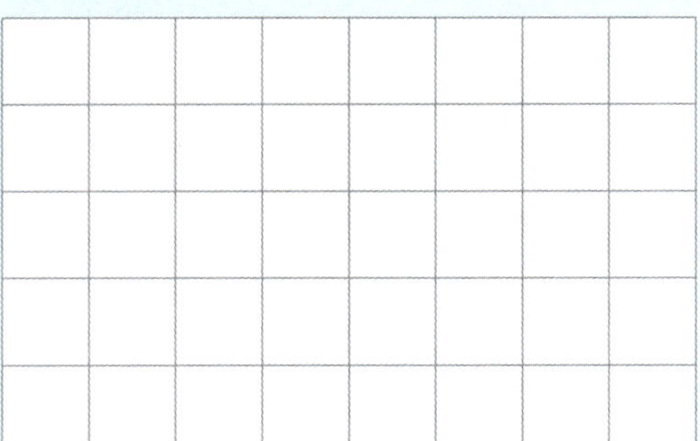

b 20 squares

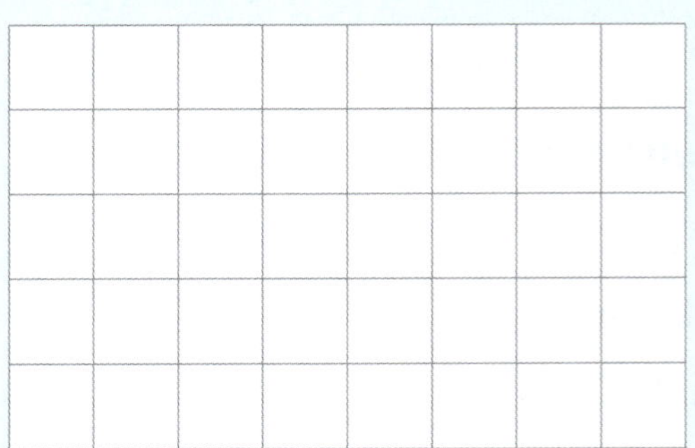

c 19 squares

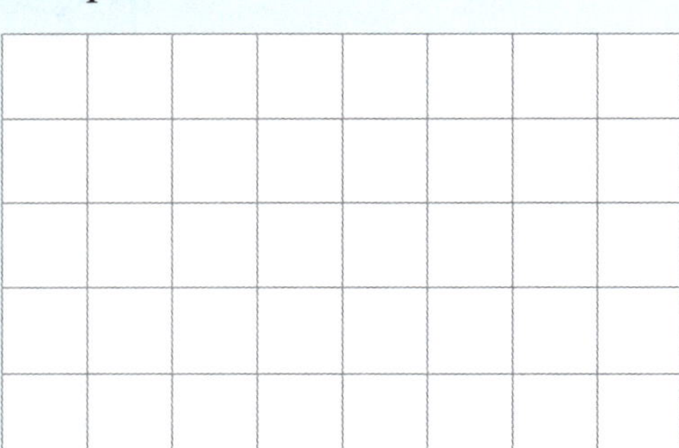

d 23 squares.

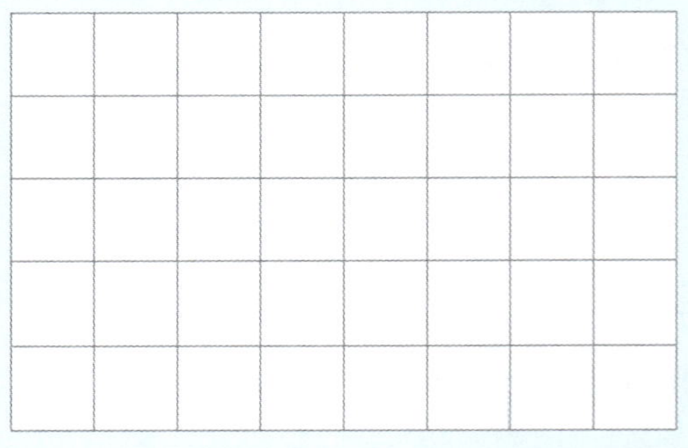

Discussion

Jamie got a phone call from his mother. "I am at the garden store. I need to know how much fertiliser to buy. Could you please measure the area of the vegetable patch for me?"

Jamie finds some square pavers in the yard.

1 How can Jamie use the pavers to measure the vegetable patch?

2 Will Jamie's measurement be helpful to his mother?

To make sure everybody understands a measurement of area, we use **standard units**. The units we use include:

- **square centimetres** (cm^2)

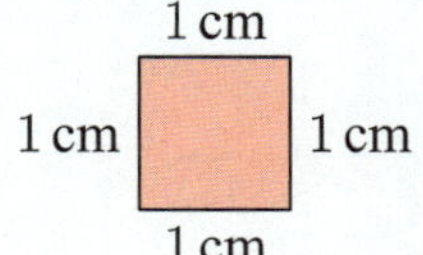

- **square metres** (m^2)

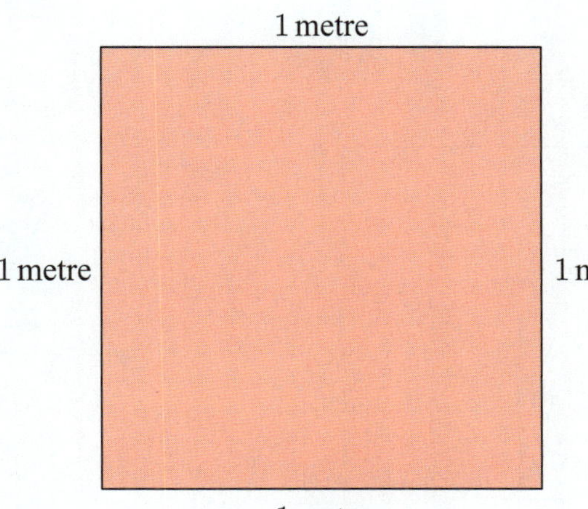

Exercise 7

These shapes are made up of squares with side length 1 cm. Write down the area of each shape in cm^2.

a

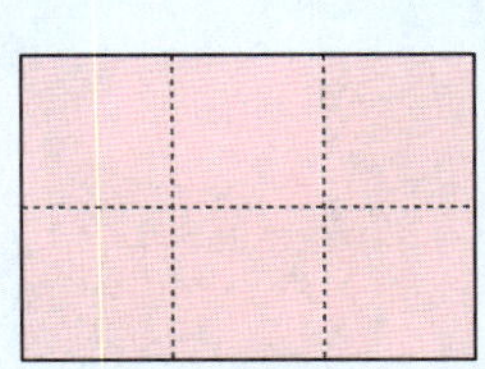

______ cm^2

b

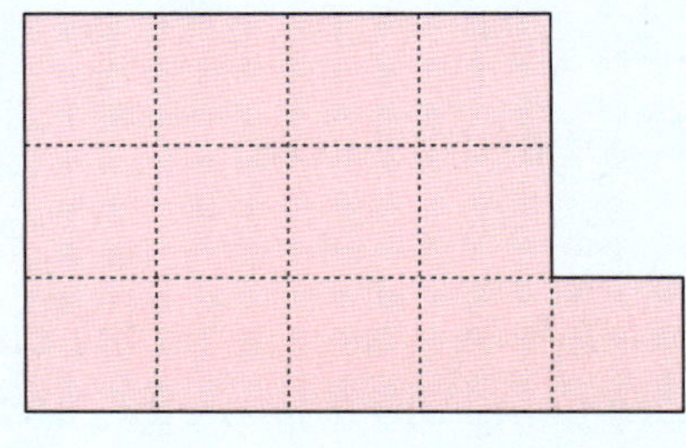

______ cm^2

c

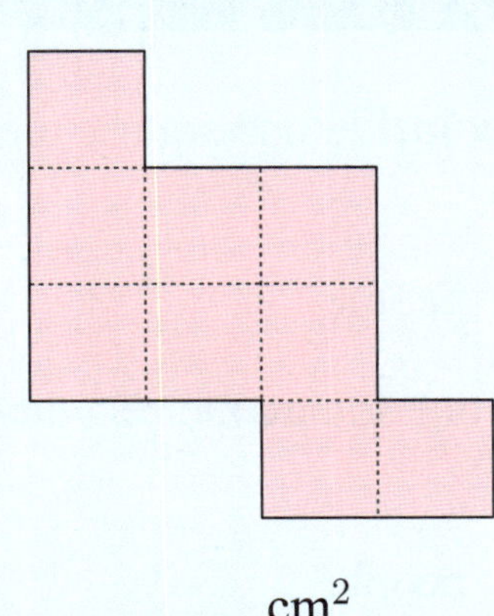

______ cm^2

d

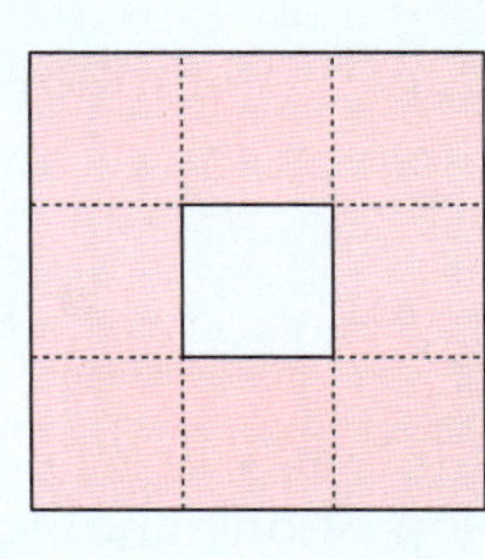

______ cm^2

Exercise 8

These shapes are made up of squares with side length 1 cm. Write down the area of each shape in cm^2.

a

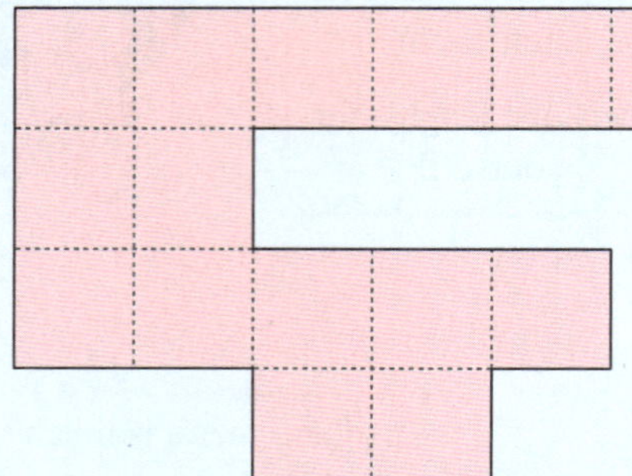

______ cm^2

b

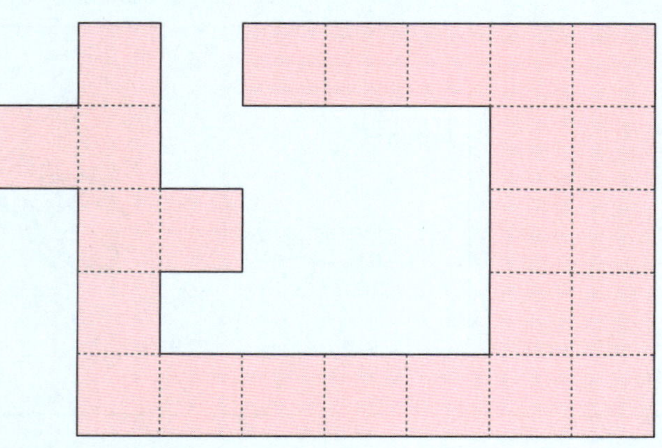

______ cm^2

Activity

This Activity will help you to understand how large 1 square metre is.

You will need: 4 metre rulers

What to do:

1 In an empty outdoor space, or on the classroom floor, make a square using the four metre rulers as shown in the picture.

2 *Estimate* how many of the following items will fit completely inside one square metre, without putting them on top of one another.

a books ______ **b** school bags ______ **c** people ______

3 How many of each item can you actually fit completely inside one square metre, without putting them on top of one another?

a books ______ **b** school bags ______ **c** people ______

Activity

Solve each problem:

a Two Grade 4 classrooms have areas $45\ m^2$ and $49\ m^2$.

The total area of the classrooms is ______ m^2.

b Sami makes 4 small pizzas. Each pizza has area $200\ cm^2$.

The total area of pizza is ______ cm^2.

c Tae places a rug with area $18\ m^2$ in a room with area $31\ m^2$.

The area of the floor that the rug does *not* cover is ______ m^2.

d Charmaine has $38\ m^2$ of material to make into curtains. Each curtain has area $4\ m^2$.

Charmaine can make ______ curtains, and there is ______ m^2 of material remaining.

For some shapes we need to count half squares.

Two half squares make a whole square.

$$\frac{1}{2}+\frac{1}{2}=1$$

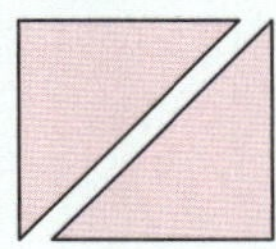

There are 9 whole squares and three half squares.

The area of this figure is

$$\begin{aligned} & 9+\frac{1}{2}+\frac{1}{2}+\frac{1}{2} \\ &= 9+1+\frac{1}{2} \\ &= 10\frac{1}{2}\ cm^2 \end{aligned}$$

Exercise 9

Circle each *half square* with area $\frac{1}{2}\ cm^2$.

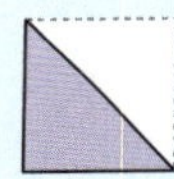 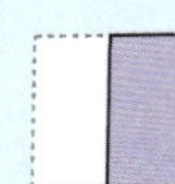 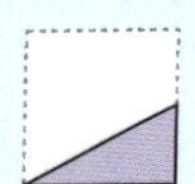 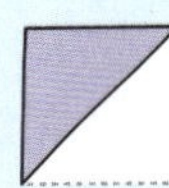 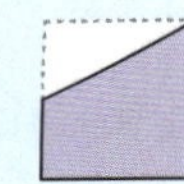  

Exercise 10

Each square in these diagrams has side length 1 cm. Write down the area of each shape in cm^2.

a 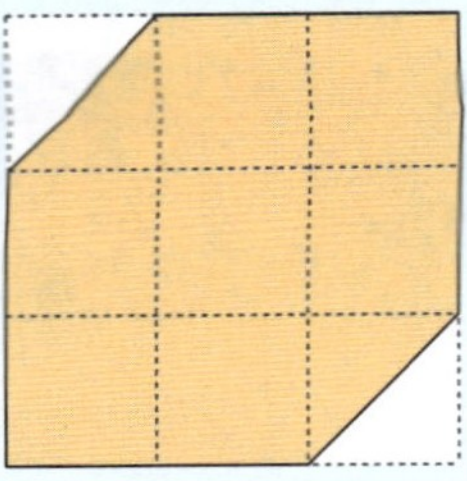

_______ cm^2

b 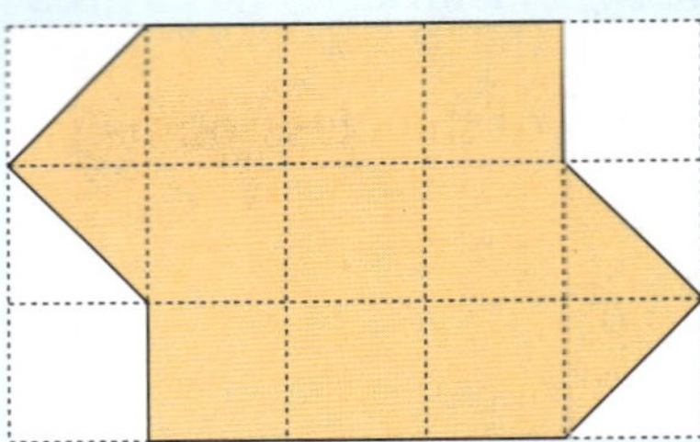

_______ cm^2

c 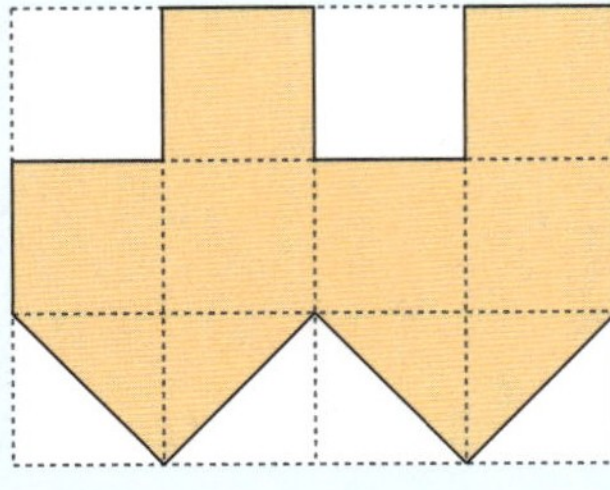

_______ cm^2

d 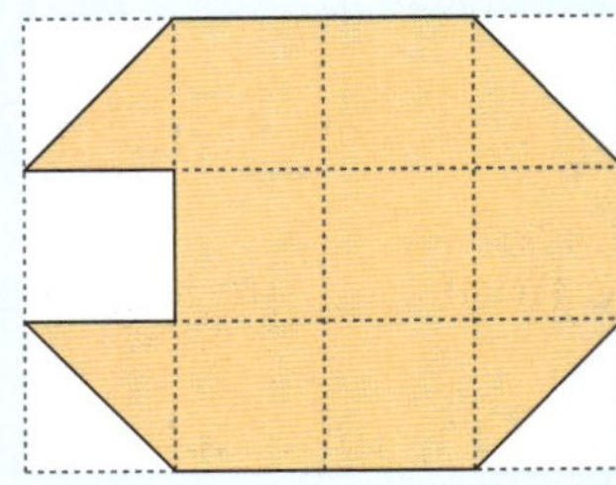

_______ cm^2

e 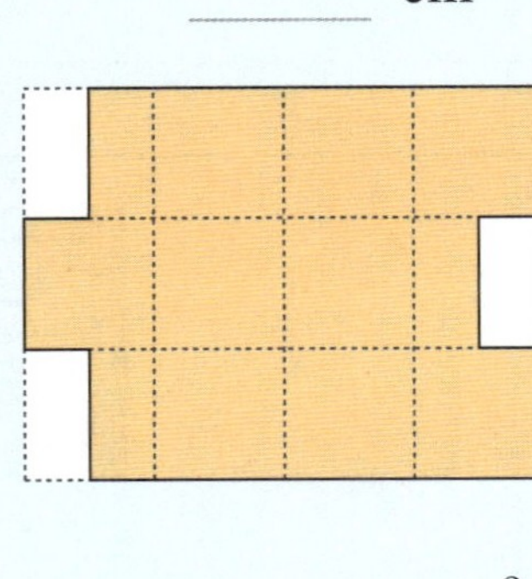

_______ cm^2

f 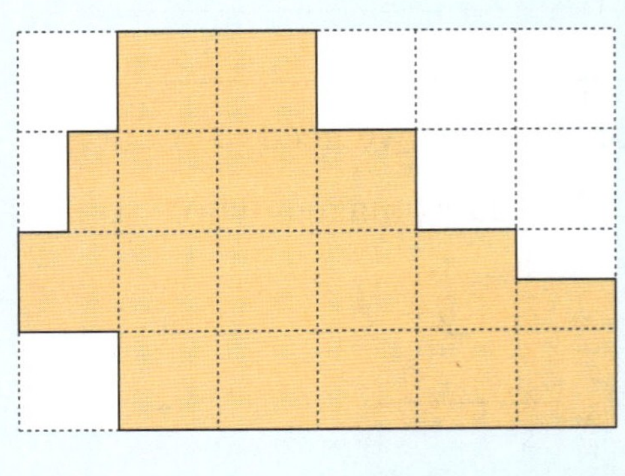

_______ cm^2

Many regions have curved boundaries, or have areas which are not an exact number of squares.

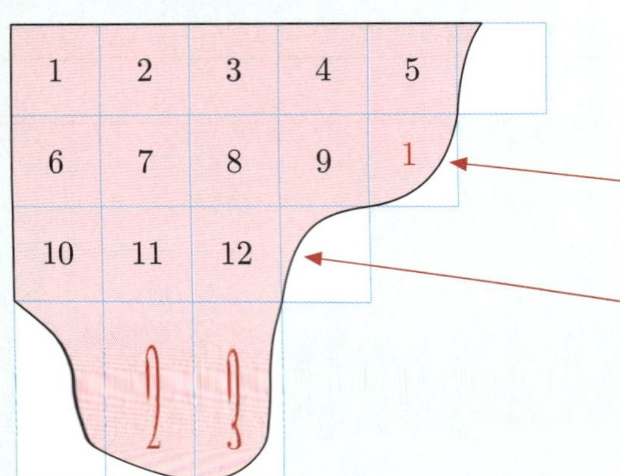

There are 12 whole squares.

Now look at the part squares.

If a square is more than half filled, we count it as one whole square.

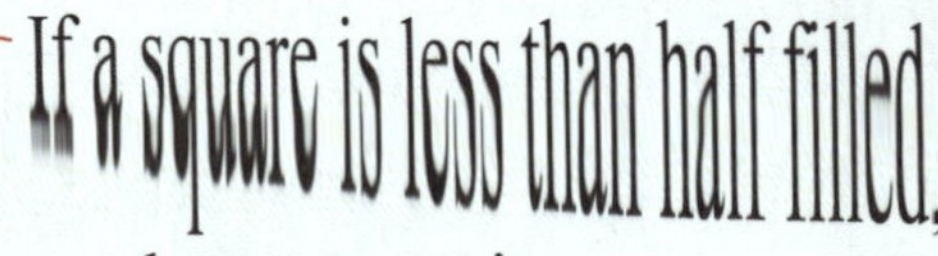

we do not count it.

There are 3 part squares which are more than half filled.

We *estimate* that the total area of the region is $12 + 3 = 15$ cm^2.

The area of this region is *approximately* 15 square centimetres.

Exercise 11

Estimate the area of each figure in cm^2:

a

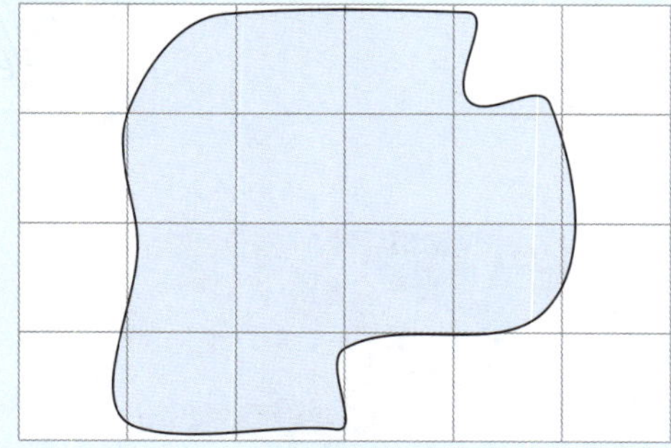

approximately ______ cm^2

b

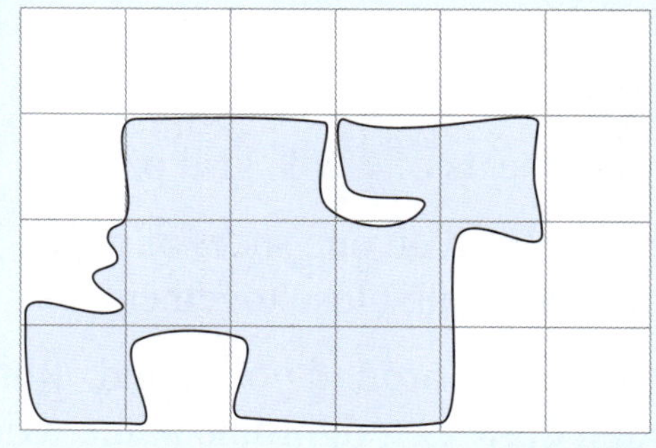

approximately ______ cm^2

c

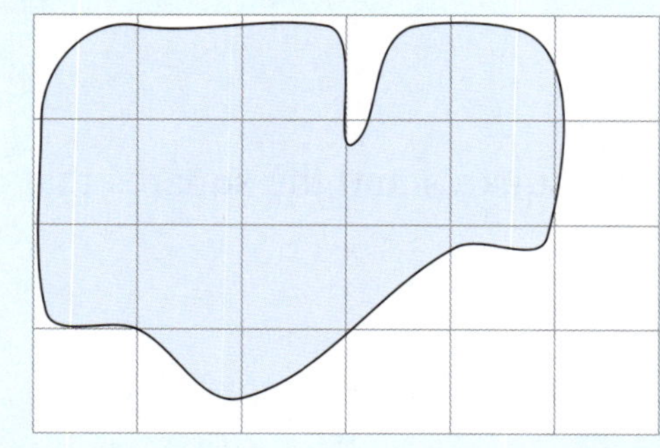

approximately ______ cm^2

d

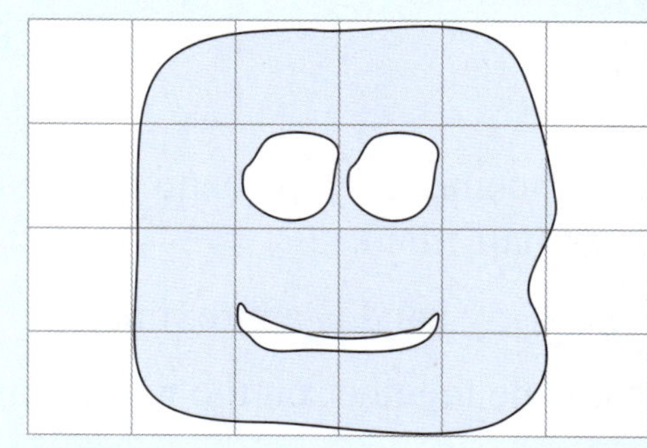

approximately ______ cm^2

e

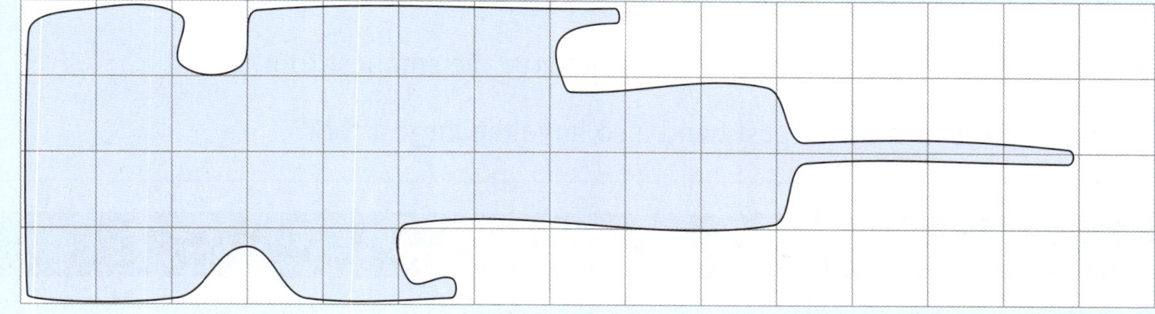

approximately ______ cm^2

f

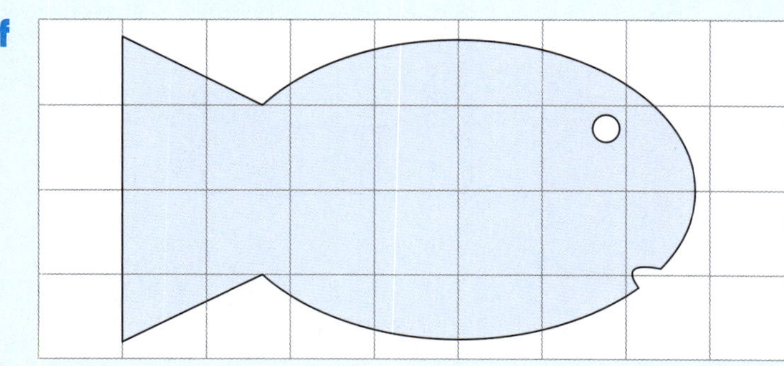

approximately ______ cm^2

Activity

You will need: two sheets of grid paper, a pencil

What to do:

1. Print two sheets of 1 cm^2 grid paper.
2. Place your hand on one sheet of paper. Keep your fingers and thumb close together.

   Carefully draw around your hand. Remove your hand and draw a straight line at the wrist.

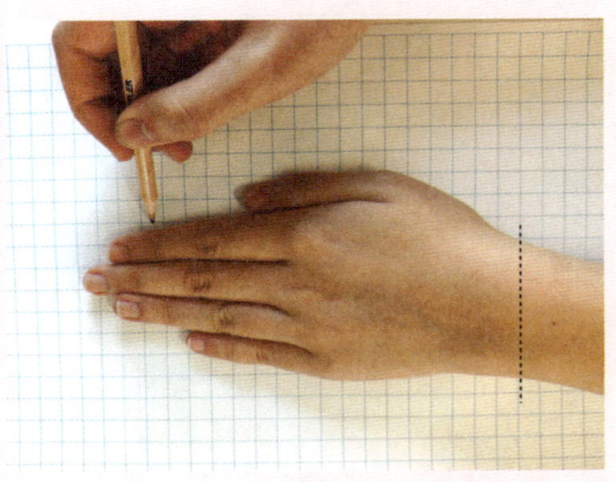

3. Estimate the area of your hand by counting the whole squares and the squares that are more than half filled.

   The area of my hand is approximately ______ cm^2.
4. Create a table together on the whiteboard to record the area of each person's hand.

   Use a spreadsheet to make a bar graph of the results.
5. Repeat the Activity to find the area of one of your feet, without a shoe.

   The area of my foot is approximately ______ cm^2.
6. Does the person with the smallest hand also have the smallest foot? ______
7. Does the person with the largest hand also have the largest foot? ______

Revision

1. Count the squares to find the area of each figure:

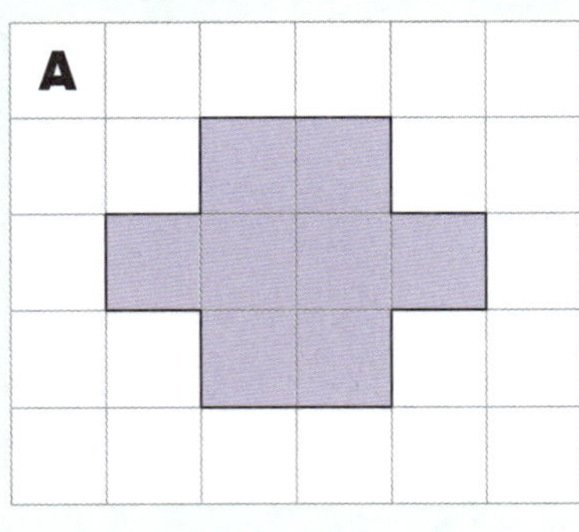

______ squares

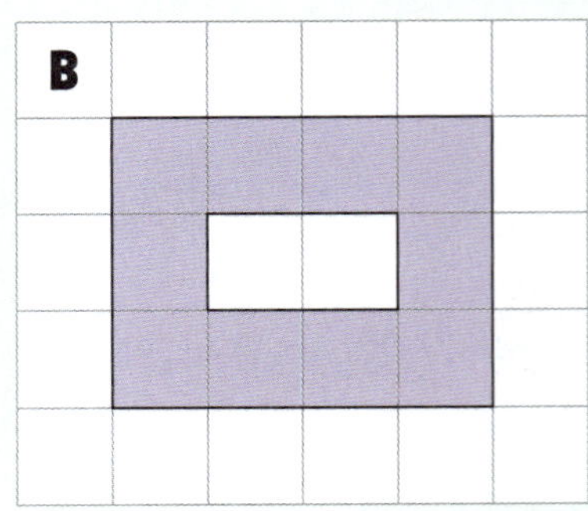

______ squares

Figure ______ has the larger area.

2 Draw a figure on the grid with area 17 squares.

3 These shapes are made up of squares with side length 1 cm.

Write down the area of each shape in cm^2.

a

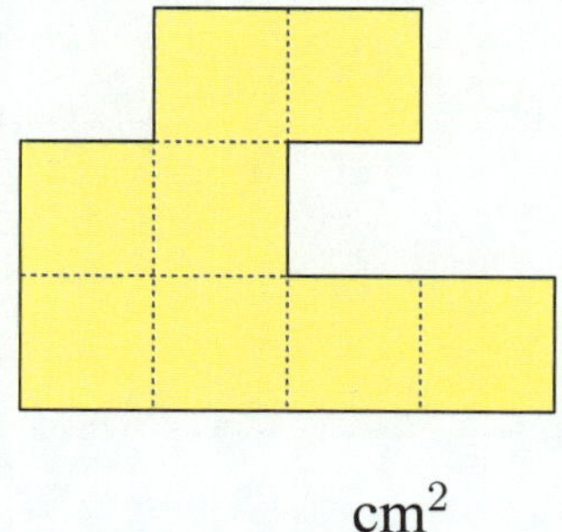

______ cm^2

b

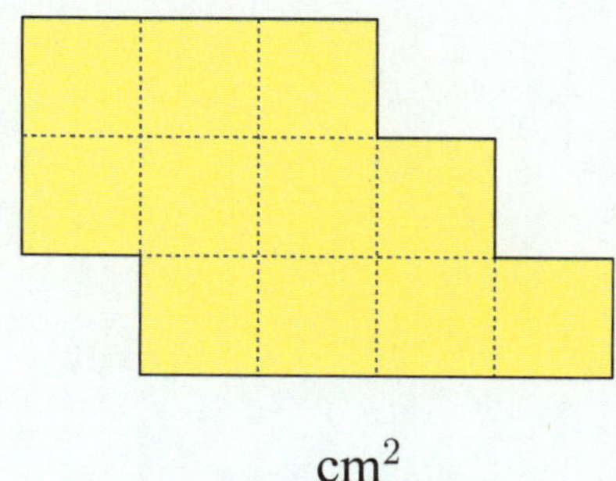

______ cm^2

4 1 square metre is the area of a square with side length ______________.

5 Carlos is buying new carpet for three rooms of his apartment.

Their areas are $12\ m^2$, $8\ m^2$, and $13\ m^2$.

The total area of carpet Carlos needs is ______ m^2.

6 Winnie is doing a jigsaw puzzle with area $372\ cm^2$.

Each piece has area $4\ cm^2$.

There are ______ pieces in the jigsaw.

7 Each square in the diagrams has side length 1 cm.

Write down the area of each shape in cm^2.

a

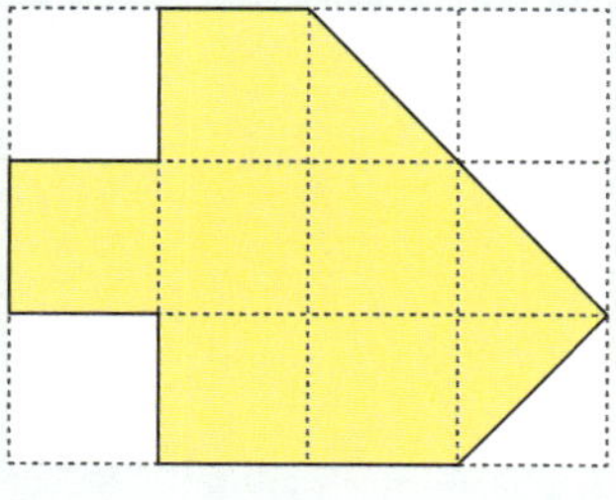

______ cm^2

b

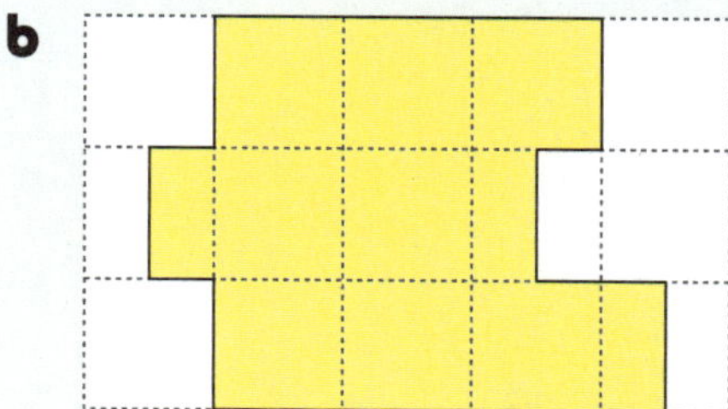

______ cm^2

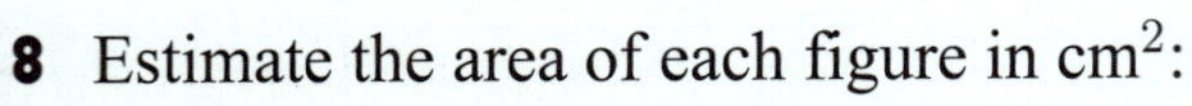

8 Estimate the area of each figure in cm^2:

a

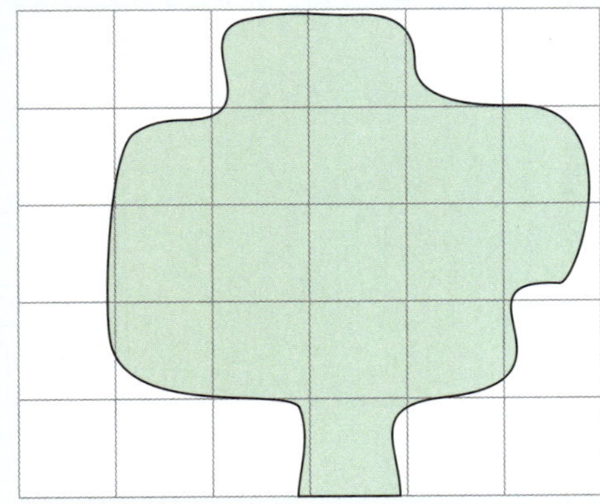

approximately ______ cm^2

b

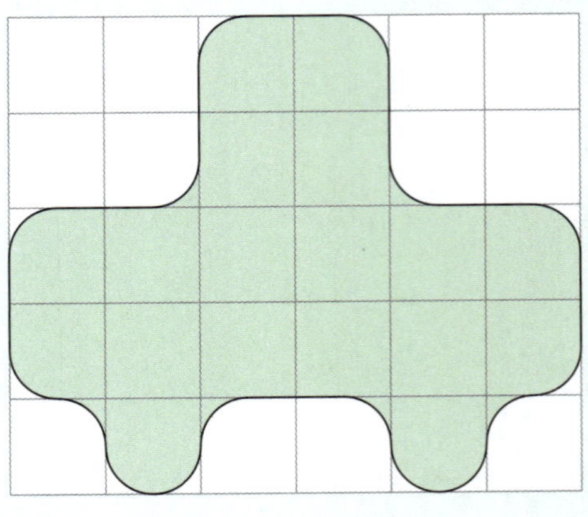

approximately ______ cm^2

c

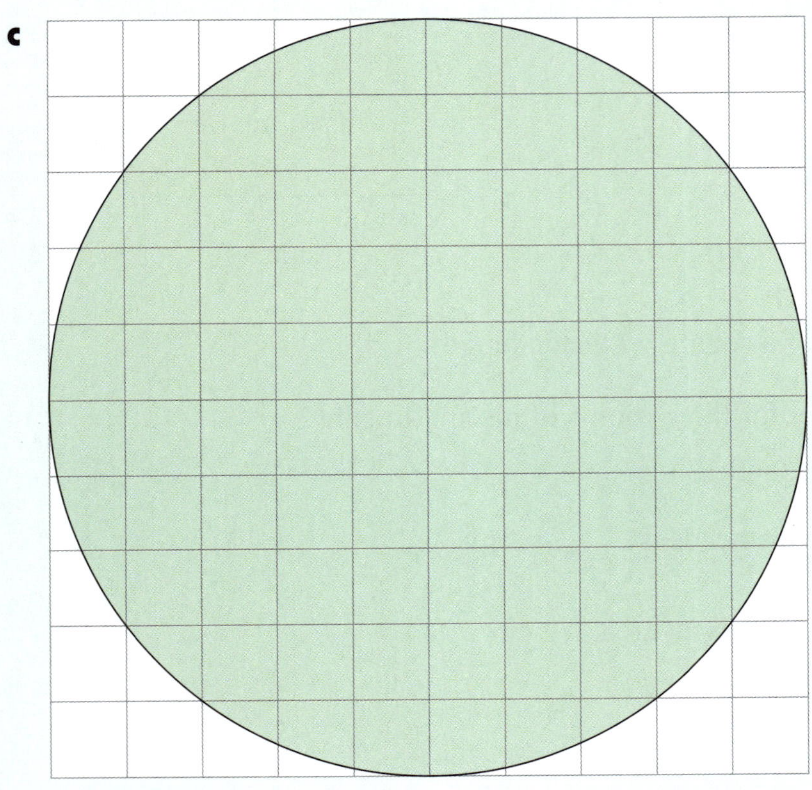

approximately ______ cm^2

CHAPTER 14: POSITION AND DIRECTION

Discussion

What words and phrases can we use to describe the **position** of a person or object?

List as many as you can.

....

....

....

....

Exercise 1

Complete each sentence describing the position of items in Molly's pantry:

a The bread bin is .... the cat food.

b The cake is .... the biscuit tin.

c The cups are to the .... of the plates.

d The soy sauce is .... the lollies.

e The carrots are .... the cat food.

f The garlic is .... the carrots and the blender.

g The olive oil is to the .... of the tomatoes.

h The mouse is in the .... of the third shelf from the ....

Exercise 2

Write a sentence to describe the position of:

a the striped shirt

b the belt

c the boots

d the cream sweater

e the suitcase

Exercise 3

a List the shapes *inside* the circle.

b List the shapes *outside* the circle.

Discussion

What do we mean by the word *adjacent*?

Exercise 4

Look at the seating arrangement at the table.

a Who is sitting opposite Zara?

b Who is sitting adjacent to Amani?

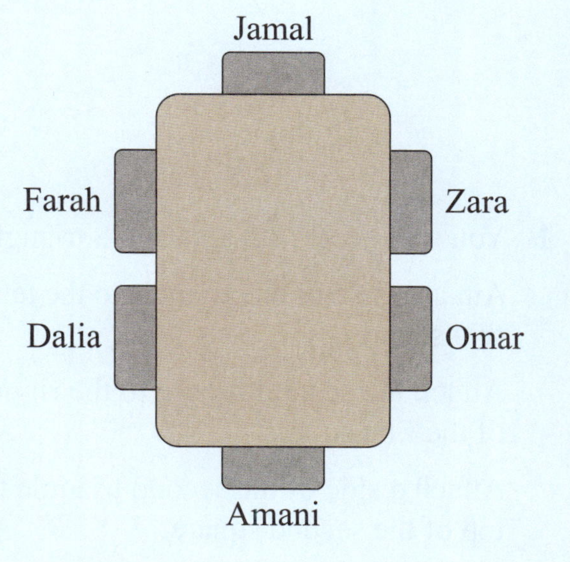

Suzie was given these instructions:

"You will make a shape using a square and two triangles.

Attach one side of a triangle to the left side of the square.

Attach one side of the second triangle to the base of the square."

Suzie used the instructions to draw this shape:

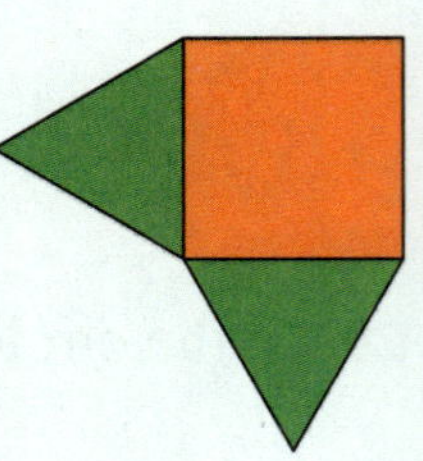

Exercise 5

Use pattern blocks to create the shape.

If you do not have pattern blocks, print this template and cut out the shapes.

Draw your shape in the space provided.

a You will need a square and 2 triangles.

Attach a side of one triangle to the top of the square.

Attach a side of the second triangle to the base of the square.

b You will need 2 squares and 2 triangles.

Attach a side of one triangle to the left side of a square.

Attach the second square to the right side of the first square.

Attach a side of the second triangle to the top of the second square.

c You will need 2 triangles, a square, and a hexagon.

Attach a side of the hexagon to the base of the square.

Attach a side of one triangle to the base of the hexagon.

Attach a side of the other triangle to the right side of the hexagon that is adjacent to the square.

Exercise 6

Write a set of instructions to make each shape:

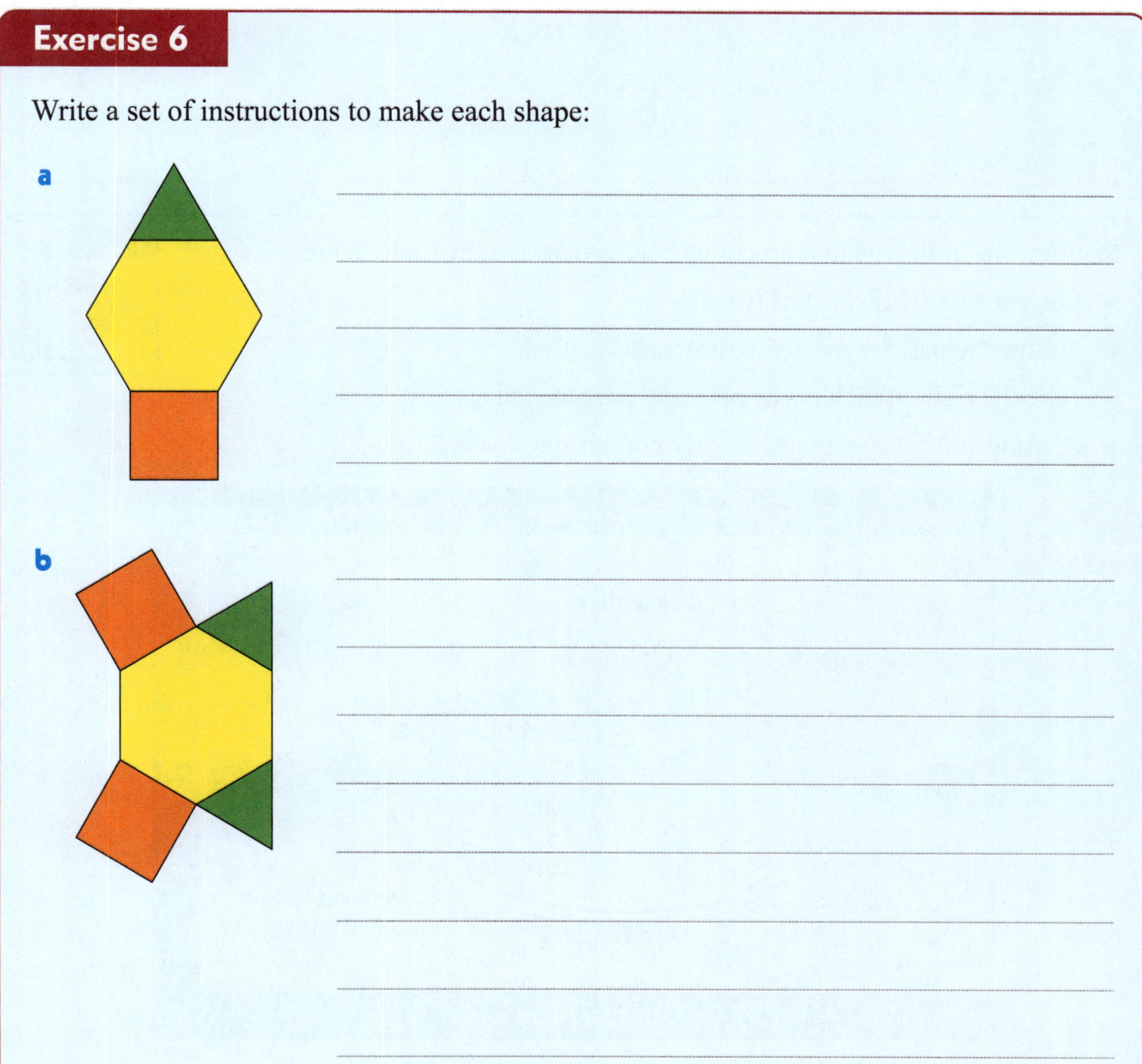

Discussion

What words and phrases can we use to give **directions** to a place or object?

List as many as you can.

Activity

Write a sentence to describe how to get from the classroom door to your desk.

Matthew has followed the directions below, marking his route in red.

- Enter the park at Newport Road.
- Follow the path between the rose gardens.
- Take the right branch of the path and pass the café.
- Continue on this path until the lily pond is on your left.

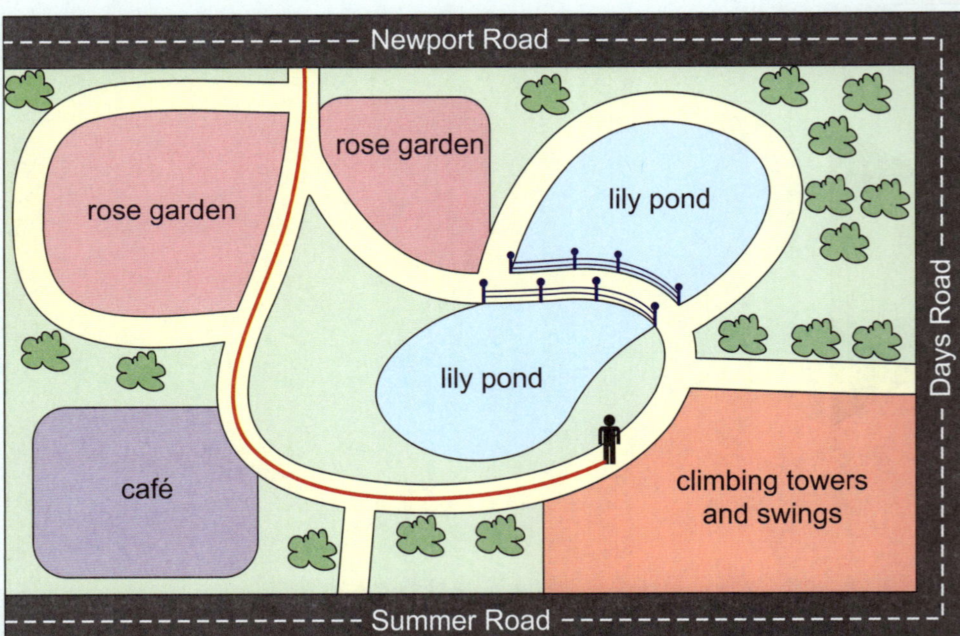

The climbing towers and swings are on Matthew's right.

Exercise 7

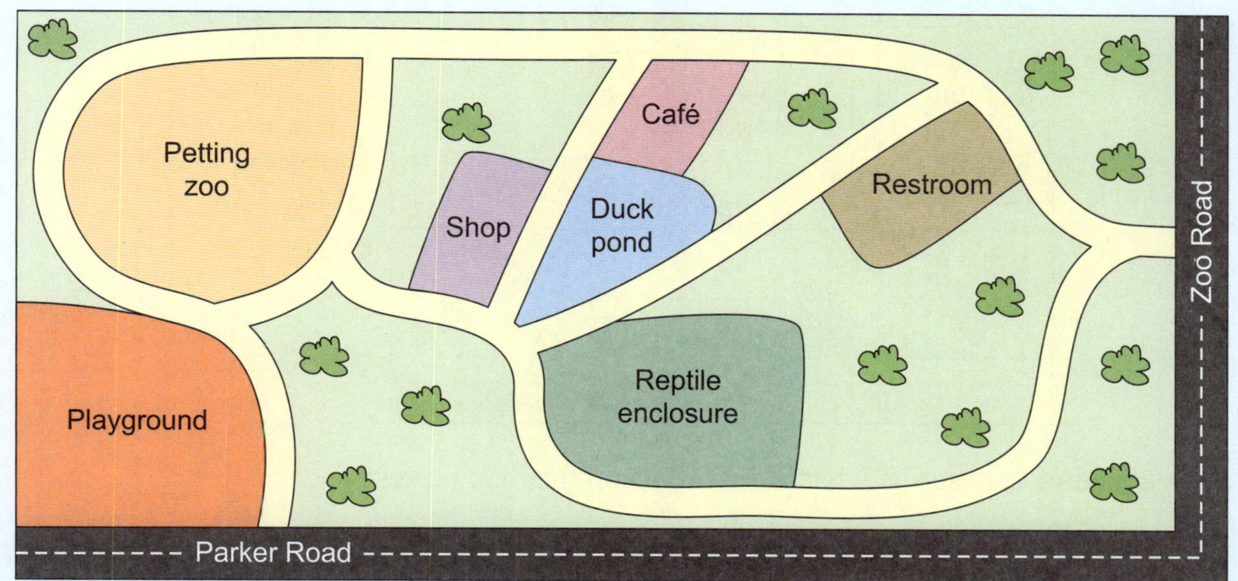

a Mark this route in red:

- Enter the park at the Zoo Road entrance.
- Follow the path to the right until you reach the café.
- Turn left at the café corner and walk past the café.

Mark your final position with 🯅.

The ______________ is to my right and the __________________ is to my left.

b Mark this route in blue:

- Enter the park at the Parker Road entrance.
- Follow the path to the left, going behind the petting zoo.
- Turn right and continue around the petting zoo, until you can turn left.
- Follow the path until you reach the shop corner, then turn right. Continue until you are almost back to Parker Road.

Mark your final position with ★.

The ____________________________ is to my left.

c Write instructions to guide a visitor from the Parker Road entrance to the restroom. Mark this route in green.

____

____

____

____

Maps of towns use small symbols to represent buildings and places of interest.

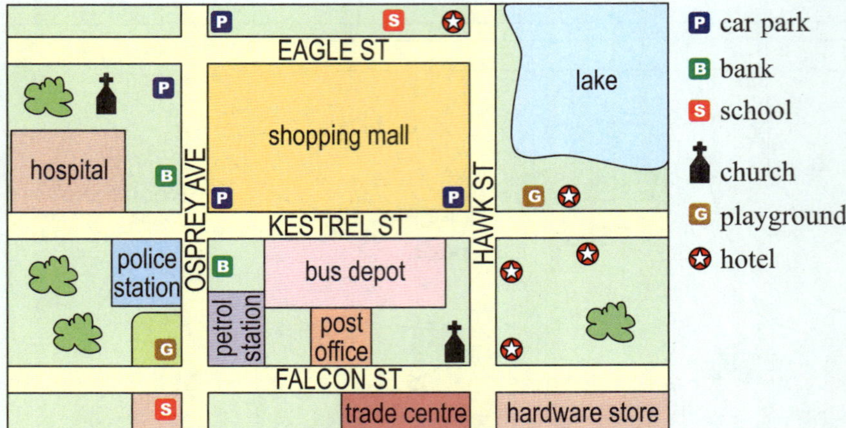

We use a **legend** to describe what the symbols on the map represent.

On this map, P shows the location of a car park.

Discussion

Why do maps show the locations of buildings and important places?

What might happen if the symbols were included, but the legend was not?

Exercise 8

Use the map above to answer these questions.

a How many hotels are marked on the map? ______

b How many churches are marked on the map? ______

c What can be found at the corner of Osprey Avenue and Kestrel Street?

______

d What buildings can be found on Eagle Street?

______

e What building is behind the hospital? ______

f What store can be found near the hotel on the corner of Falcon Street and Hawk Street?

______

g What place of interest is opposite the petrol station on Osprey Avenue?

______

When we draw a map, we need to draw each object much smaller than it really is.

We adjust the size of every object according to a **scale** so that we are still able to *compare* the sizes of objects.

The scale of this map is "1 unit = 100 metres".

This means that the distance between the grid lines represents 100 m of actual distance.

Scale: 1 unit = 100 metres

We can use the scale to calculate actual distances.

- Cove Terrace is 4 grid units long, so its actual length is 400 m.
- Beach Street is 7 grid units long, so its actual length is 700 m.
- Mary is 5 grid units from the car park, so Mary is actually 500 m from the car park.

Exercise 9

The scale of a map is "1 unit = 100 m". Find the actual distance represented by:

a 2 grid units ....

b 3 grid units ....

c 6 grid units ....

d 8 grid units ....

e $\frac{1}{2}$ grid unit ....

Exercise 10

The scale of a map is "1 unit = 20 m". Find the actual distance represented by:

a 10 grid units ....

b 2 grid units ....

c 4 grid units ....

d 7 grid units ....

e $\frac{1}{2}$ grid unit ....

Exercise 11

This map shows a school.

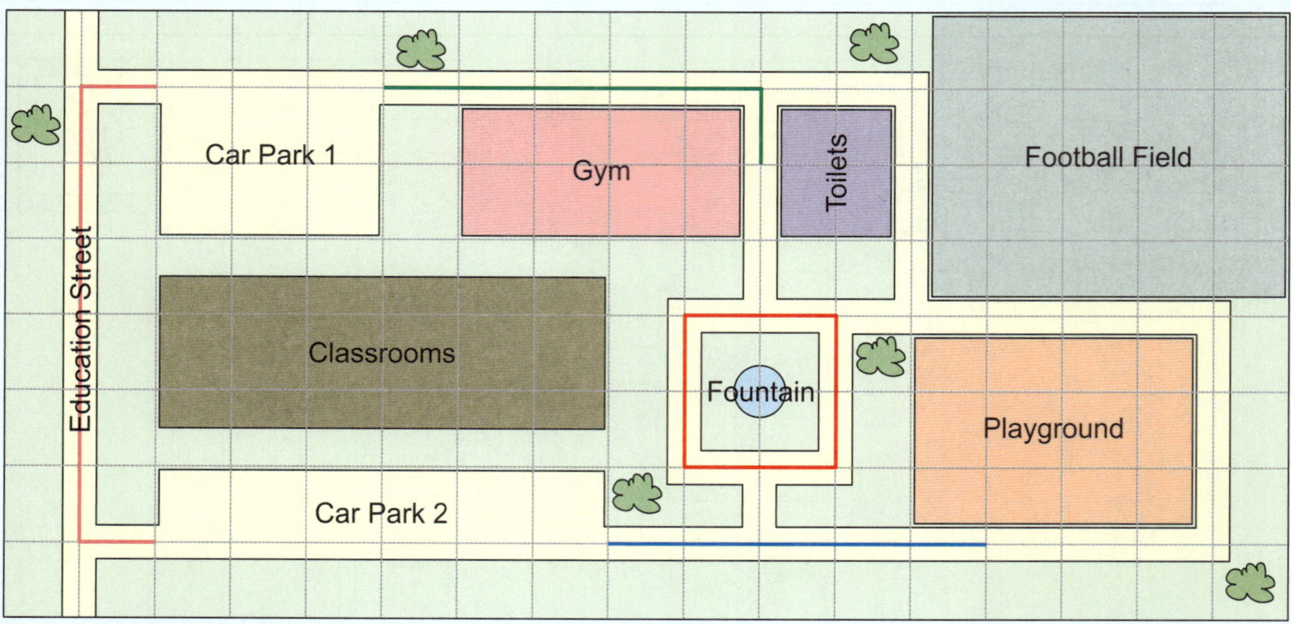

Scale: 1 unit = 20 metres

a Complete:

On this map, 1 grid unit = ______ metres.

On this map, 2 grid units = ______ metres.

Car Park 1 is ______ metres wide.

Car Park 2 is ______ metres wide.

b Circle the correct answers.

The total length of the blue route from the playground to Car Park 2 is:

- 80 metres
- 100 metres
- 120 metres

The total length of the pink route from Car Park 1 to Car Park 2 is:

- 160 metres
- 180 metres
- 200 metres

c If you walk the green route from Car Park 1 to the gym, how far will you walk?

______

d If you walk the red route around the fountain, how far will you walk?

______

Exercise 12

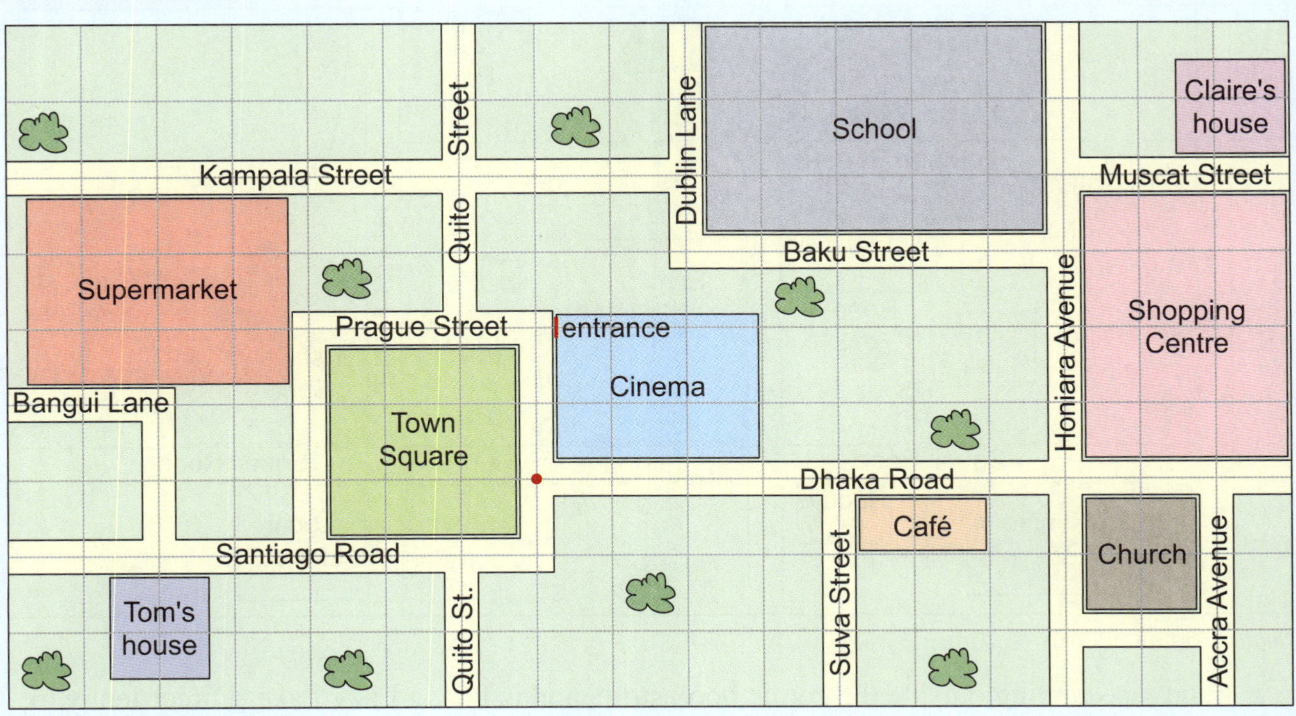

Scale: 1 unit = 10 m

a You are at the red dot on the edge of the town square. You walk 40 m along Dhaka Road, then turn right. Which street are you now in?

b Claire has just left her house and is on Muscat Street. How would you direct her to the church?

c Billy is standing at the intersection of two streets, and is facing the town square. Looking to his left, at the end of the road he can see the school. Where is Billy?

d Tom is going to the cinema. He is standing outside his house opposite Bangui Lane. Tom walks to the end of Santiago Road, then turns left and walks to the cinema's entrance at the top corner of Prague Street. How far did Tom walk?

e Estimate how far you would need to walk to get from the shopping centre to the supermarket.

Exercise 13

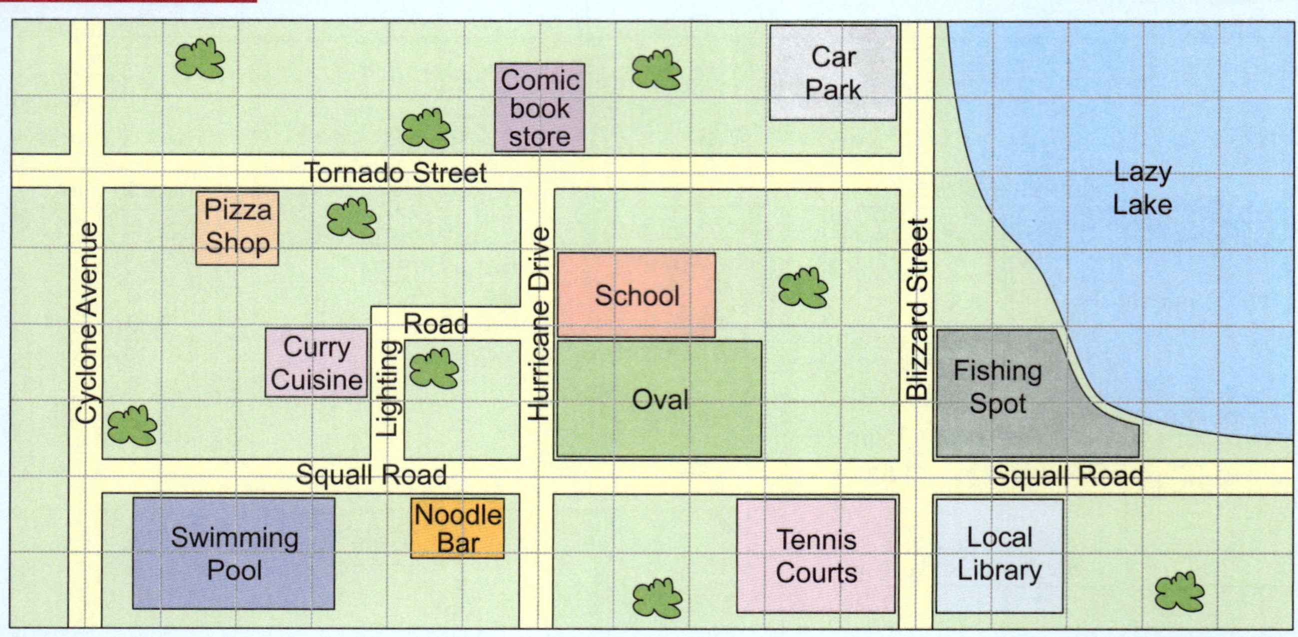

Scale: 1 unit = 20 m

a Jenny is standing outside the comic book store, and is facing Lazy Lake. Direct Jenny to the tennis courts.

....

....

....

b Name the two landmarks at the corner of Squall Road and Hurricane Drive.

....

c Create a legend for this map by drawing appropriate symbols in these boxes.

Bus stop	☐	Pedestrian crossing	☐
Traffic lights	☐	Post box	☐

d Add symbols to the map to show:

- a pedestrian crossing outside the Noodle Bar on Squall Road
- a bus stop on Hurricane Drive between the school and oval
- traffic lights at the intersection of Tornado Street and Cyclone Avenue
- a post box half way between Lightning Road and Cyclone Avenue, but opposite the swimming pool.

e How long is Squall Road? ....

f How long is Blizzard Street? ....

Revision

1 Look at this doll's house.

Write a sentence to describe the position of the:

a cupboard

b mirror

c table

d doll

2 **a** List the shapes *inside* the square.

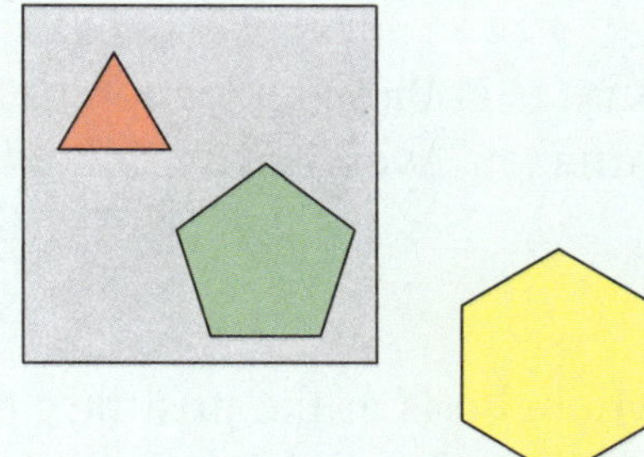

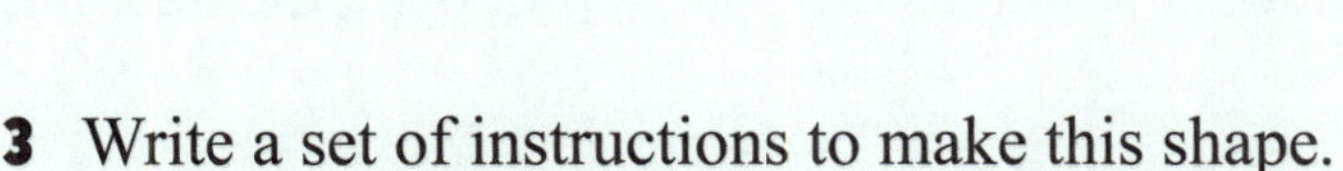

b Draw a circle *outside* the square.

3 Write a set of instructions to make this shape.

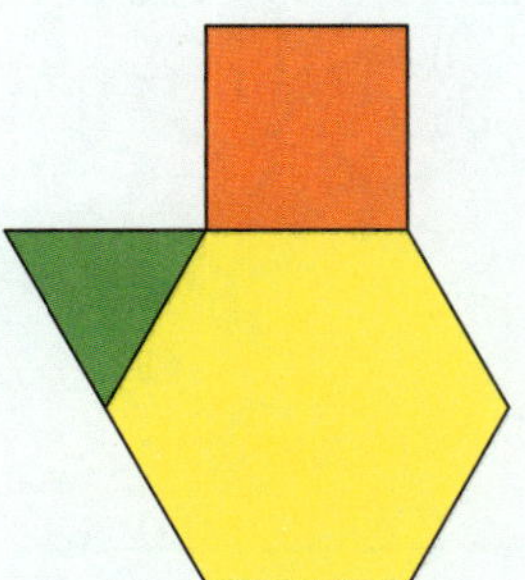

4

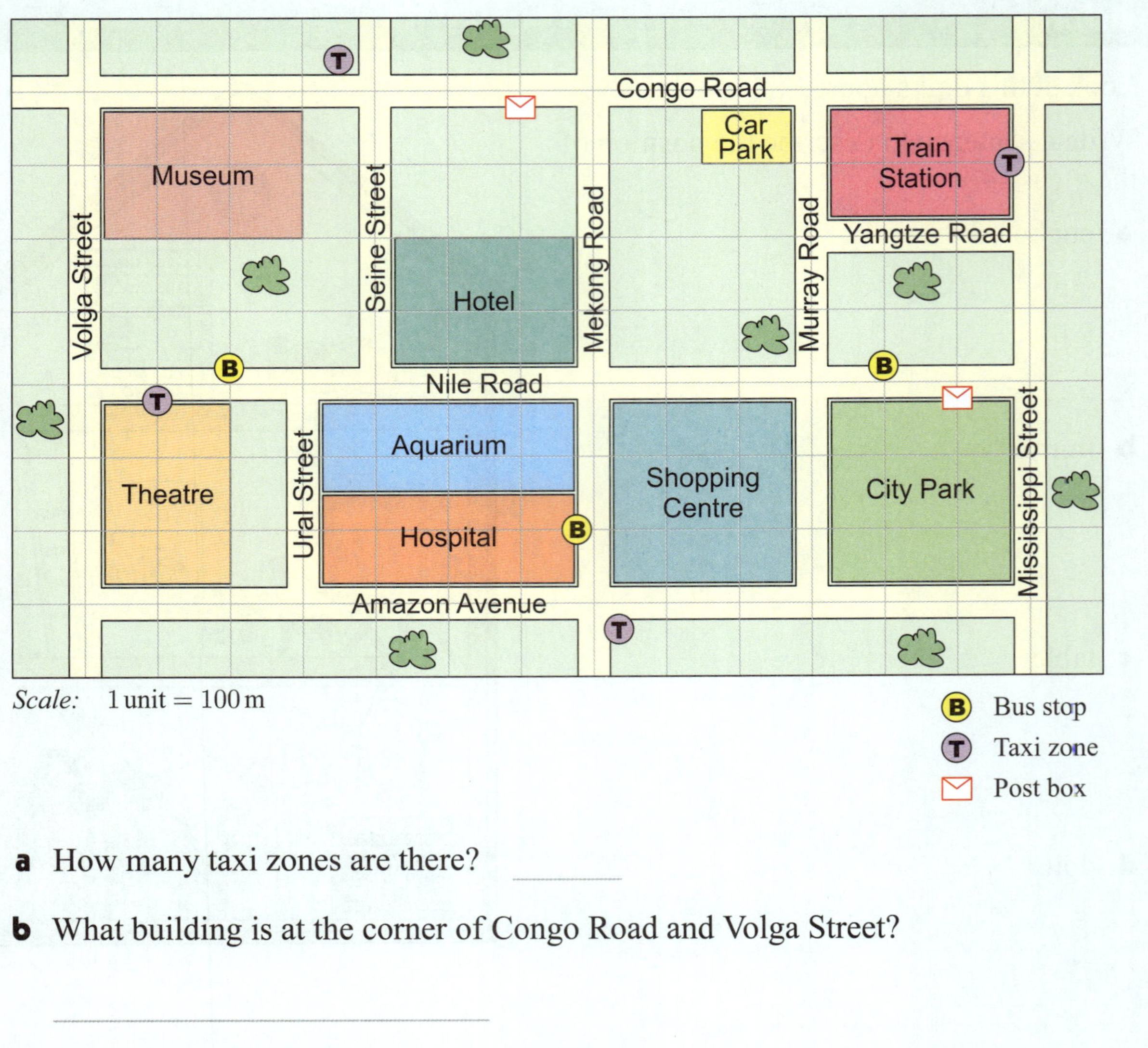

a How many taxi zones are there? ....

b What building is at the corner of Congo Road and Volga Street?

....

c How far apart are the bus stops on Nile Road? ....

d Kim is at the intersection of Amazon Avenue and Mississippi Street. He walks along Amazon Avenue for 400 metres, then looks to his right. What can he see?

....

e Michelle is at the post box near the city park. She walks down Nile Road until she reaches Mekong Road. She turns left, then walks to the bus stop outside the hospital.

How far has she walked? ....

f David is at the taxi zone near the theatre. How would you direct him to the car park?

....

....

....

CHAPTER 15: SOLIDS

A **solid** is a 3-dimensional shape which takes up **space**.

We can *draw* solids as 2-dimensional shapes on a page.

Any flat surface of a solid is called a **face**.

Any line where two faces meet is called an **edge**.

Any point where three faces meet is called a **corner**.

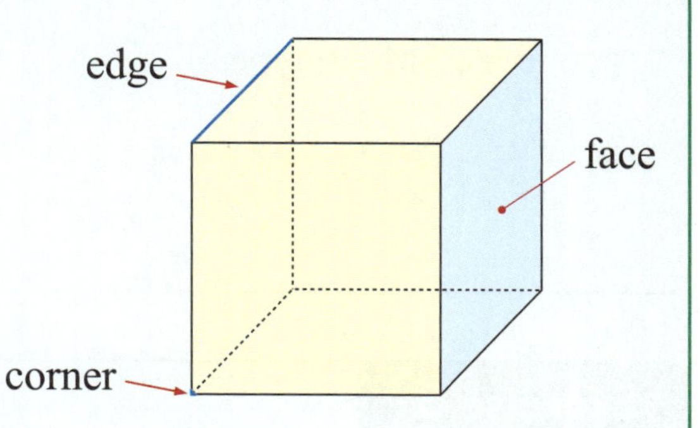

Exercise 1

Here are some solids whose surfaces are all flat.

Complete the table by counting the faces, edges, and corners.

Solid	*Diagram*	*Faces*	*Edges*	*Corners*
rectangular prism				
triangular prism				
square-based pyramid				
triangular-based pyramid				

Exercise 2

This is a **cube**.

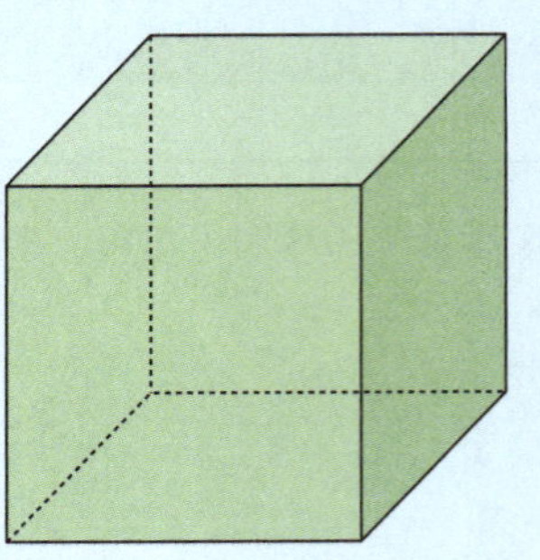

A cube has 6 square ________ .

A cube is a special type of

________________________ .

Exercise 3

Here are some solids which include a *curved* surface.

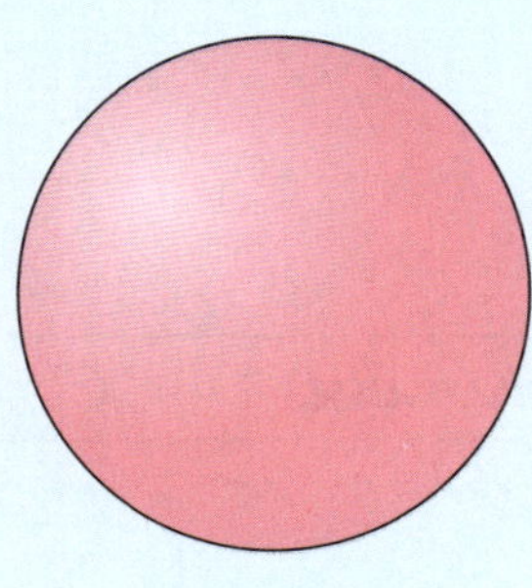
sphere

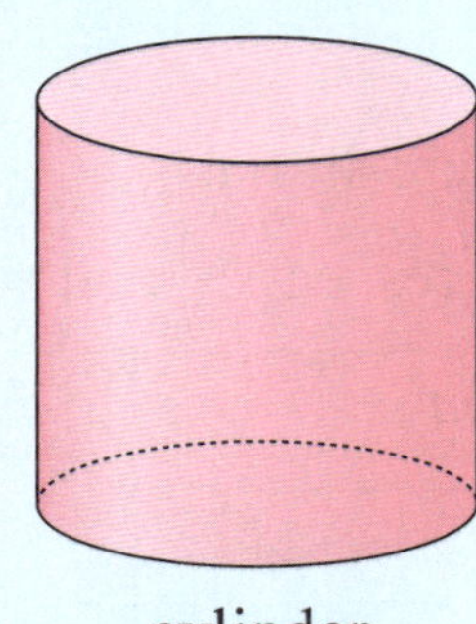
cylinder

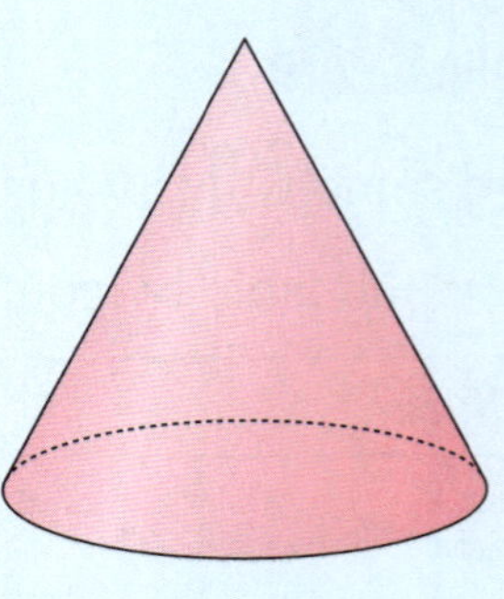
cone

Complete each sentence:

a A ____________ has two circular faces and one curved surface.

b A ____________ has one curved surface only.

c A ____________ has one circular face and one curved surface.

Discussion

What do we mean by the *apex* of a pyramid or cone?

Draw an arrow to indicate the apex on each diagram.

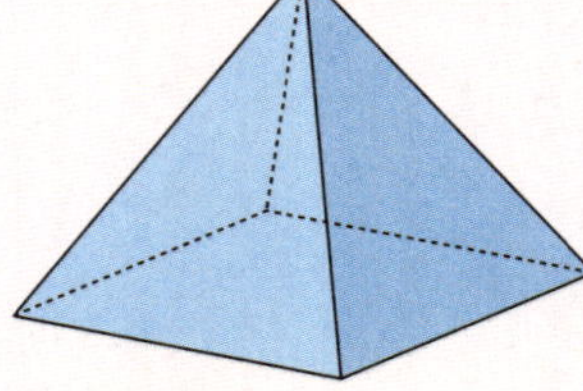

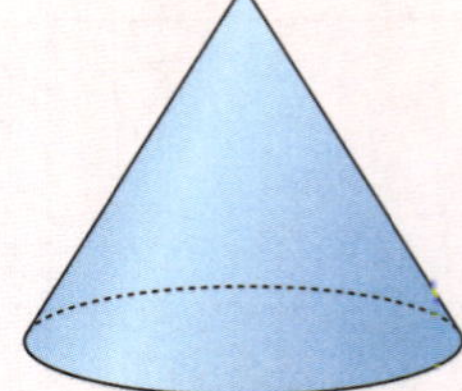

Activity

Draw each solid using the steps given:

a Rectangular prism

1 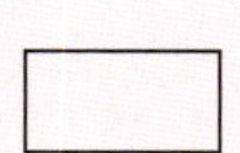

2 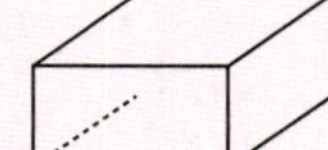

3 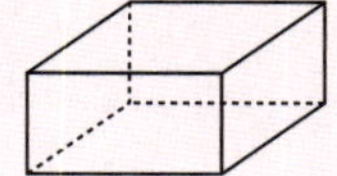

b Sphere

1 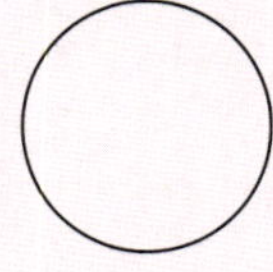

2 

3 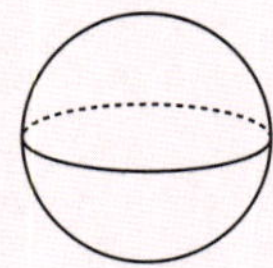

c Cone

1 

2 

3 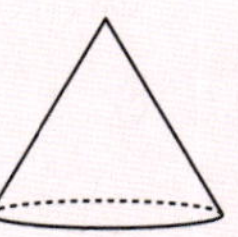

d Cylinder

1 

2 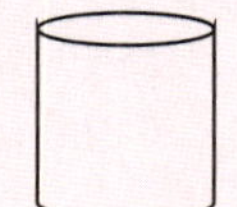

3 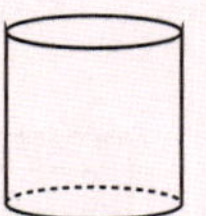

e Square-based pyramid

1

2 

3 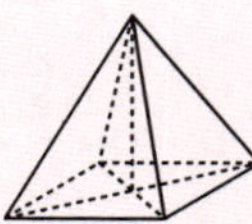

4 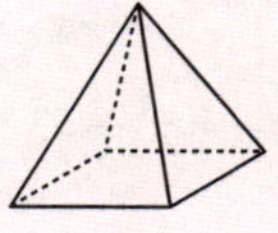

Exercise 4

Name the solid. Copy the diagram in the space alongside.

a

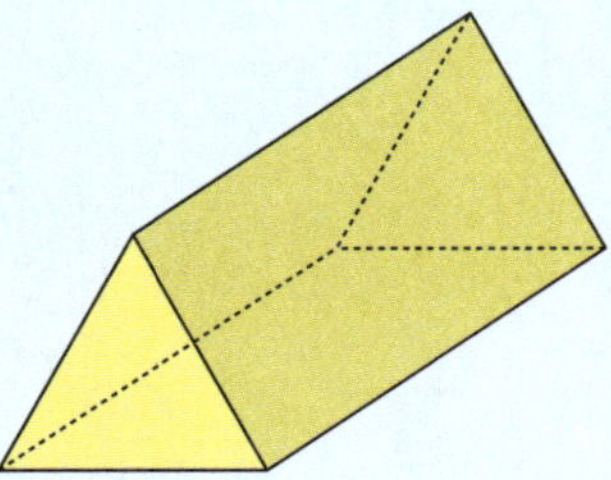

....

b

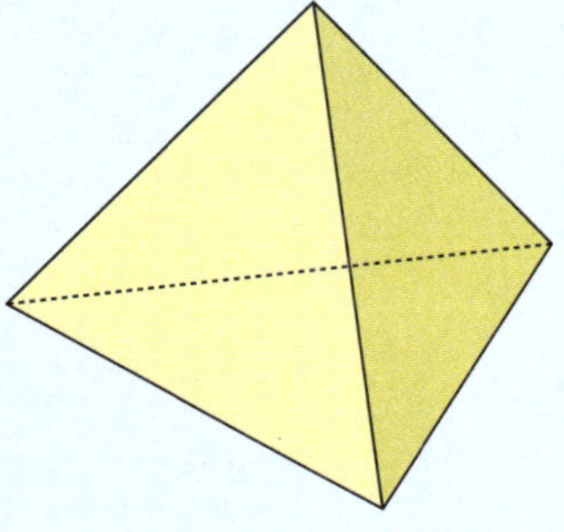

....

A **net** is a 2-dimensional shape which can be folded to make a 3-dimensional solid.

For example, when this net is folded along the dotted lines, we form a square-based pyramid.

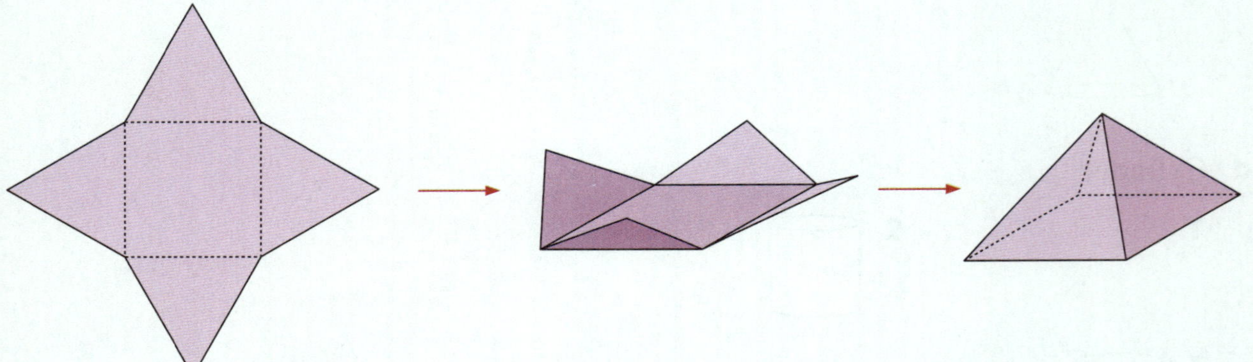

Click the icon to watch how the net is folded into the pyramid.

Activity — Making solids

You will need: printed sheets, coloured pencils, scissors, tape

What to do:

1. Print out the nets.
2. Colour each surface of each solid a different colour.
3. Cut around the *thick* black lines.
4. Fold along each red line.
5. Use tape to join the sides.

Exercise 5

Match each net with its solid:

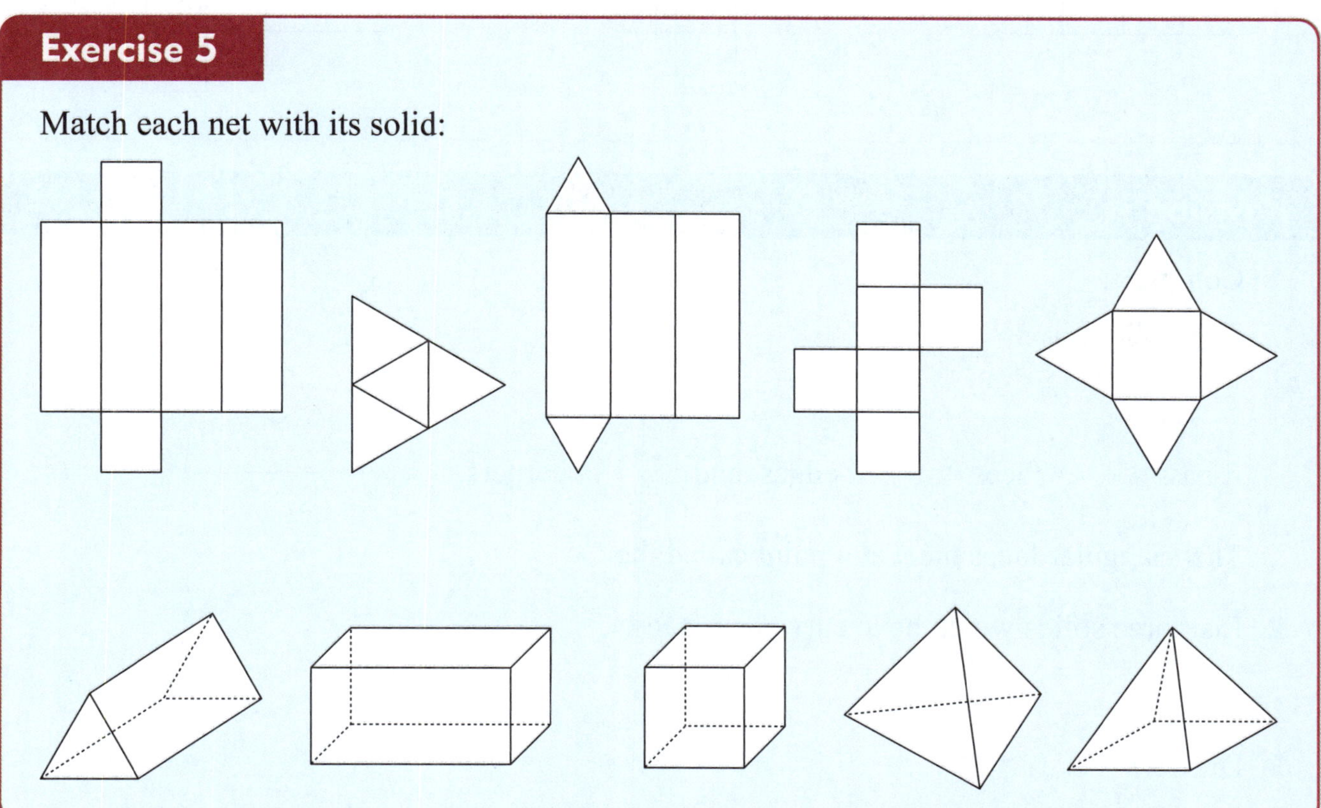

Discussion

What solids do you think could be made with these nets?

a **b**

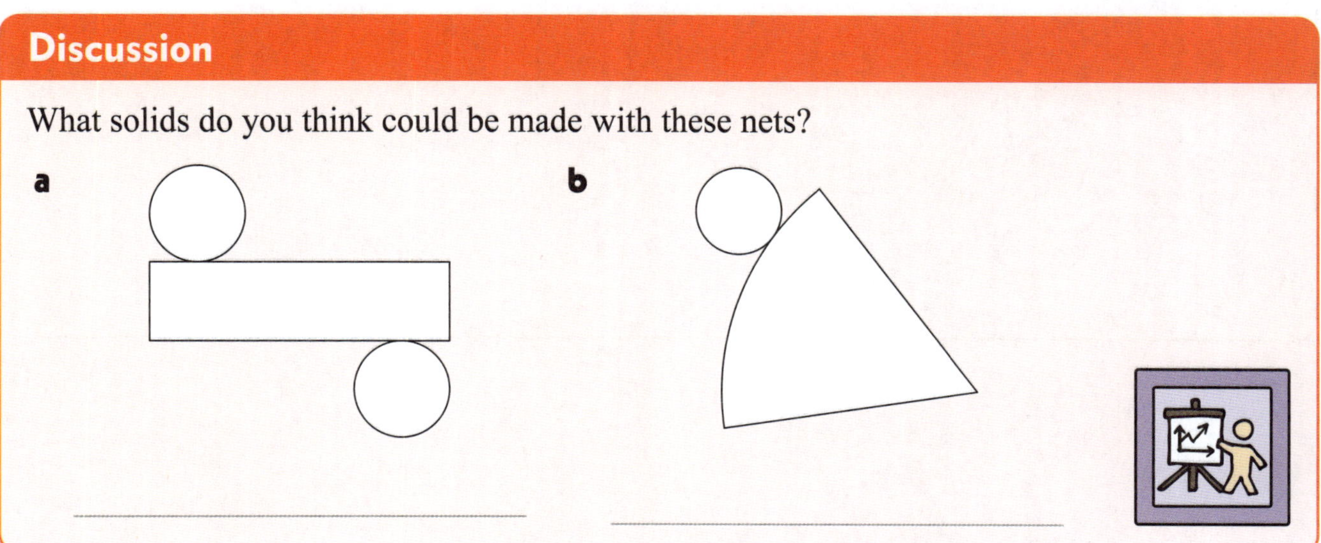

Activity

Some of the nets below will *not* make a cube.

Draw a circle around each net that will not make a cube.

A

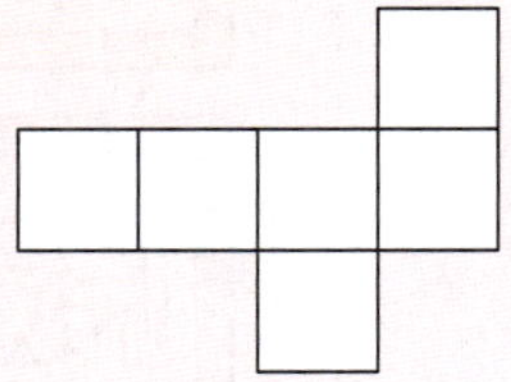

B

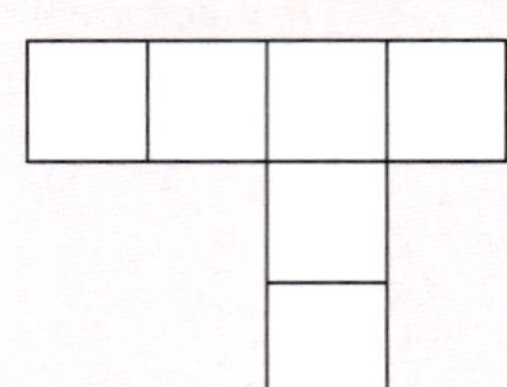

C

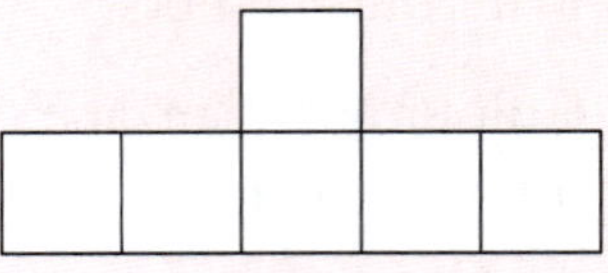

D

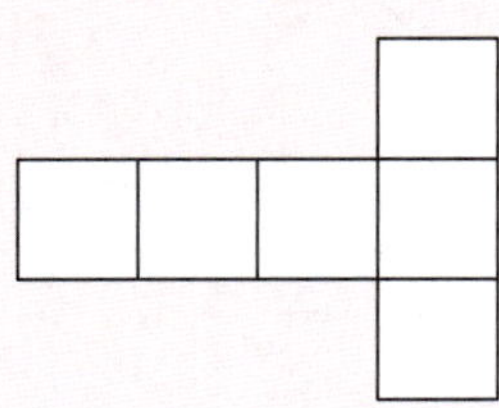

E

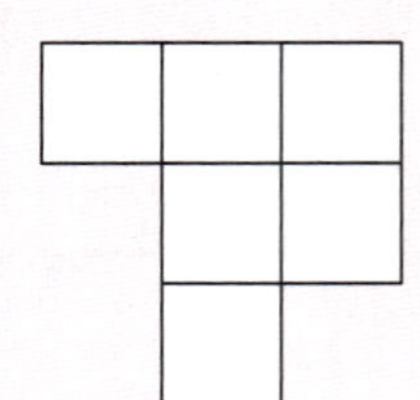

F

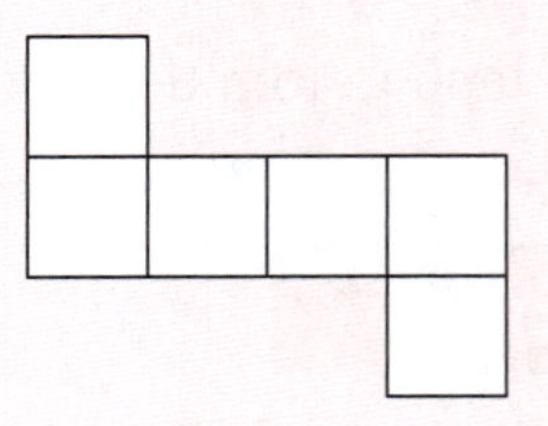

Revision

1 Complete:

This solid is a

____________________ .

It has _____ faces, _____ edges, and _____ corners.

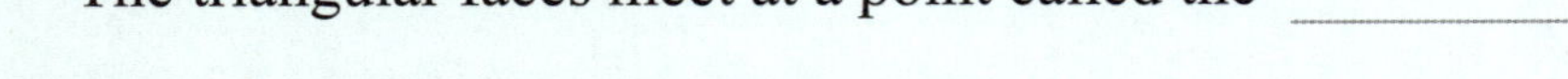

The triangular faces meet at a point called the __________ .

2 List three solids which have curved surfaces.

____________ ____________ ____________

3 Draw:

a a triangular prism

b a cylinder.

CHAPTER 16: VOLUME

Discussion

Look at these solids.

Which solid is larger?

How do you know it is larger?

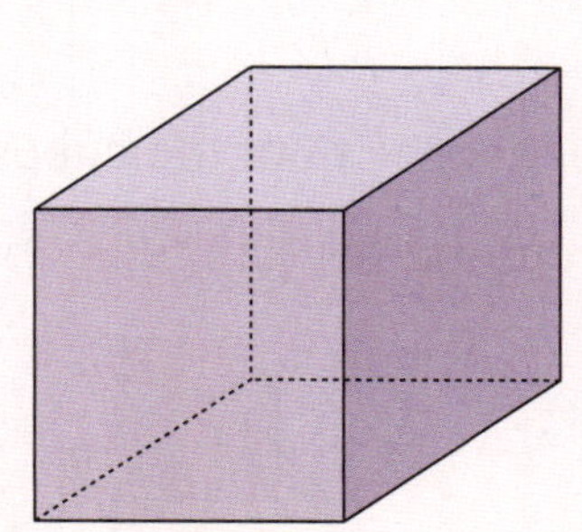

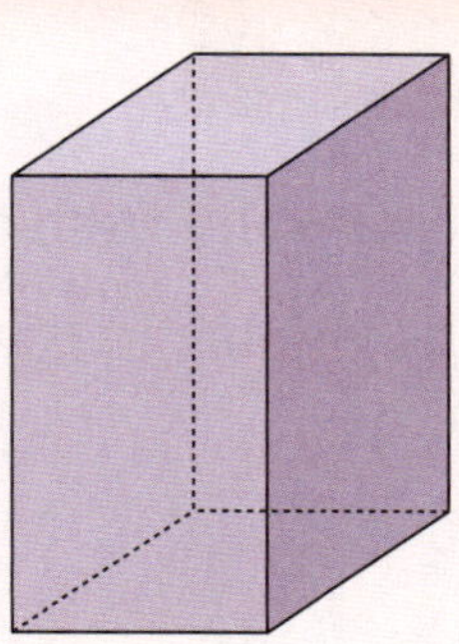

When we talk about the size of a solid, we are usually interested in how much *space* it takes up.

The **volume** of a solid is the amount of space it takes up.

To measure the volume of a solid, we could build it out of cubes.

There are 6 cubes on the first layer.

There are 2 cubes on the second layer.

The solid has volume $6 + 2 = 8$ cubes.

Discussion

Look at the solid above with volume 8 cubes.

Can you *see* all of the 6 cubes on the first layer?

How did we *know* there were 6 cubes on the first layer?

Exercise 1

Count the cubes to find the volume of each solid:

a

.... cubes

b

.... cubes

c

.... cubes

A **cubic centimetre** is the volume of a cube whose edges all have length 1 cm.

We write a cubic centimetre as **cm³**.

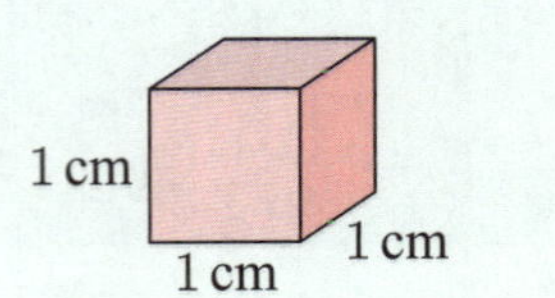

Exercise 2

Build each solid from centicubes. Count the cubes to find the volume of the solid.

Write its volume in cubic centimetres (cm³).

a ______ cm³

b ______ cm³

c ______ cm³

d ______ cm³

e ______ cm³

f ______ cm³

g ______ cm³

h ______ cm³

i ______ cm³

j ______ cm³

k ______ cm³

Exercise 3

Write the volume of the solid in cm^3.

Only build the solid if you need to.

a

______ cm^3

b

______ cm^3

c

______ cm^3

d

______ cm^3

Game

Click the icon to practise finding the volume of solids.

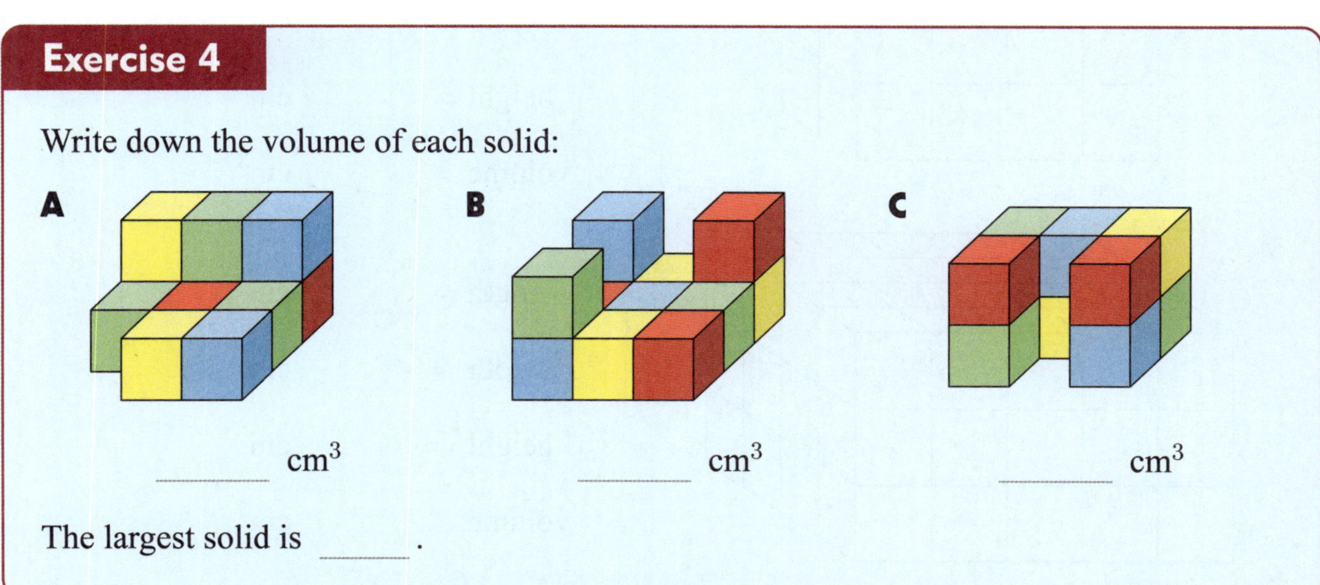

Exercise 4

Write down the volume of each solid:

A

______ cm^3

B

______ cm^3

C

______ cm^3

The largest solid is ______.

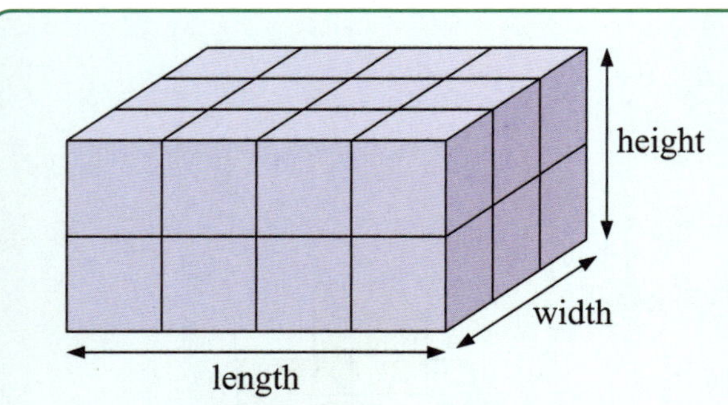

length = 4 cm
width = 3 cm
height = 2 cm
volume = 24 cm^3

Exercise 5

Complete the measurements.

a

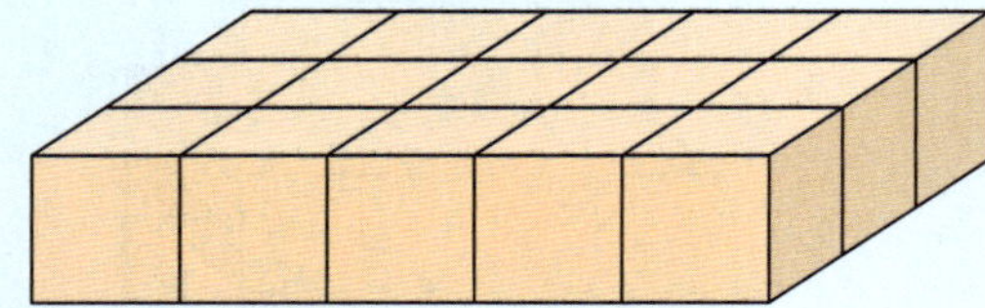

length = ______ cm

width = ______ cm

height = ______ cm

volume = ______ cm^3

b

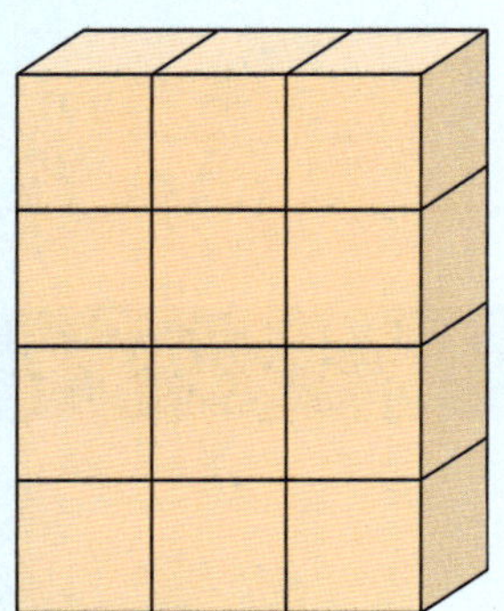

length = ______ cm

width = ______ cm

height = ______ cm

volume = ______ cm^3

c

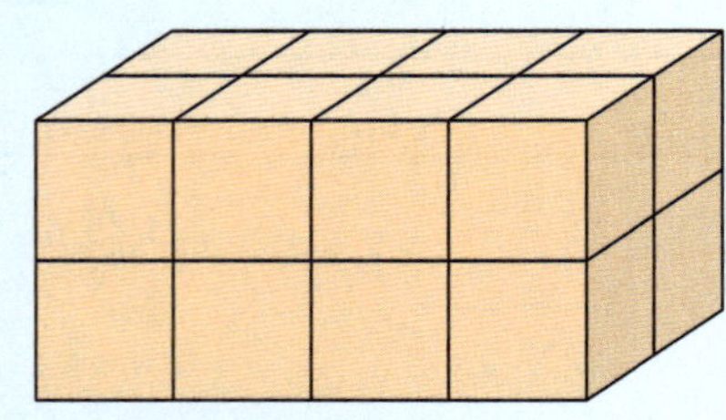

length = ______ cm

width = ______ cm

height = ______ cm

volume = ______ cm^3

d

length = ______ cm

width = ______ cm

height = ______ cm

volume = ______ cm^3

Exercise 6

A B C

Write the solids in ascending order of:

a length ______ , ______ , ______

b width ______ , ______ , ______

c height ______ , ______ , ______

d volume ______ , ______ , ______

Exercise 7

A basketball has volume 7470 cm^3.

A volleyball has volume 4850 cm^3.

How much more volume does the basketball have?

__________ cm^3

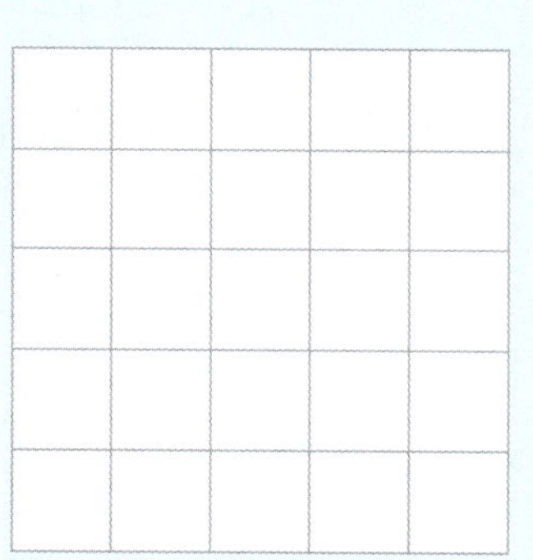

Activity

In this Activity you should work in pairs.

You will need: lots of centicubes, a can

What to do:

1. Use your centicubes to build a model of your can.
2. Count the centicubes to *estimate* the volume of the can.

   ______ cm^3

3. Compare your model and answer with other pairs of students.

Revision

1 Build each solid from centicubes. Write its volume in cubic centimetres.

a

______ cm^3

b

______ cm^3

2 Write the volume of the solid in cm^3. Only build the solid if you need to.

a

______ cm^3

b

______ cm^3

3 Complete the measurements.

length = ______ cm

width = ______ cm

height = ______ cm

volume = ______ cm^3

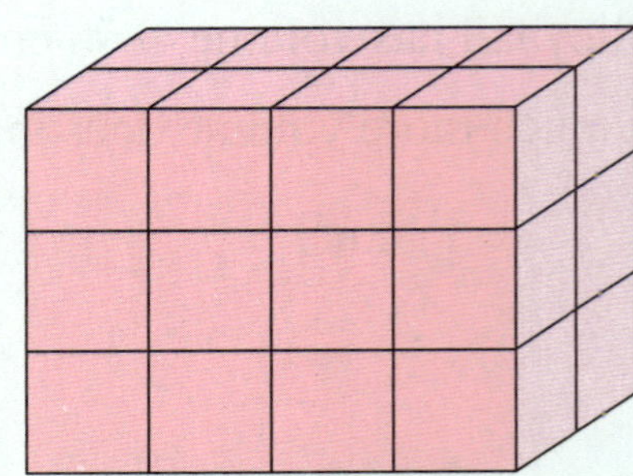

4 A sculptor used 175 cm^3 of clay to make a model and 48 cm^3 of clay to make a base for the model. How much clay did the sculptor use in total?

The sculptor used ________ cm^3 of clay in total.

CHAPTER 17: CAPACITY

The **capacity** of a container is the amount of liquid it can hold.

Exercise 1

Arrange these containers in order of capacity.

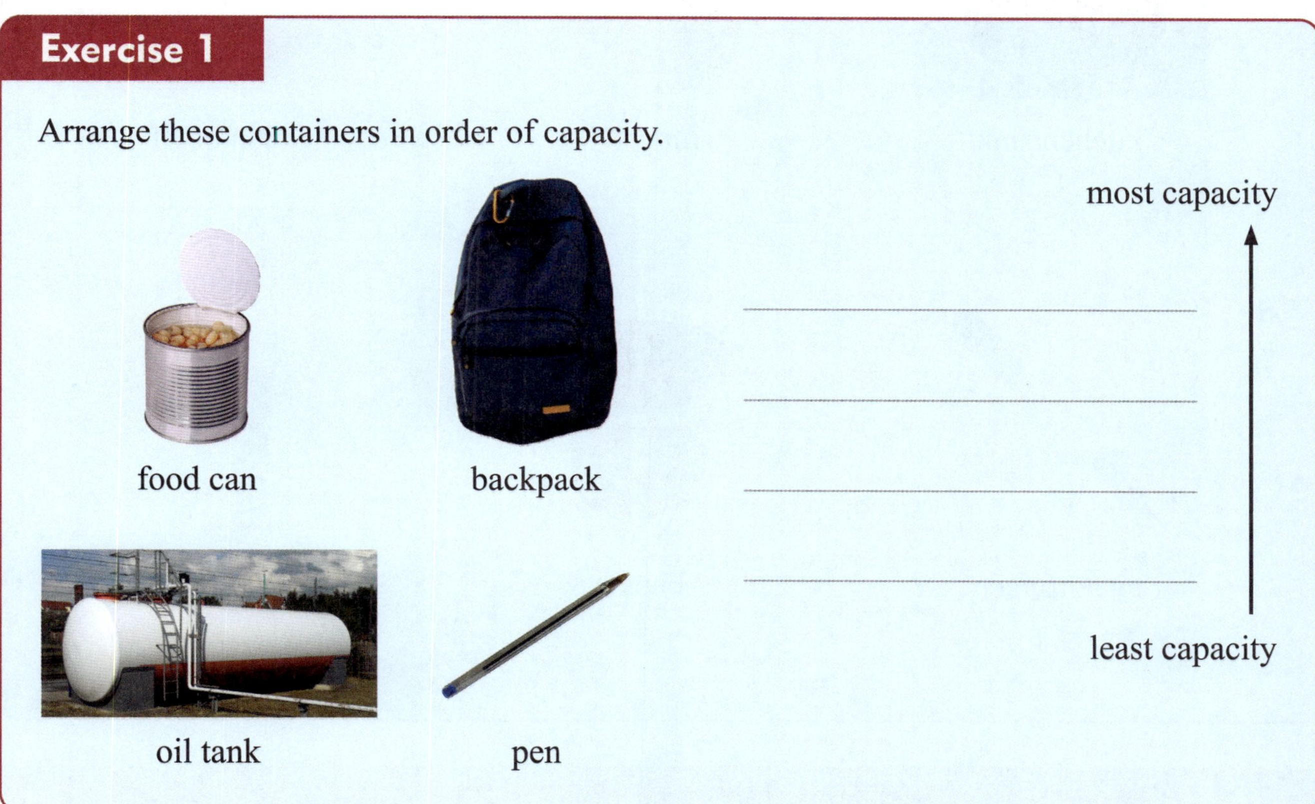

food can backpack oil tank pen

We measure capacity using **millilitres** (mL) or **litres** (L).

$$1 \text{ L} = 1000 \text{ mL}$$

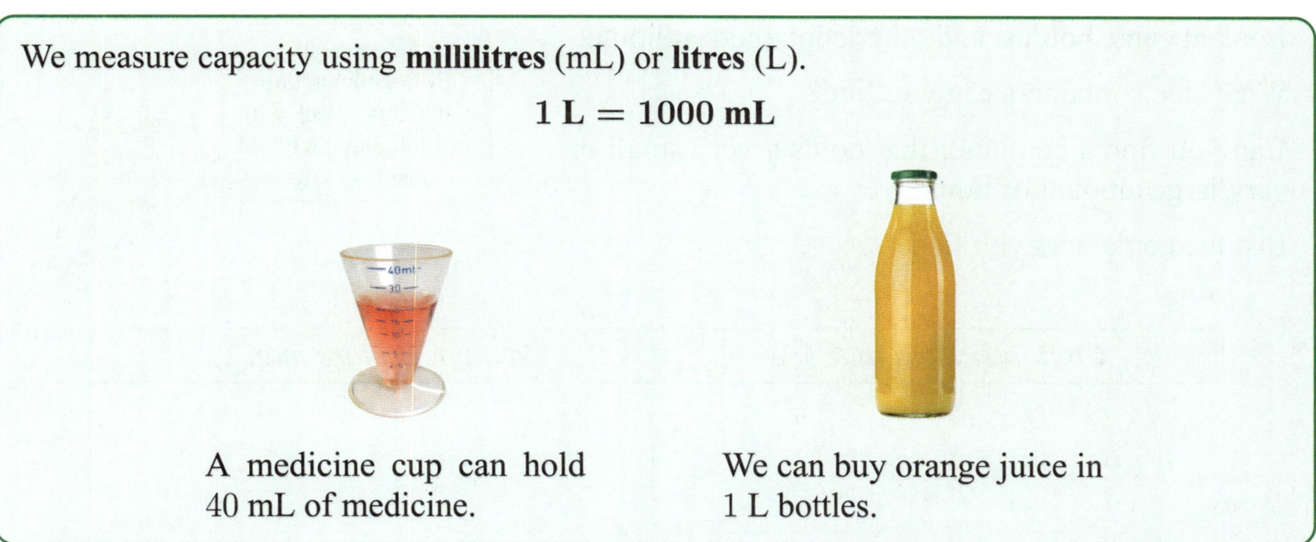

A medicine cup can hold 40 mL of medicine.

We can buy orange juice in 1 L bottles.

Discussion

We have seen millimetres and millilitres.

What does *milli*- mean at the start of a word?

Exercise 2

Decide whether each container holds *more* or *less* than 1 L.

a 

kitchen sink

b 

mug

c 

fish bowl

d 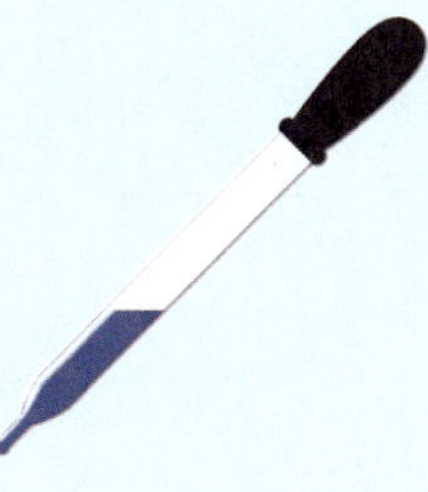

eye dropper

e 

bucket

f 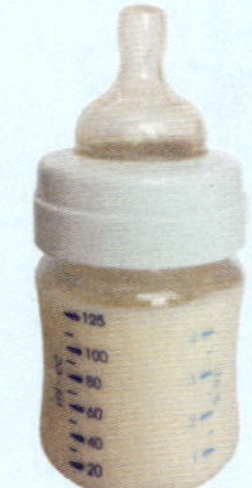

baby bottle

Activity

Look at cans, bottles, and other containers at home.

What size containers can you find?

Can you find a container that holds a very small or very large amount of liquid?

List the containers you find.

Containers less than 1 L	*Containers more than* 1 L

Exercise 3

Complete:

1 L = 1000 mL	6 L = ______ mL
______ L = 2000 mL	7 L = ______ mL
3 L = ______ mL	______ L = 8000 mL
______ L = 4000 mL	9 L = ______ mL
______ L = 5000 mL	______ L = 10 000 mL

We also use millilitres and litres to measure the *amount* of liquid in a container.

We use a **scale** to read the amount of liquid.

Discussion

Look at this container.

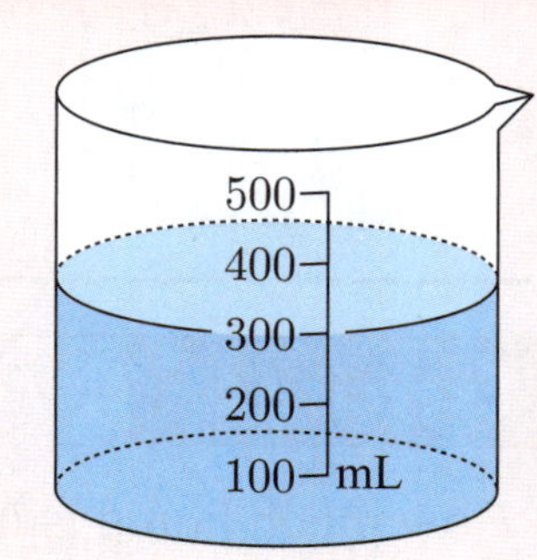

What *units* does it measure in? ______

How much liquid fits between each mark on the scale?

______

What is the *capacity* of the container? ______

How much liquid is in the container? ______

Exercise 4

How much liquid is in the container?

a

b

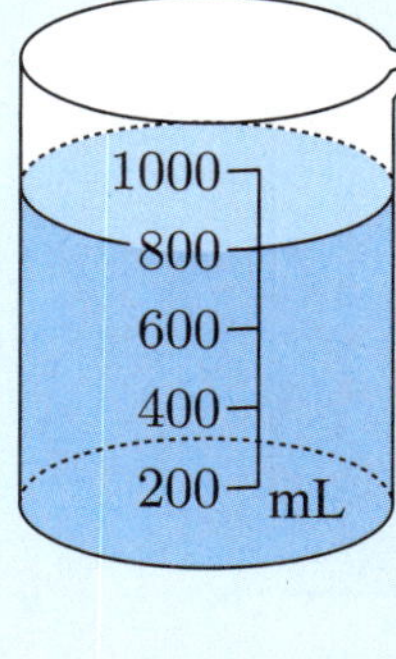

c

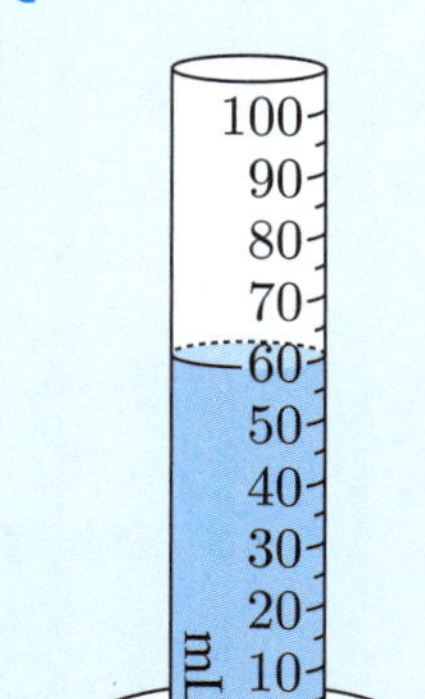
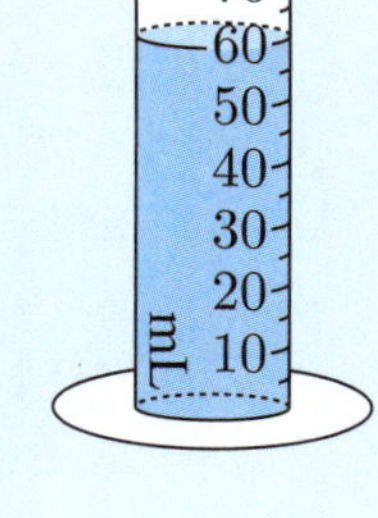

d

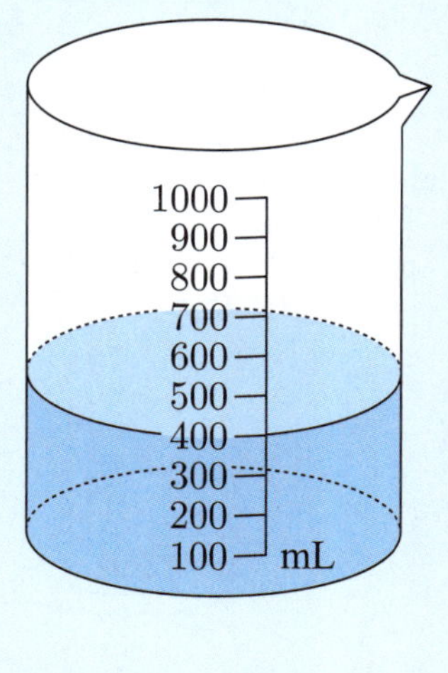

______ ______ ______ ______

Exercise 5

Draw the given amount of liquid in the container.

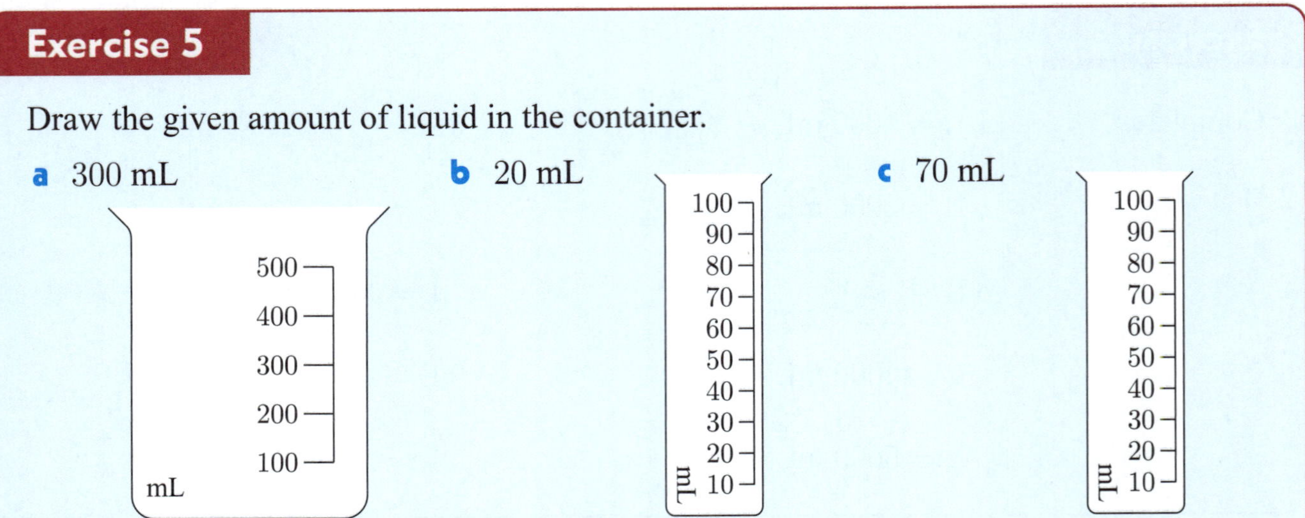

500
400
300
200
100
mL

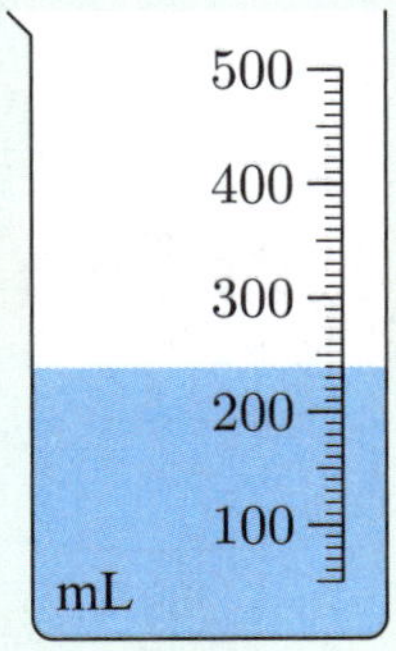

Each major division represents 100 mL.

Each minor division represents 10 mL.

There is 240 mL of liquid in this container.

Exercise 6

How much liquid is in the container?

a 500 400 300 200 100 mL ....

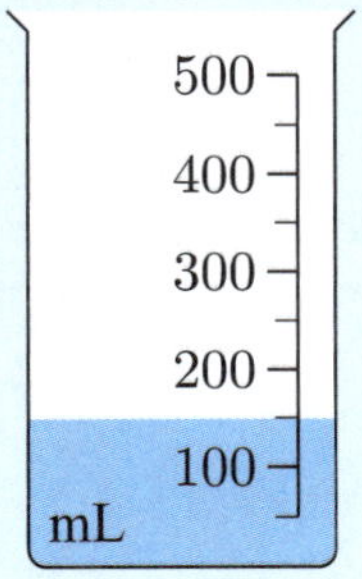

b 100 90 80 70 60 50 40 30 20 10 mL ....

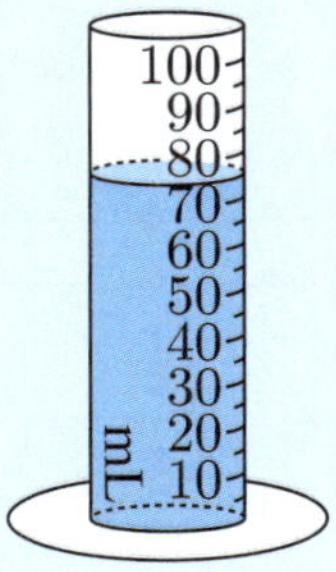

c 500 400 300 200 100 mL ....

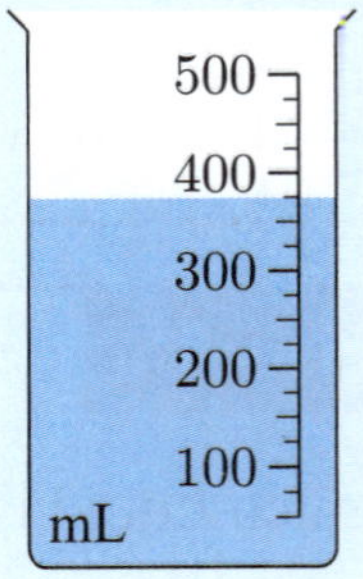

d 100 90 80 70 60 50 40 30 20 10 mL ....

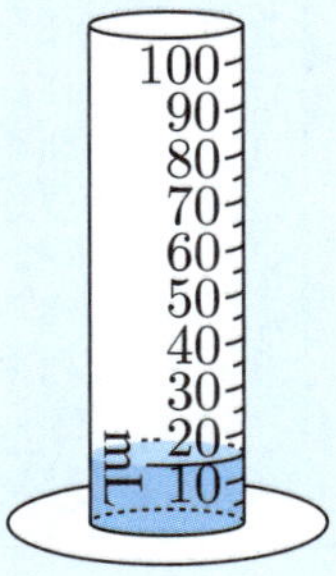

e 50 L 40 L 30 L 20 L 10 L ....

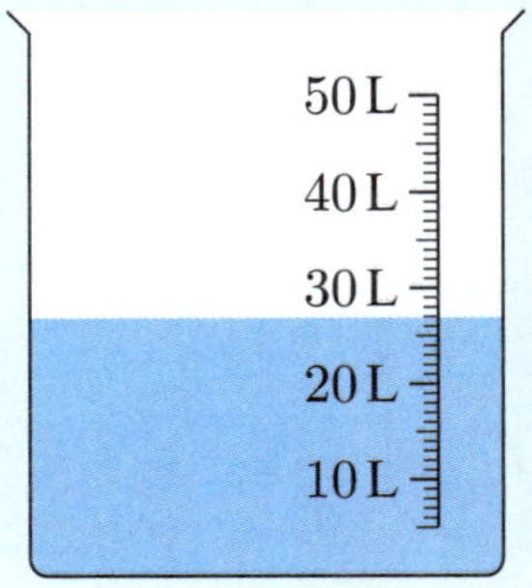

f 500 400 300 200 100 mL ....

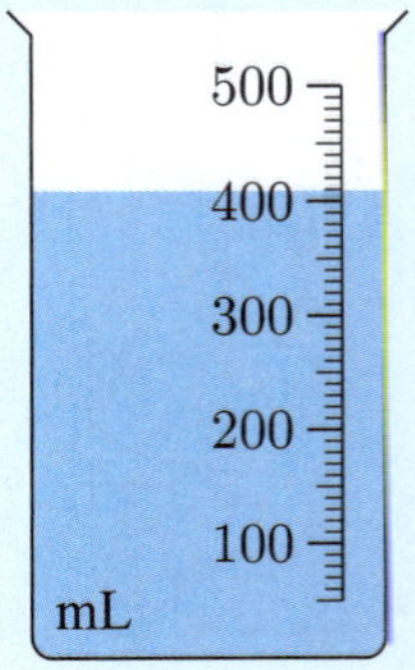

Exercise 7

Draw the given amount of liquid in the container.

a 350 mL

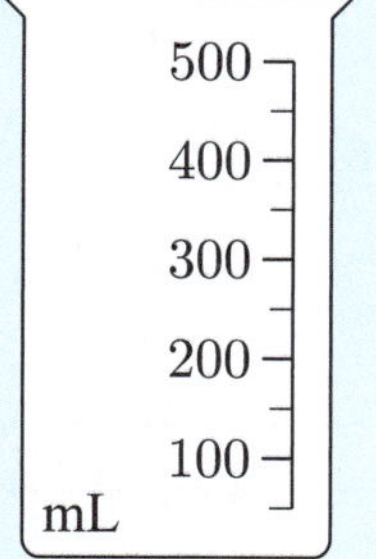

b 55 mL

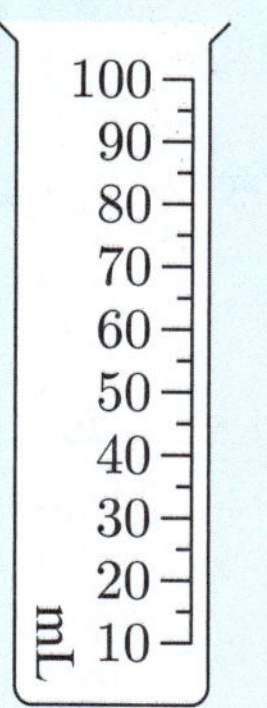

c 250 mL

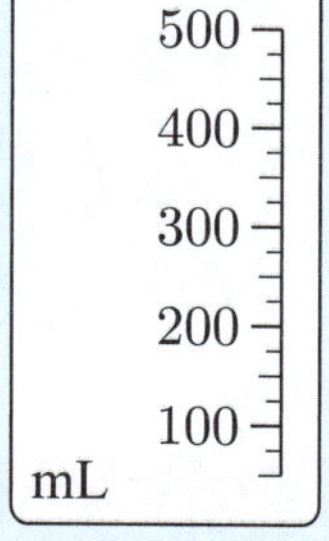

d 16 L

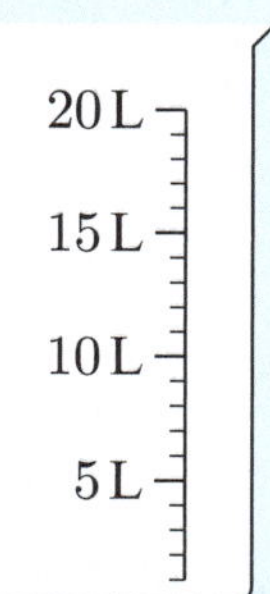

e 190 mL

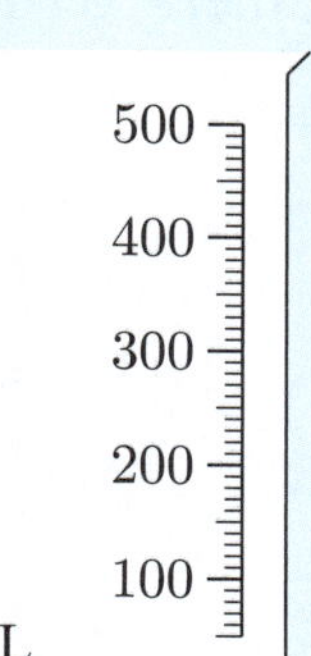

f 475 mL

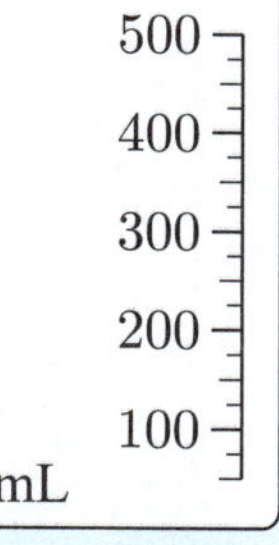

Exercise 8

Describe how much liquid is in each container:

a

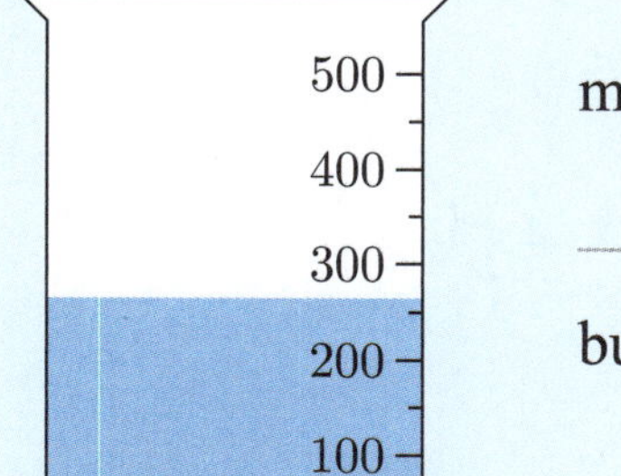

more than ....

but less than ....

b

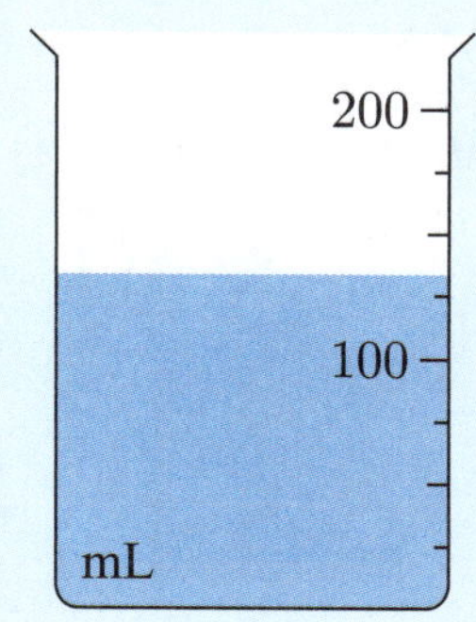

more than ....

but less than ....

Exercise 9

Estimate how much liquid is in the container. Describe how you made your estimate.

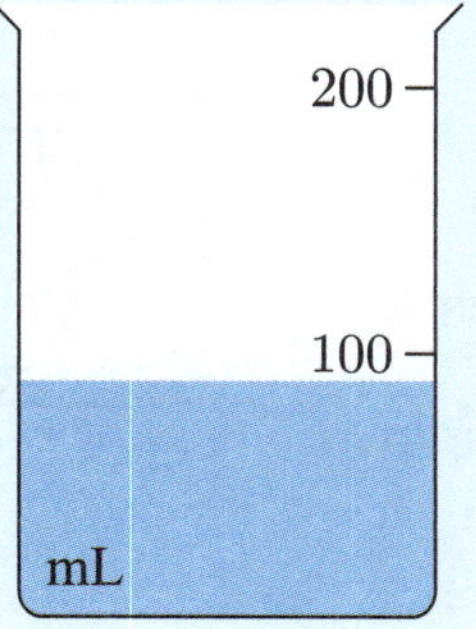

....

....

....

On this scale, each whole litre is divided into quarters.

There is $4\frac{1}{4}$ L of liquid in the container.

Exercise 10

How much liquid is in the container? Write your answer as a mixed number of litres.

a

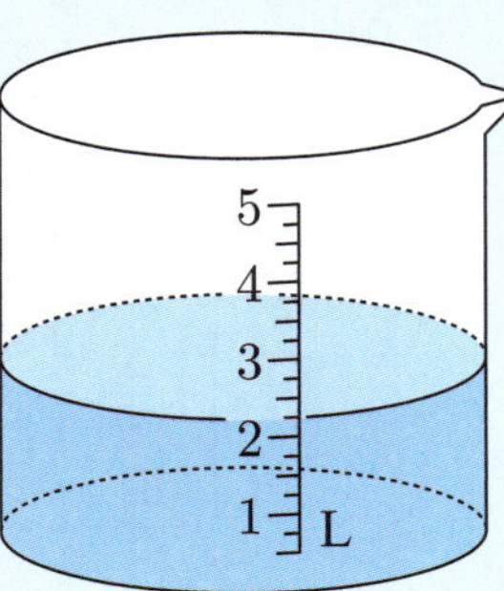

b

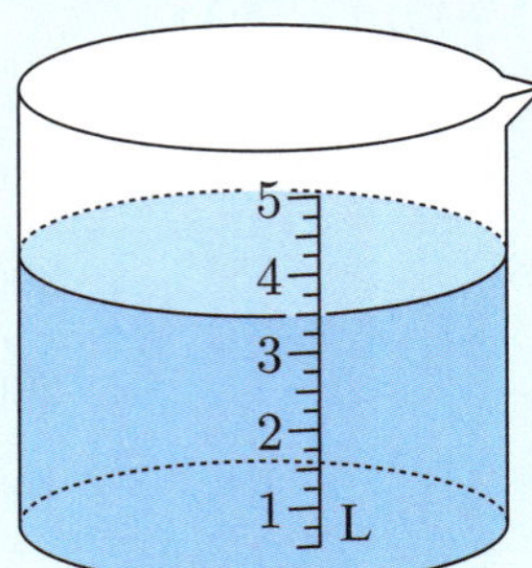

c

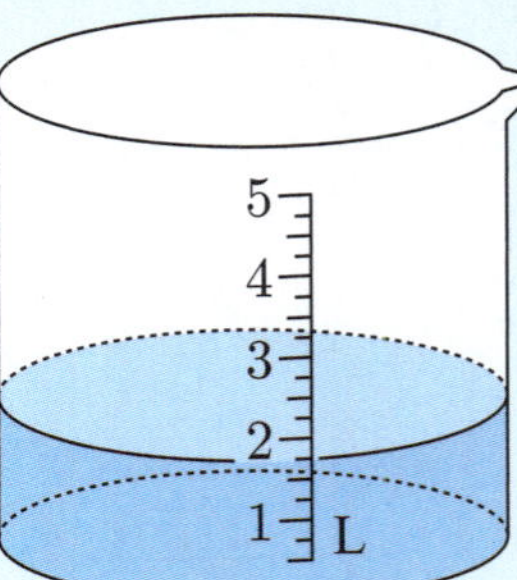

Exercise 11

Draw the given amount in the container.

a $3\frac{1}{4}$ L

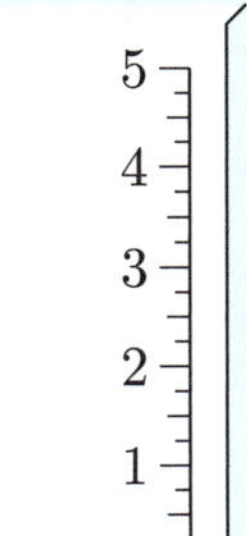

b $4\frac{3}{4}$ L

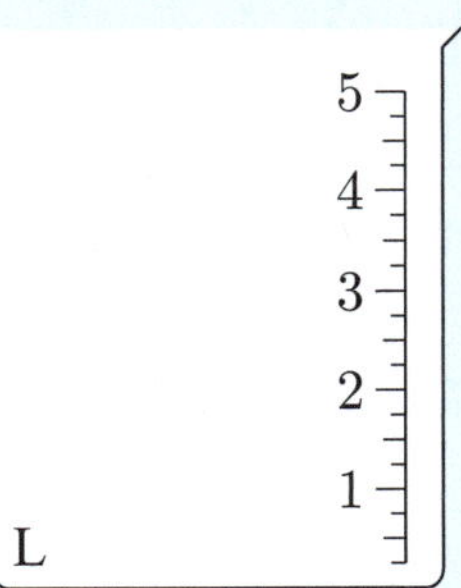

c $1\frac{1}{2}$ L

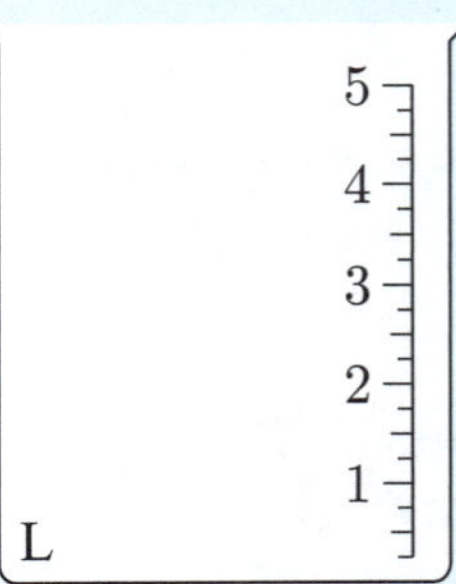

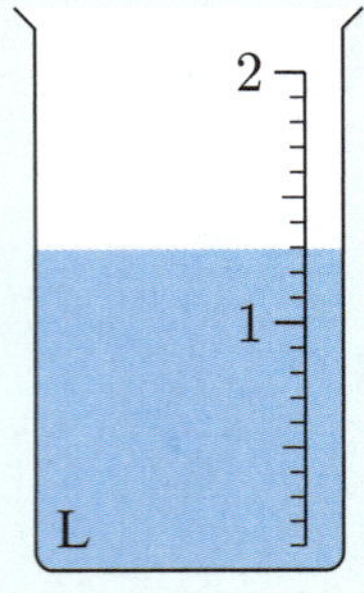

On this scale, each whole litre is divided into tenths.

There is $1\frac{3}{10}$ L or 1.3 L of liquid in the container.

Exercise 12

How much liquid is in the container? Write your answer as a decimal number of litres.

a

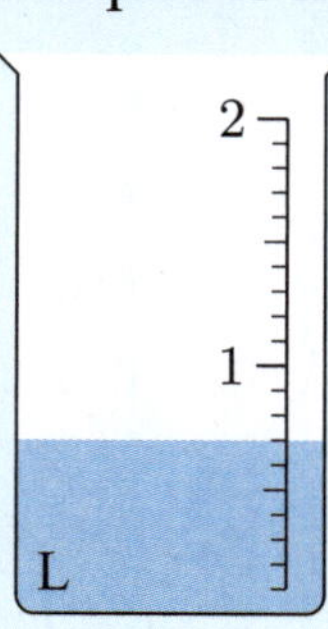

____

b

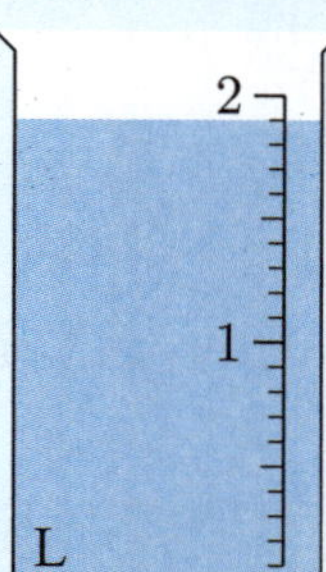

____

c

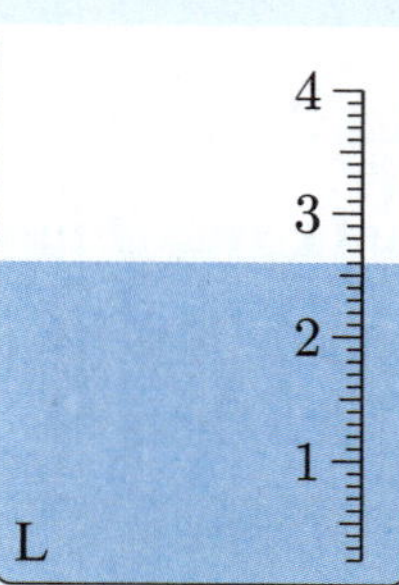

____

Exercise 13

Draw the given amount of liquid in the container.

a 1.4 L

b 0.6 L

c 1.7 L

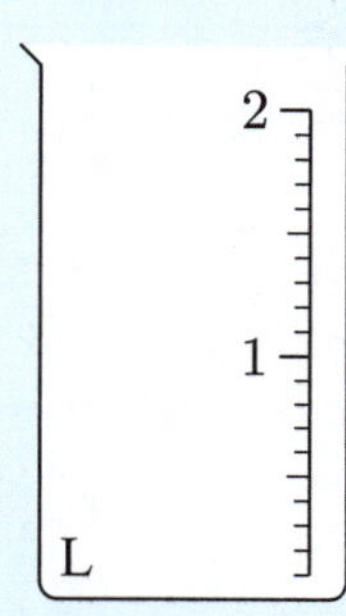

d 2.3 L

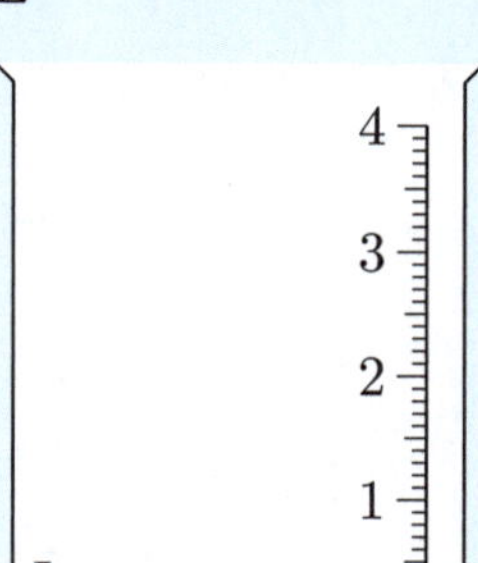

e 3.8 L

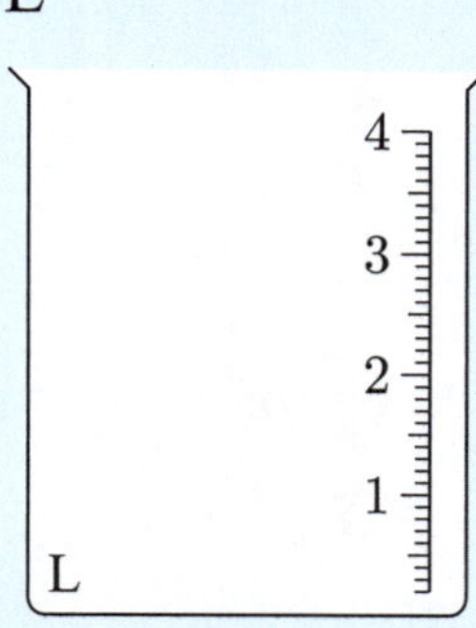

f 2.9 L

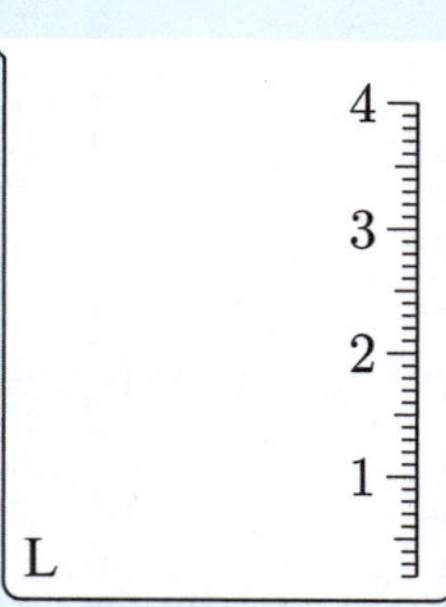

Exercise 14

Estimate the level of 3.2 L of water in this bucket.

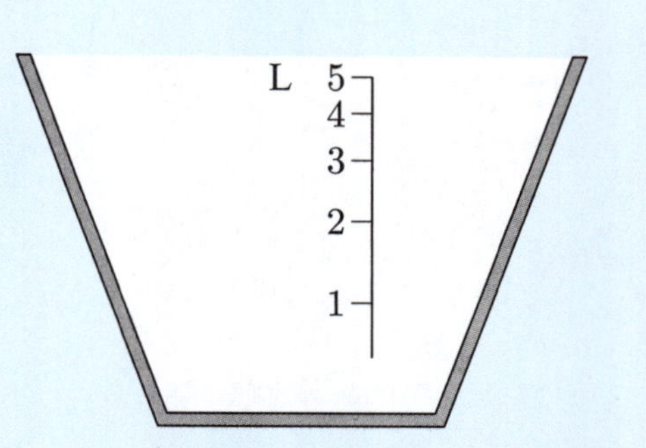

Activity — Measuring capacity in litres

You will need: a jug or container that measures 1 litre, an empty ice cream tub, a saucepan, a bucket, and a beaker

What to do:

1. Gather your equipment near a tap or water tray.
2. *Estimate* how many 1 litre jugs of water will fill each container. Write your estimates in the table below.

Container	*Estimate*	*Number of 1 litre jugs to fill the container*	*Capacity*
ice cream tub			
saucepan			
bucket			
beaker			

3. **a** If we only use *full* jugs of water, will we get an accurate measurement of the capacity of the container?

   ....

   b What do we need to do in order to get a more accurate measurement?

   ....

   ....

4. Find the actual capacity of each container and complete the table.

5. Write the containers in order:

   .... ↑ most capacity

   ....

   ....

   .... least capacity

Revision

1 **a** Arrange these containers in order.

tub

glass

most capacity

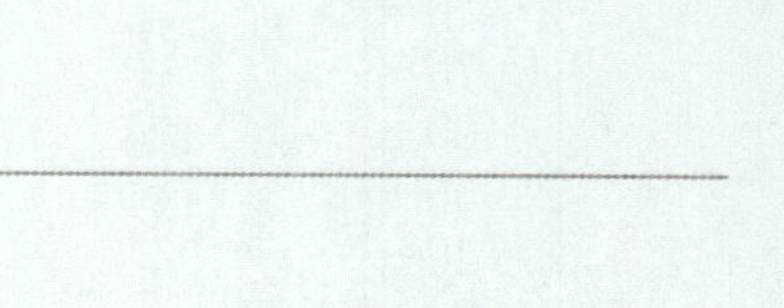

least capacity

toothpaste tube skip bin

b The containers which would hold less than 1 L are the

________________ and the ________________.

2 Complete:

1 L = ________ mL 5 L = ________ mL

______ L = 3000 mL ______ L = 7000 mL

3 How much liquid is in the container?

a

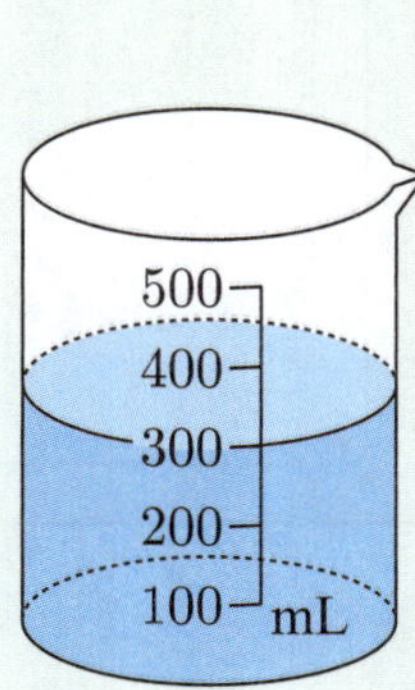

b

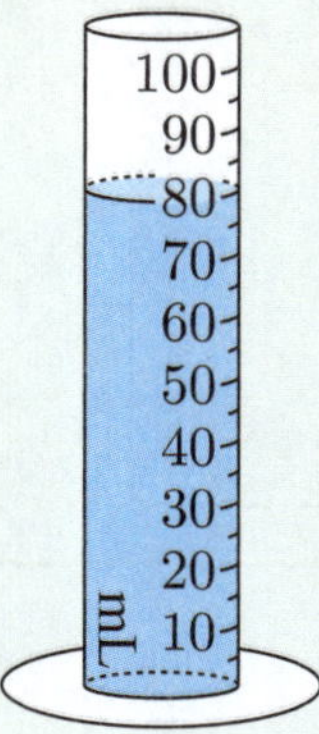

c

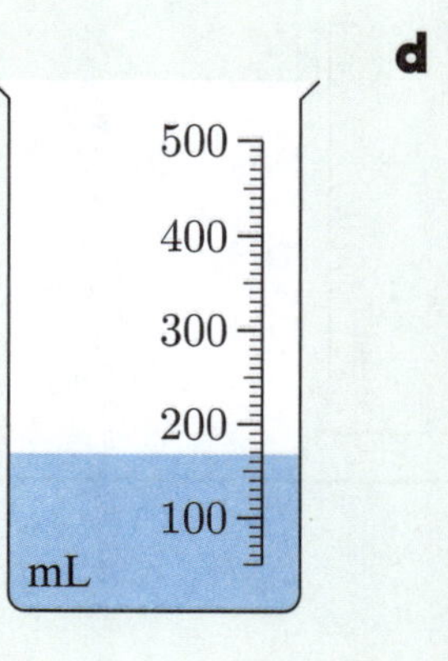

d

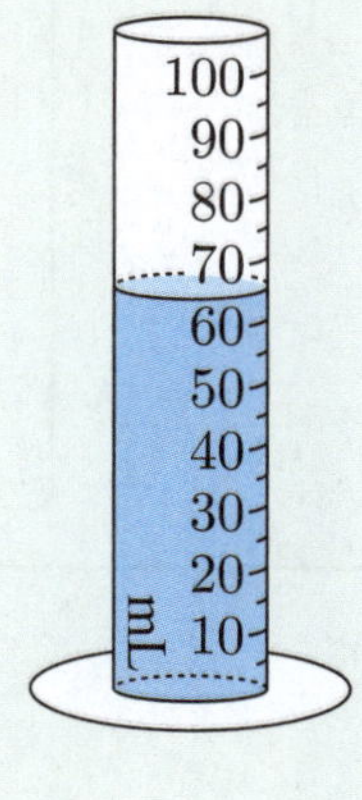

________ ________ ________ ________

4 Describe how much liquid is in the container.

more than ________

but less than ________

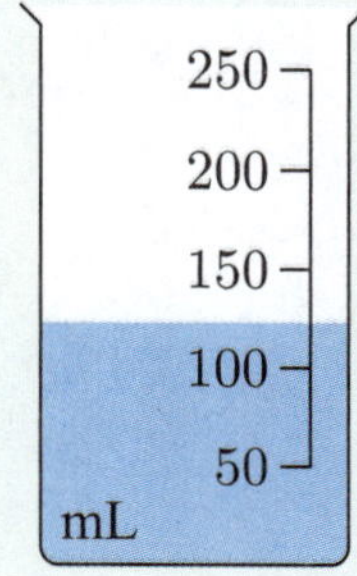

5 Draw the given amount of liquid in the container.

a 30 mL

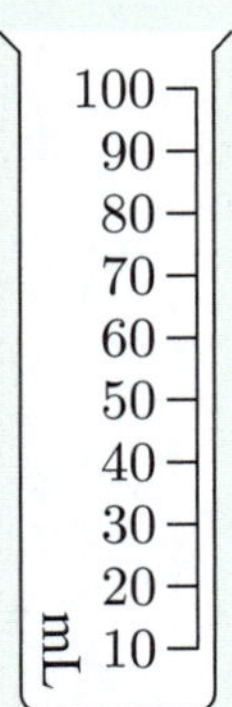

b 180 mL

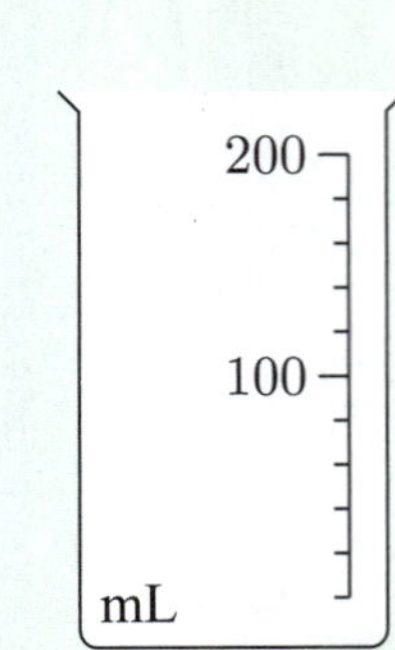

c 430 mL

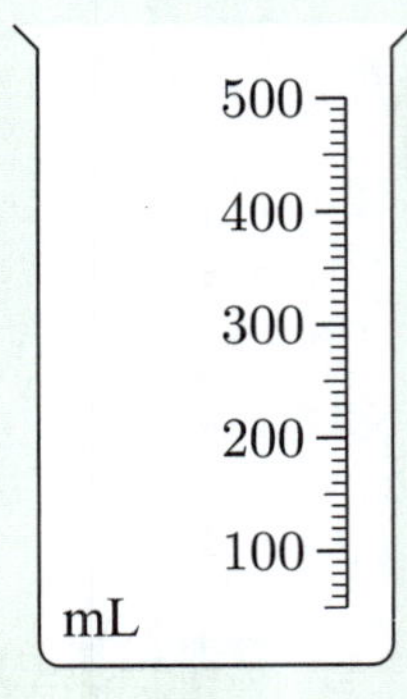

d 125mL

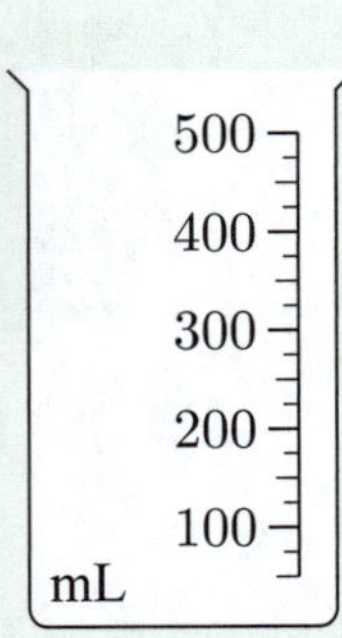

6 How much liquid is in the container?

a

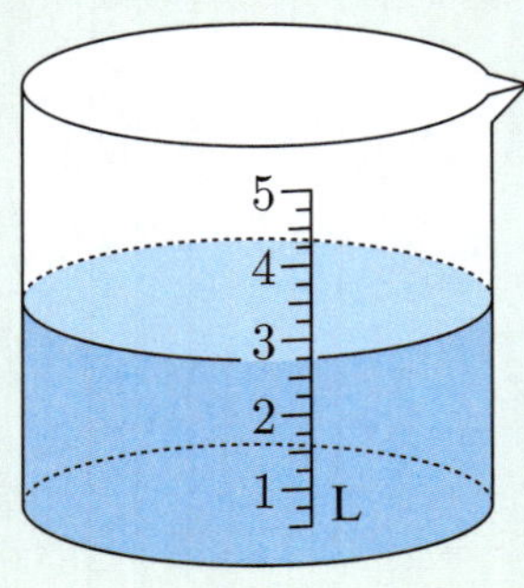

....

b

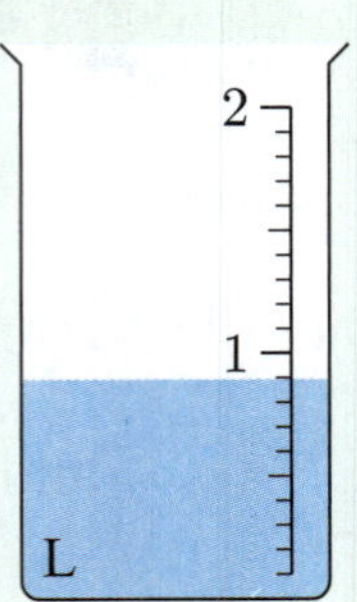

....

c

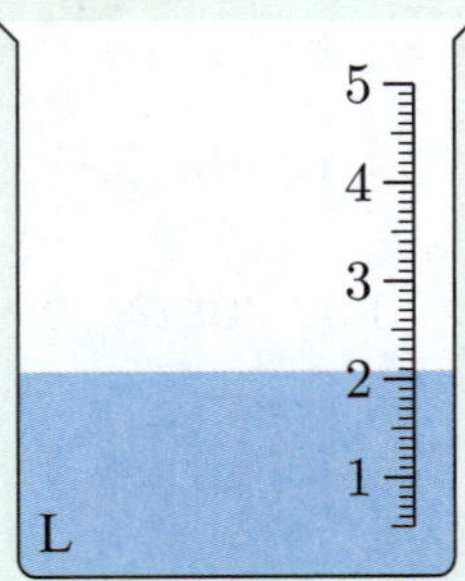

....

7 Draw the given amount of liquid in the container.

a 1.6 L

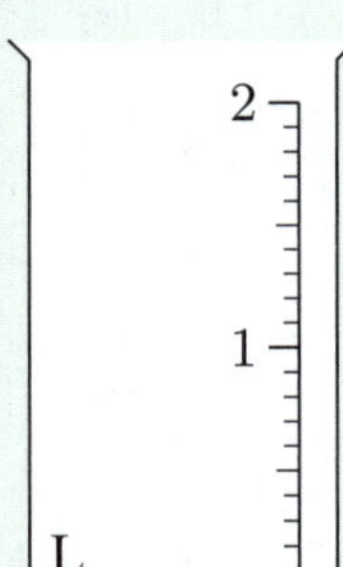

b 0.8 L

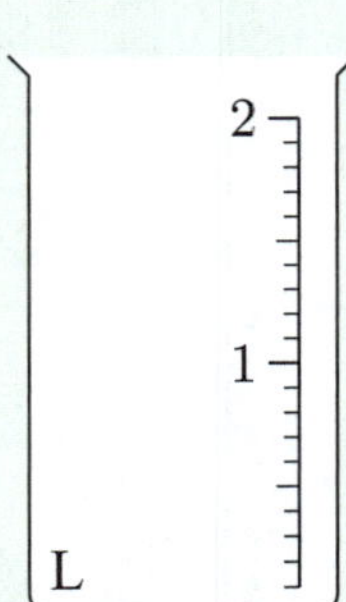

c 3.4 L

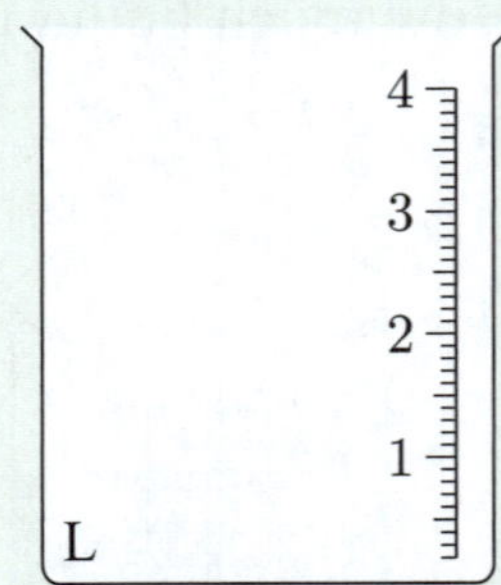

CHAPTER 18: MASS

The **mass** of an object is a measure of how *heavy* it is.

Exercise 1

Use *heavier* or *lighter* to complete each sentence:

a A table is ______________ than a plate.

b A bee is ______________ than a bird.

c A tennis ball is ______________ than a racquet.

d A trombone is ______________ than a flute.

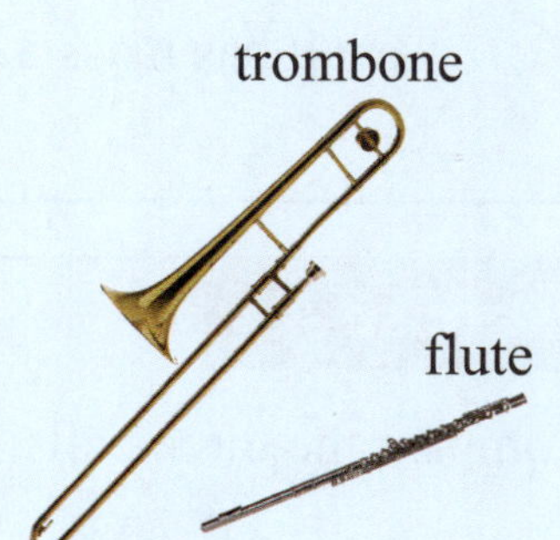

Exercise 2

Complete this sentence comparing the masses of these cats.

A ______________ is heavier

than a ______________,

but a ______________ is heaviest.

Exercise 3

Write a sentence to compare the masses of these birds.

Use the words *lighter* and *lightest*.

____

____

____

____

We measure mass using **grams** (g), **kilograms** (kg), or **tonnes** (t).

$$1\ \mathbf{kg} = 1000\ \mathbf{g}$$
$$1\ \mathbf{t} = 1000\ \mathbf{kg}$$

A snail has mass 5 g.

1 litre of water has mass 1 kg.

A bull has mass 1 t.

Exercise 4

Compare the three balls:

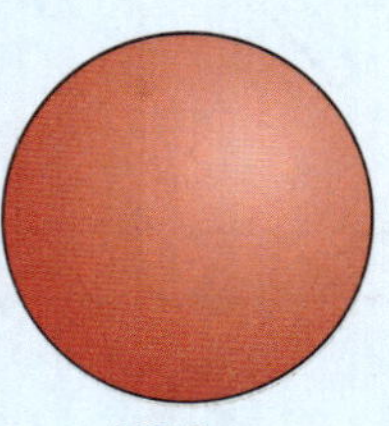

750 g

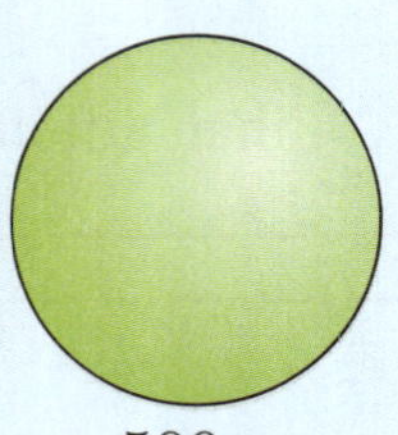

500 g

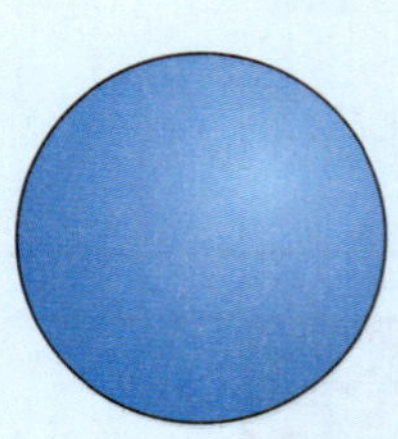

700 g

a Which ball is the lightest? ____________________

b Is the blue ball heavier than the red ball? ________

c Which ball is the heaviest? ____________________

Exercise 5

Compare the three books:

350 g

680 g

530 g

a Is the blue book heavier or lighter than the yellow book? ________________

b Is the pink book heavier or lighter than the blue book? ________________

c Which book is the heaviest? ____________________

d Which book is lighter than the yellow book? ____________________

Exercise 6

Culin collects 80 g of parsley and 35 g of basil from his garden.

In total, Culin collected ____________ g of herbs.

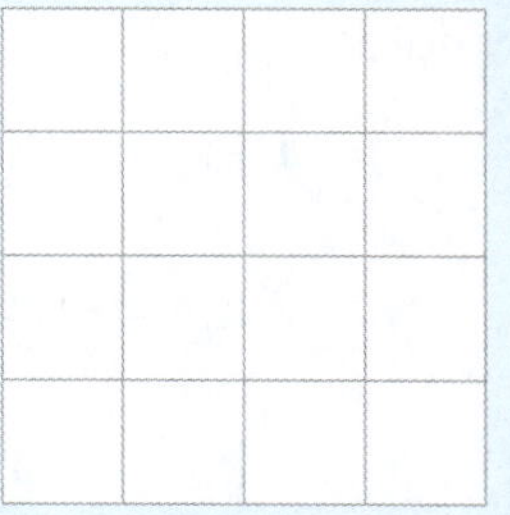

Exercise 7

broccoli

cauliflower

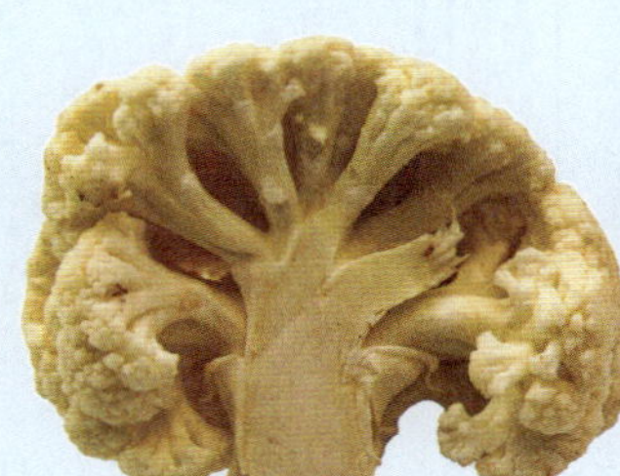

420 g

sweet potato

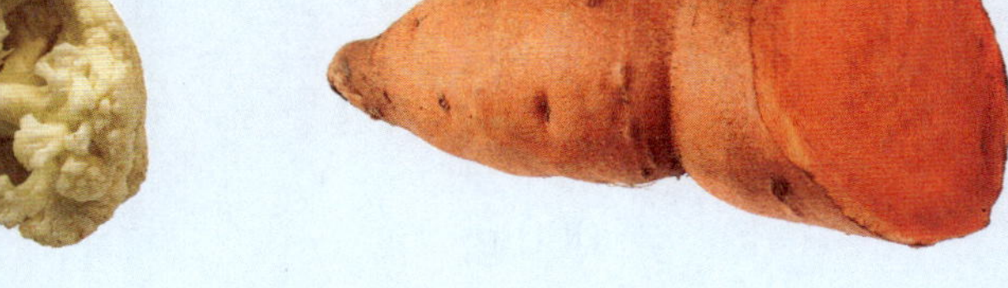

a The cauliflower is 60 g heavier than the broccoli.
How heavy is the broccoli?

The broccoli has mass ____________ .

b The sweet potato has double the mass of the broccoli.
How heavy is the sweet potato?

The sweet potato has mass ____________ .

Exercise 8

Decide whether you would use grams, kilograms, or tonnes to measure the mass of each object.

a a sheep ________________

b a mobile phone ________________

c a drink bottle ________________

d a truck ________________

e a mouse ________________

f a lion ________________

g a tree ________________

h a grape ________________

Exercise 9

Complete:

$1\text{ kg} = 1000\text{ g}$

$2\text{ kg} = \underline{\qquad\qquad}\text{ g}$

$\underline{\qquad}\text{ kg} = 3000\text{ g}$

$4\text{ kg} = \underline{\qquad\qquad}\text{ g}$

$\underline{\qquad}\text{ kg} = 5000\text{ g}$

$\underline{\qquad}\text{ kg} = 6000\text{ g}$

$7\text{ kg} = \underline{\qquad\qquad}\text{ g}$

$8\text{ kg} = \underline{\qquad\qquad}\text{ g}$

$\underline{\qquad}\text{ kg} = 9000\text{ g}$

$10\text{ kg} = \underline{\qquad\qquad}\text{ g}$

Exercise 10

Complete:

$1\text{ t} = 1000\text{ kg}$

$\underline{\qquad}\text{ t} = 2000\text{ kg}$

$3\text{ t} = \underline{\qquad\qquad}\text{ kg}$

$\underline{\qquad}\text{ t} = \underline{\qquad\qquad}\text{ kg}$

$\underline{\qquad}\text{ t} = 5000\text{ kg}$

$6\text{ t} = \underline{\qquad\qquad}\text{ kg}$

$7\text{ t} = \underline{\qquad\qquad}\text{ kg}$

$\underline{\qquad}\text{ t} = 8000\text{ kg}$

$9\text{ t} = \underline{\qquad\qquad}\text{ kg}$

$\underline{\qquad}\text{ t} = 10\,000\text{ kg}$

$\frac{1}{2}$ of $1000\text{ g} = 500\text{ g}$ so $\mathbf{\frac{1}{2}\text{ kg} = 500\text{ g}}$

$\frac{1}{4}$ of $1000\text{ g} = 250\text{ g}$ so $\mathbf{\frac{1}{4}\text{ kg} = 250\text{ g}}$

$\frac{3}{4}$ of $1000\text{ g} = 750\text{ g}$ so $\mathbf{\frac{3}{4}\text{ kg} = 750\text{ g}}$

Exercise 11

Complete:

$1\frac{1}{2}\text{ kg} = 1000\text{ g} + 500\text{ g} = 1500\text{ g}$

a $2\frac{1}{2}\text{ kg} = 2000\text{ g} + \underline{\qquad\qquad} = \underline{\qquad\qquad}$

b $2\frac{1}{4}\text{ kg} = \underline{\qquad\qquad} + \underline{\qquad\qquad} = \underline{\qquad\qquad}$

c $3\frac{3}{4}\text{ kg} = \underline{\qquad\qquad} + \underline{\qquad\qquad} = \underline{\qquad\qquad}$

d $1\frac{1}{4}\text{ kg} = \underline{\qquad\qquad} + \underline{\qquad\qquad} = \underline{\qquad\qquad}$

e $2\frac{3}{4}\text{ kg} = \underline{\qquad\qquad} + \underline{\qquad\qquad} = \underline{\qquad\qquad}$

Exercise 12

Complete:

$3250\text{ g} = 3000\text{ g} + 250\text{ g} = 3\frac{1}{4}\text{ kg}$

a $4500\text{ g} =$ ______ + ______ = ______

b $1750\text{ g} =$ ______ + ______ = ______

c $4750\text{ g} =$ ______ + ______ = ______

d $6250\text{ g} =$ ______ + ______ = ______

$\frac{1}{10}$ of $1000\text{ g} = 100\text{ g}$

so $\frac{1}{10}\text{ kg} = 100\text{ g}$

or $0.1\text{ kg} = 100\text{ g}$

Exercise 13

Complete:

a $0.2\text{ kg} =$ ______ g

b $0.4\text{ kg} =$ ______ g

c $0.7\text{ kg} =$ ______ g

d $300\text{ g} =$ ______ kg

e $900\text{ g} =$ ______ kg

f $600\text{ g} =$ ______ kg

Exercise 14

Complete:

$3.1\text{ kg} = 3000\text{ g} + 100\text{ g} = 3100\text{ g}$

a $2.3\text{ kg} = 2000\text{ g}$ + ______ = ______

b $1.8\text{ kg} =$ ______ + ______ = ______

c $4.9\text{ kg} =$ ______ + ______ = ______

d $10.4\text{ kg} =$ ______ + ______ = ______

Exercise 15

Complete:

2100 g = 2000 g + 100 g = 2.1 kg

a 3200 g = 3000 g + ________ = ________

b 5400 g = ________ + ________ = ________

c 4700 g = ________ + ________ = ________

d 8500 g = ________ + ________ = ________

To find the mass of an object, we use **scales**.

Balance scales

Spring balance

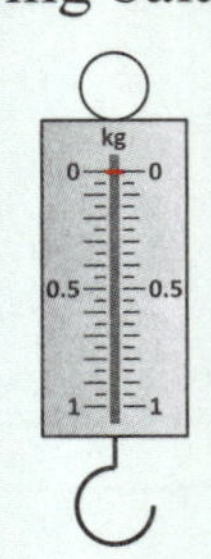

Kitchen scales

Digital scales

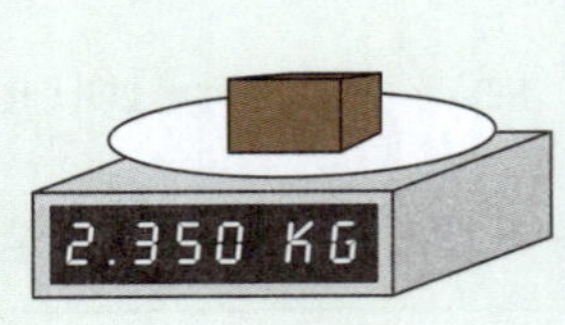

Discussion

Can you see the "scale" on the spring balance and kitchen scales?

Do the balance scales have a "scale" we can see? If not, how do we use the balance scales to measure mass?

Exercise 16

Read the mass on each digital scale:

a

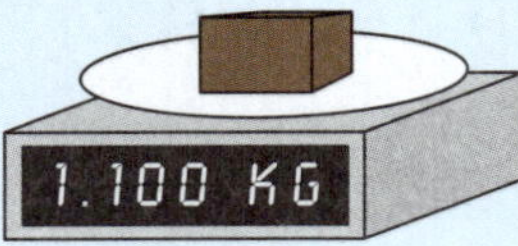

____

b

____

c

____

d

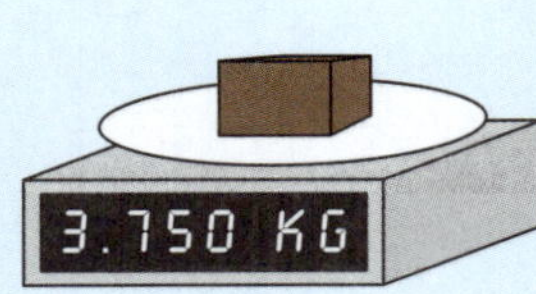

____

e

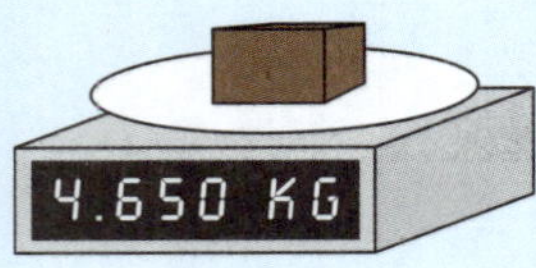

____

f

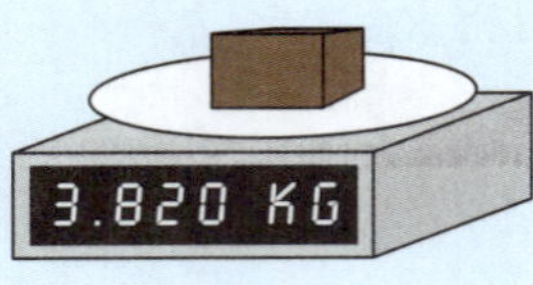

____

Watch the video to help you correctly read kitchen scales.

Exercise 17

Read the mass on the kitchen scales.

a ....

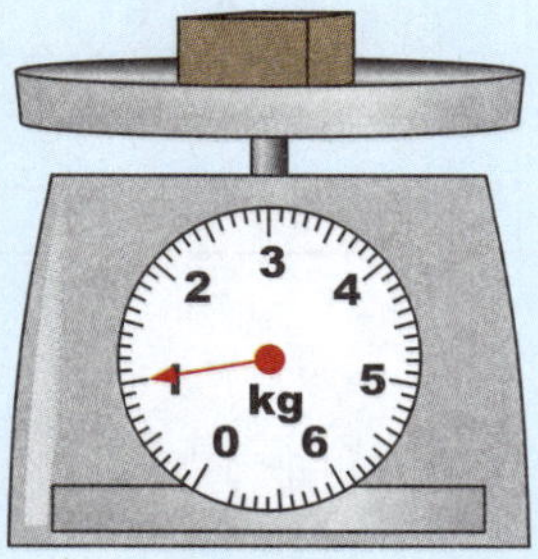

b ....

c ....

d ....

e ....

f ....

g ....

h ....

i ....

Discussion

Do you think kitchen scales are more accurate if they have a digital display?

Suppose you are buying fruit, vegetables, or meat at the supermarket. Is is more *fair* to use digital scales?

Exercise 18

Show the mass on the scale.

a 1.4 kg

b 5.5 kg

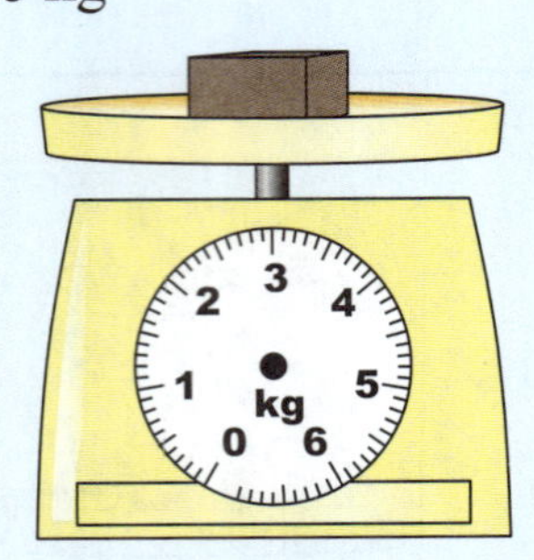

c 3.7 kg

Exercise 19

Anastasia uses her kitchen scales to measure the masses of three books.

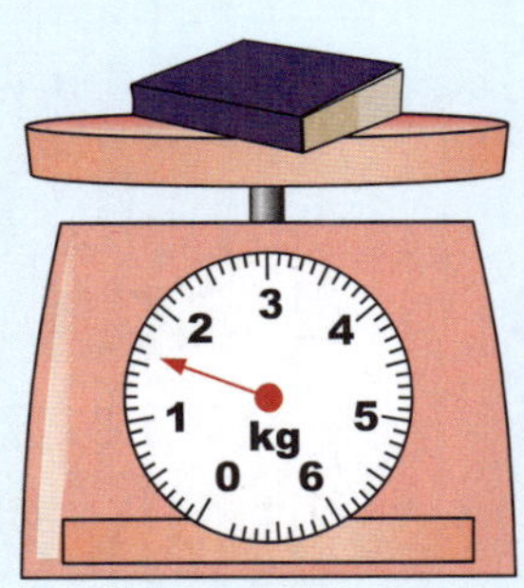

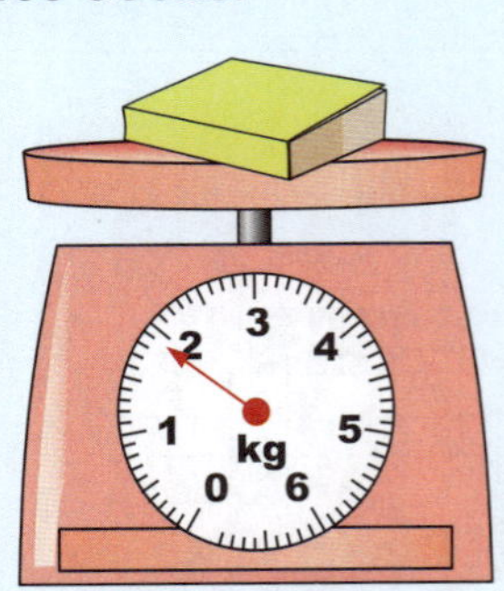

a Which book is the heaviest? ____________________

b How much heavier is the yellow book than the purple book?

____

Revision

1 Use *lightest*, *lighter*, *heavier*, or *heaviest* to compare the masses of these objects.

phone

car

pin

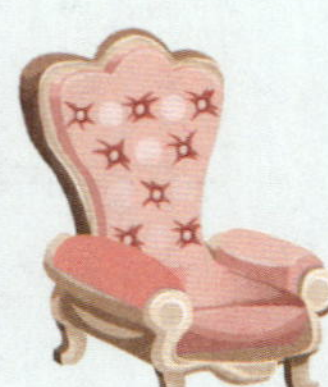
armchair

The car is the ________________ object.

The armchair is ________________ than the phone.

The ________________ object is the pin.

2 Compare the three guinea pigs:

Luna	Larna	Lexi

820 g	950 g	780 g

a The lightest guinea pig is ________________.

b Is Larna lighter or heavier than Luna? ________________

c Larry is 90 g heavier than Luna.

Larry has mass ______________.

3 Decide whether you would use grams, kilograms, or tonnes to measure the mass of each object.

a a rhinoceros ________________

b a spider ________________

c a refrigerator ________________

4 Complete:

a

1 kg = 1000 g

_____ kg = 2000 g

3 kg = ________ g

6 kg = ________ g

_____ kg = 8000 g

b

1 kg = 1000 g

$\frac{1}{2}$ kg = ________ g

_____ kg = 250 g

$\frac{3}{4}$ kg = ________ g

5 Write each mass in grams:

a $3\frac{1}{2}$ kg = 3000 g + ____________ = ____________

b $1\frac{3}{4}$ kg = ____________ + ____________ = ____________

c $6\frac{1}{4}$ kg = ____________ + ____________ = ____________

d $10\frac{1}{2}$ kg = ____________ + ____________ = ____________

6 Write each mass as a mixed number of kg:

a 5250 g = 5000 g + ________ = ________

b 6750 g = ________ + ________ = ________

7 Complete:

a 0.8 kg = ________ g

b 0.3 kg = ________ g

c 700 g = ________ kg

d 200 g = ________ kg

8 Write each mass in grams:

a 3.7 kg = 3000 g + ________ = ________

b 6.2 kg = ________ + ________ = ________

9 Write each mass as a decimal number of kg:

a 4800 g = 4000 g + ________ = ________

b 1300 g = ________ + ________ = ________

10 Read the mass on the kitchen scales.

a

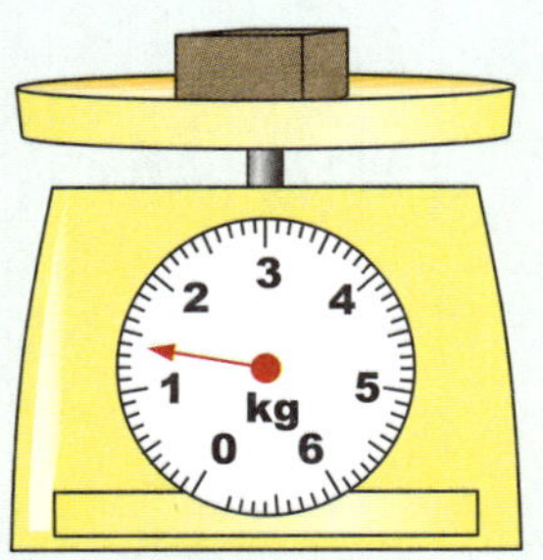

____

b

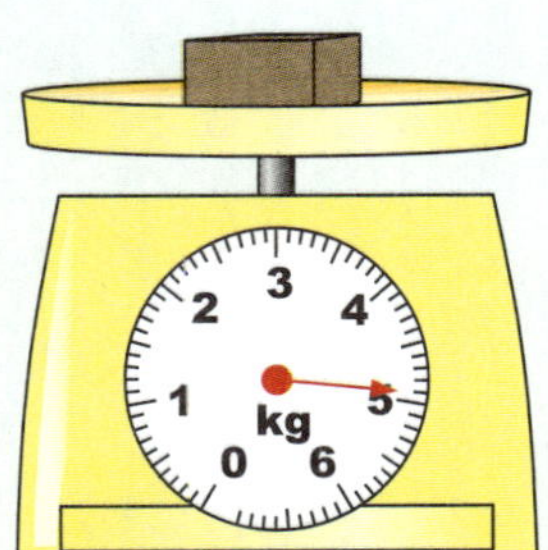

____

c

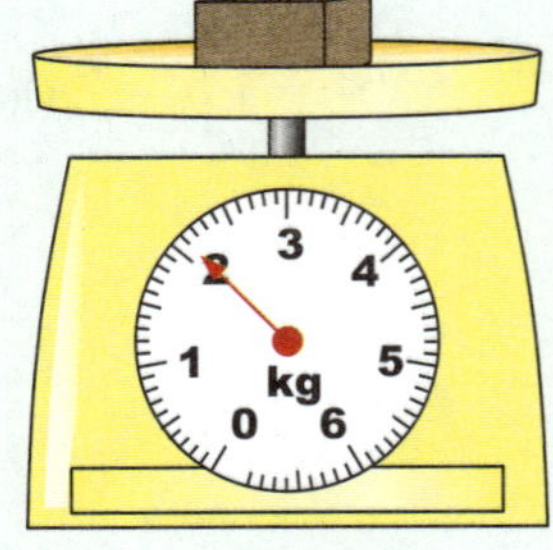

____

11 Show the mass on the scale.

a 3.8 kg

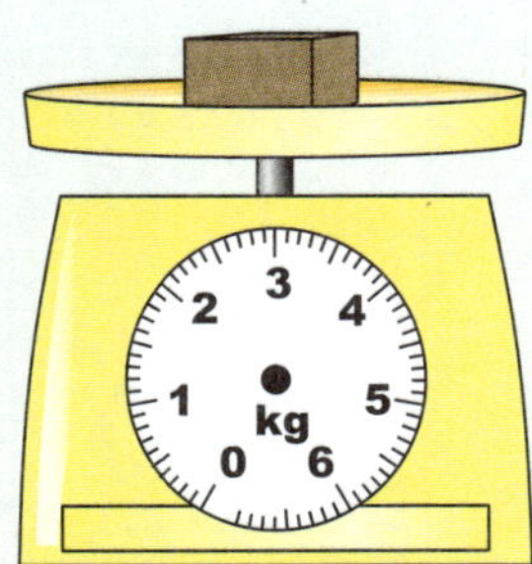

b 1.2 kg

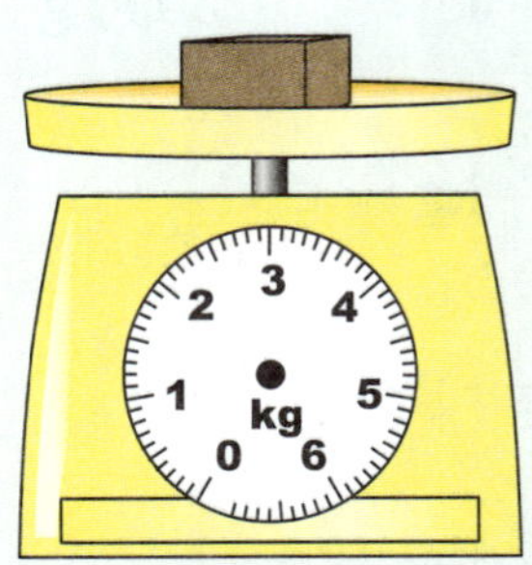

CHAPTER 19: TEMPERATURE

Temperature is a measure of how hot or cold something is.

Discussion

What words do we use to talk about temperature?

hot words	cold words

To measure temperature we use a **thermometer**.

This thermometer measures temperature in **degrees Celsius** or **°C**.

- Each major division represents 10 degrees Celsius.
- Each minor division represents 1 degree Celsius.

This thermometer shows 18°C.

We read this as "eighteen degrees Celsius".

The greater the number shown on the thermometer, the warmer the temperature.

Exercise 1

Write each temperature in words:

a 24°C ....

b 41°C ....

c 16.4°C ....

d 37.2°C ....

Exercise 2

Write each temperature in numerals:

a nineteen degrees Celsius ....

b sixty two degrees Celsius ....

c thirty eight point four degrees Celsius ....

d three point nine degrees Celsius ....

e four point five degrees Celsius ....

Exercise 3

What temperature is shown on each thermometer?

a

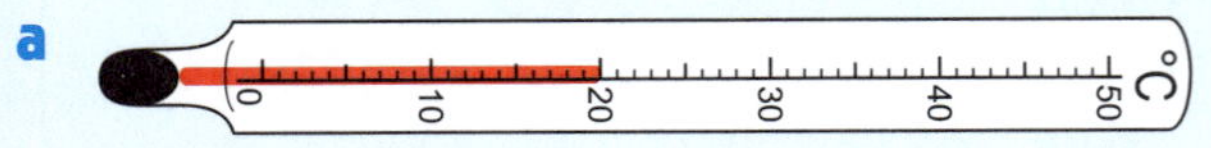

....

b

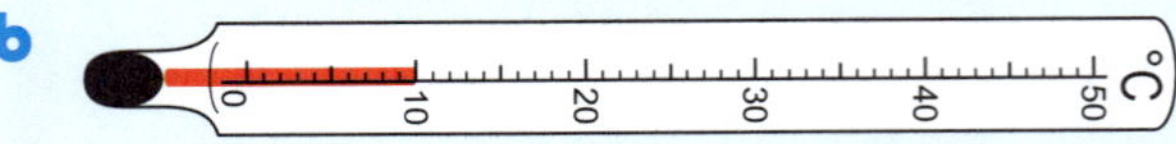

....

c

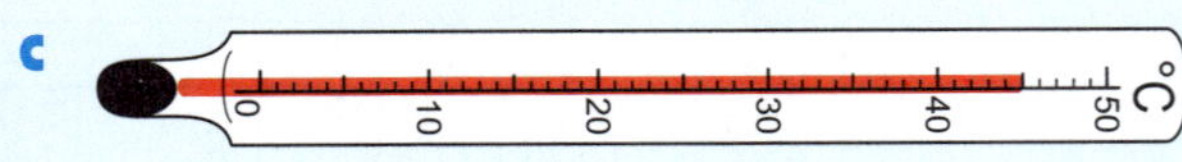

....

d

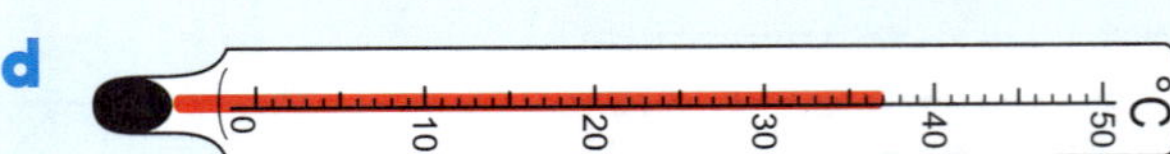

....

e

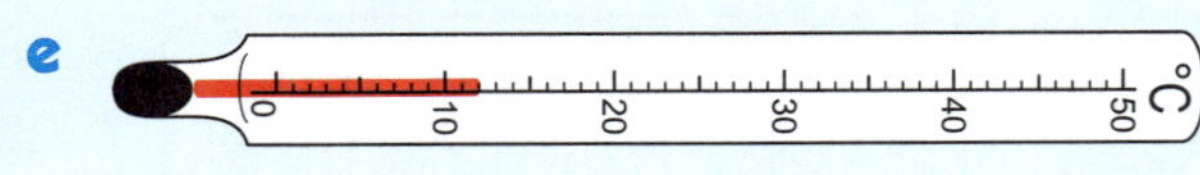

....

f

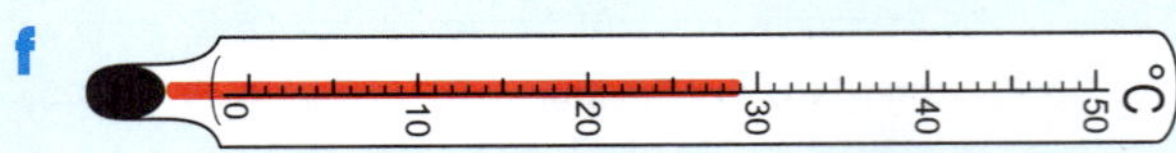

....

Exercise 4

Show the temperature on the thermometer.

a 30°C

b 40°C

c 25°C

d 18°C

e 22°C

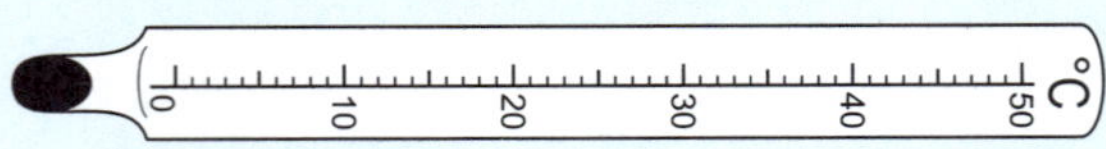

f 39°C

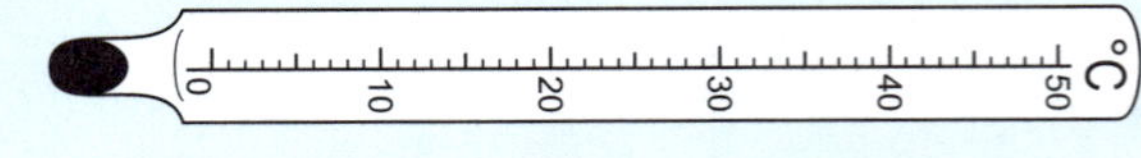

Exercise 5

The temperature of Ian's coffee is 58°C.

The temperature of Jessica's tea is 83°C.

How much warmer is Jessica's tea than Ian's coffee?

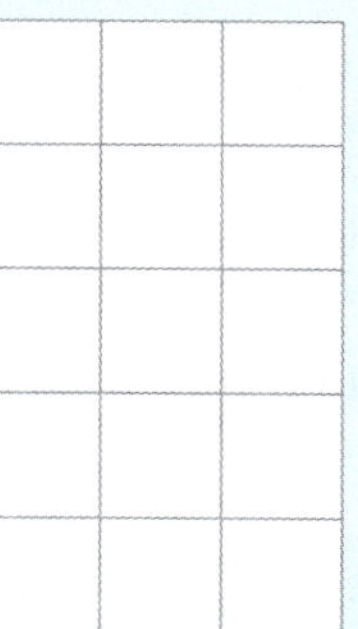

______

Discussion

This is a thermometer for measuring body temperature.

- What is a normal body temperature for a person?
- What happens to your temperature if you have a "fever"?
- Where in your school would you find a thermometer like this?

Exercise 6

What temperature is shown on each thermometer?

a 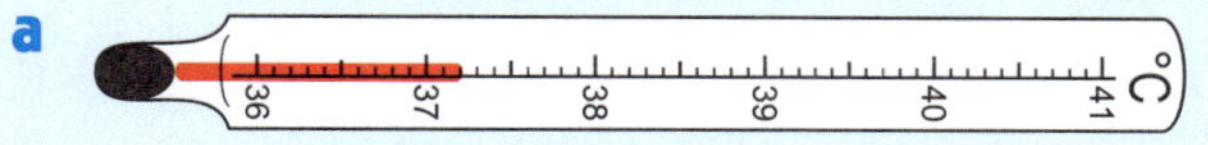

______

b 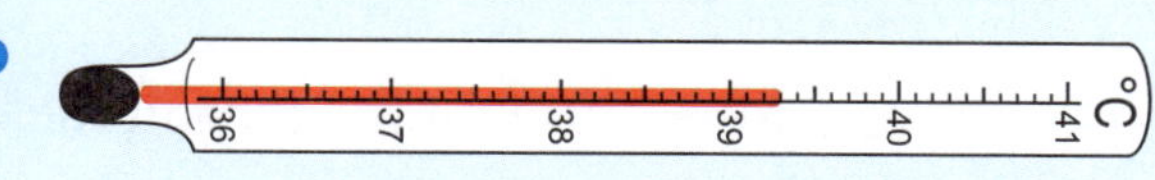

______

Exercise 7

The temperature overnight was 1.8°C.

It has risen by 0.5°C.

What is the temperature now? ______

Discussion

At what temperature does water freeze? ______

How do we describe the temperature if it is colder than "freezing"?

Discussion

What is the temperature where you are right now? ____________

What is a particularly hot temperature where you are? ____________

What is a particularly cold temperature where you are? ____________

Revision

1 Write each temperature in words:

a 32°C ____________

b 18.7°C ____________

2 Write each temperature in numerals:

a twenty four point six degrees Celsius ____________

b forty two point eight degrees Celsius ____________

3 What temperature is shown on each thermometer?

a ____________

b 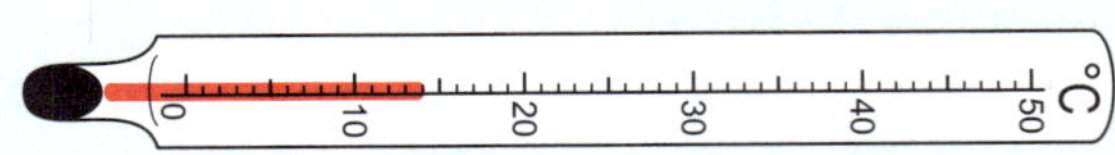

____

4 Show the temperature on the thermometer.

a 43°C

b 19°C

CHAPTER 20: MONEY

We use **money** to buy items.

Different types of money are used for different countries.

The money used in a country is called its **currency**.

Most currencies have a *large* unit and a *small* unit. The large unit usually has 100 times the value of the small unit.

In this Chapter, we will mainly use **dollars** and **cents**, with 1 dollar = 100 cents. The symbol for *dollar* is \$, and the symbol for *cent* is c.

We will work with this "play" set of coins and notes:

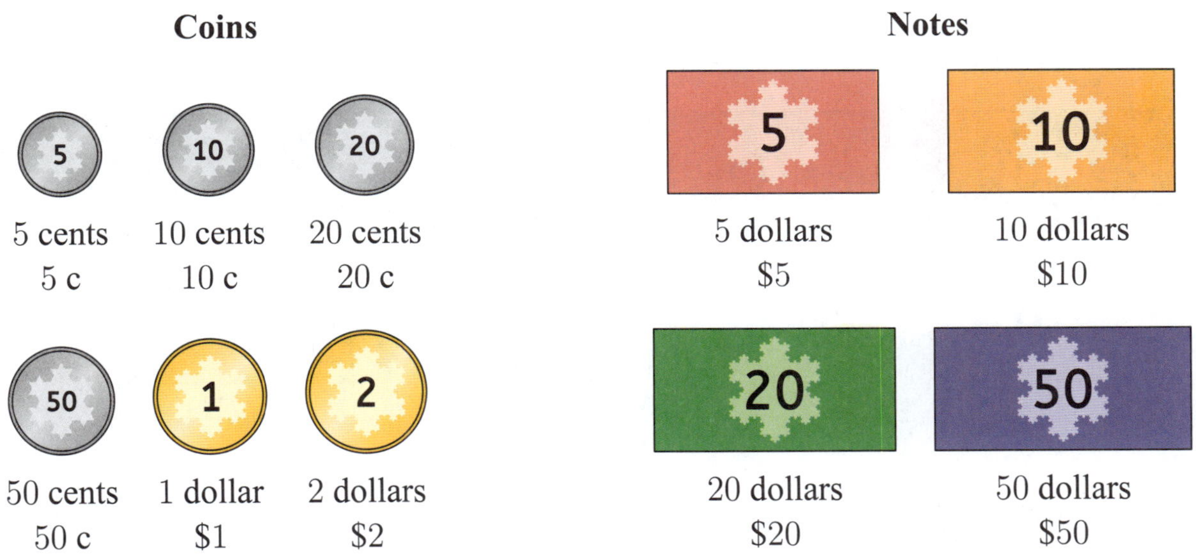

This set of coins and notes is quite different to the set of coins and notes used in the United States of America.

If you prefer, you can click the icon and complete a set of Exercises using United States currency.

Exercise 1

Here is 35 cents or 35 c.

$20\text{ c} + 10\text{ c} + 5\text{ c} = 35\text{ c}$

How much money is shown?

a   

_______ c

b  

_______ c

c  

_______ c

d   

_______ c

Exercise 2

How many dollars are shown?

a 

$ _______

b 

$ _______

c  

$ _______

d  

$ _______

e  

$ _______

Exercise 3

How much money is in each bag?

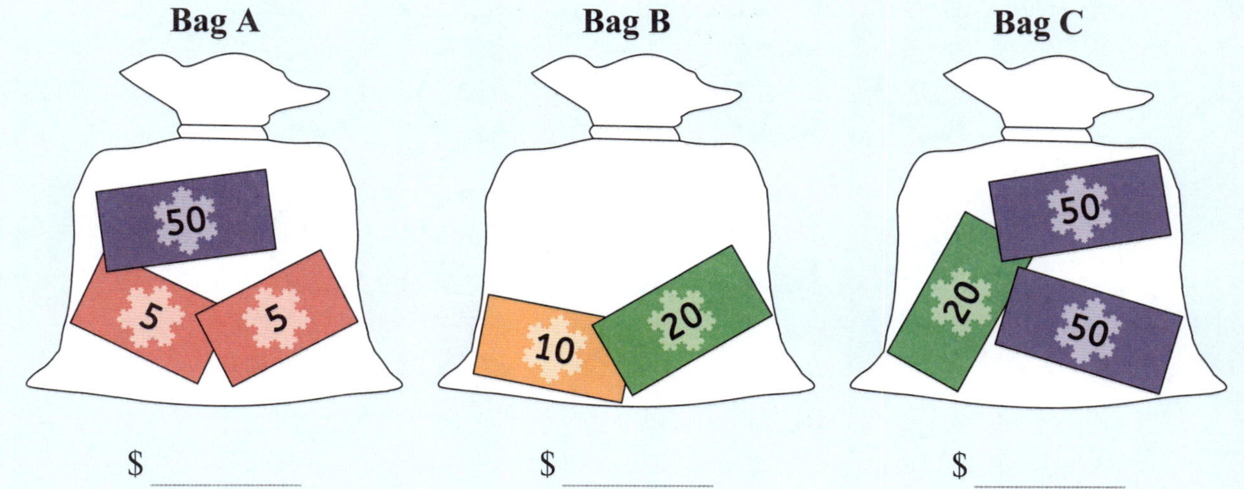

$ ________ $ ________ $ ________

The amount in Bag A is double the amount in Bag ______ .

The amount in Bag C is double the amount in Bag ______ .

Exercise 4

Decide whether each amount is *equal* to, *more* than, or *less* than 1 dollar.

Write *equal*, *more*, or *less*.

1 dollar = 100 cents!

a 

____

b 

____

c  

____

d 

____

Puzzle

By drawing "fences" along the grid lines, divide this grid into 5 regions so that each region contains 1 dollar.

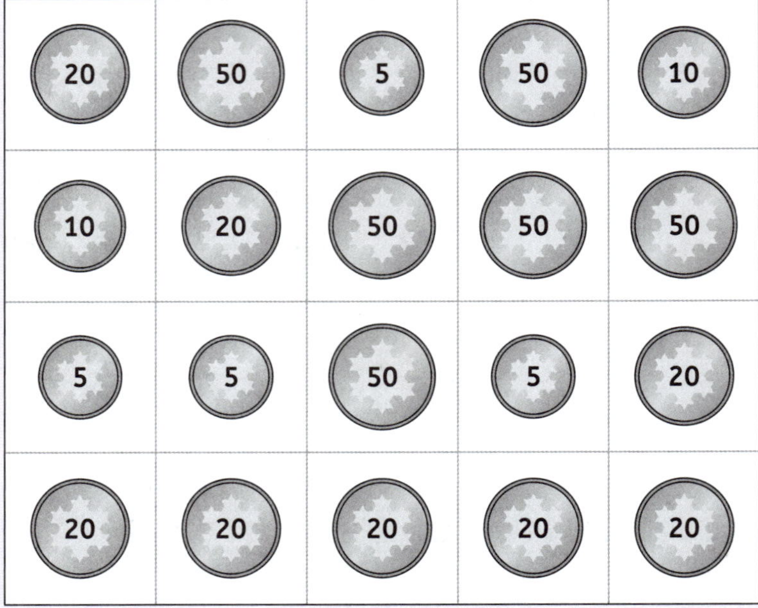

Exercise 5

$1 = 100 cents

$2 = ______ cents

$3 = ______ cents

$ ______ = 400 cents

$5 = ______ cents

$ ______ = 600 cents

$ ______ = 700 cents

$8 = ______ cents

$ ______ = 900 cents

$10 = ______ cents

Exercise 6

Jenni has ten 50 cent coins.

In total, Jenni has 10×50 c = ______ c

= $ ______

Challenge

1 How many 10 c coins are needed to make $100? ....

2 How many 5 c coins are needed to make $200? ....

Activity — The Egyptian pound

The Egyptian pound uses **pounds** and **piastres**. 1 pound = 100 piastres.

The symbol for pound is £, and the symbol for piastre is PT.

Coins

25 PT

50 PT

£1

Notes

£5

£10

£20

£50

£100

£200

What to do:

1 How much money is shown?

a   

£ ....

b 

£ ....

c    

£ ....

d     

£ ....

2 **a** How many  are needed to make  ? ....

b How many  are needed to make  ? ....

Since 100 cents = 1 dollar,

1 cent = $\frac{1}{100}$ dollar.

We can therefore write an amount of dollars and cents as a **decimal number** of dollars.

$2 35 c

"Two dollars and thirty five cents."

We write $ 2 . 3 5

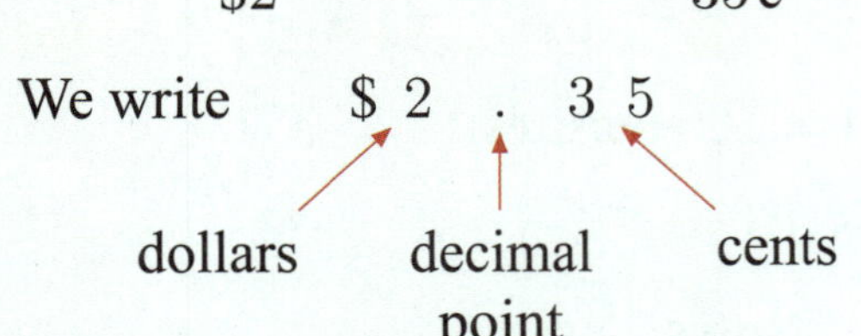

dollars decimal point cents

Exercise 7

Match the equal amounts.

three dollars and fifteen cents	$4.50
four dollars and fifty cents	$5.40
forty five cents	$3.15
five dollars and forty cents	$4.05
three dollars	$0.45
four dollars and five cents	$3.00

2 50 50 50 5

$2 $1 55 c

Group as many whole dollars together as you can!

I have 3 whole dollars, and 55 cents left over.

In total, I have $3.55.

dollars cents

Exercise 8

How much money is shown?

a

$ ______ . ______

b

$ ______ . ______

c

$ ______ . ______

d

$ ______ . ______

e

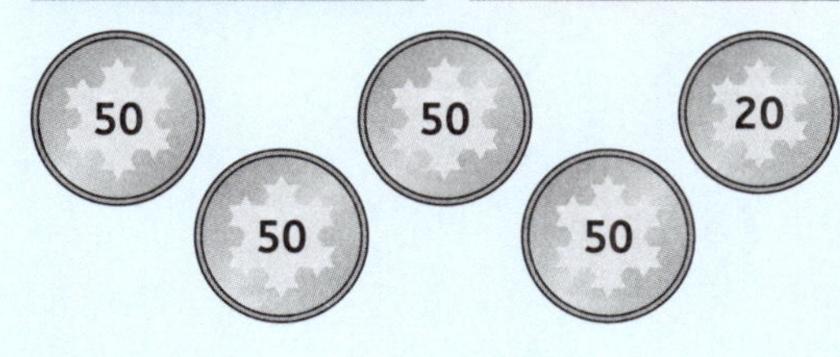

$ ______ . ______

f

$ ______ . ______

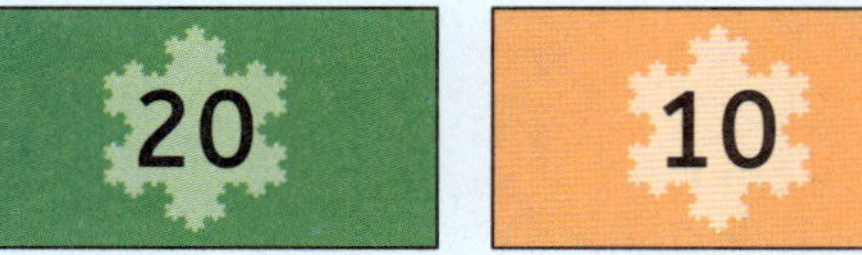

Exercise 9

dollars cents

$$\$2.85 = 200 \text{ c} + 85 \text{ c}$$
$$= 285 \text{ c}$$

Write in cents.

a $1.95 = ______ c + ______ c

= ______ c

b $2.10 = ______ c + ______ c

= ______ c

c $3.05 = ______ c

d $4.80 = ______ c

Exercise 10

$$\begin{aligned} 370\text{ c} &= \overset{\$3}{\overset{\downarrow}{300\text{ c}}} + 70\text{ c} \\ &= \$3.70 \end{aligned}$$

Write in \$ as a decimal number.

a 185 c = ______ c + ______ c
= \$ ______ . ______

b 230 c = ______ c + ______ c
= \$ ______ . ______

c 160 c = \$ ______ . ______

d 215 c = \$ ______ . ______

e 340 c = \$ ______ . ______

f 565 c = \$ ______ . ______

Exercise 11

$$70\text{ c} + 40\text{ c} = 110\text{ c} = \$1.10$$

Complete:

a 80 c + 50 c = ______ c
= \$ ______ . ______

b 60 c + 60 c = ______ c
= \$ ______ . ______

c 90 c + 70 c = ______ c
= \$ ______ . ______

d 70 c + 80 c = ______ c
= \$ ______ . ______

Exercise 12

Write as cents, then as a decimal number of dollars.

a

______ c

\$ ______ . ______

b

______ c

\$ ______ . ______

Exercise 13

I have \$2.40, which is \$2 and 40 c.

If I add 

, I have \$2 and 90 c, which is \$2.90 .

Complete:

a \$3.10 + (20) = \$ ______ . ______

b \$2.65 + (5) = \$ ______ . ______

c \$4.15 + (5) (50) = \$ ______ . ______

d \$1.60 + (20) (20) = \$ ______ . ______

Exercise 14

Draw the missing coin.

a \$1.20 + = \$1.30

b \$2.30 + = \$2.50

c \$3.50 + = \$4.00

d \$4.95 + = \$5.00

Listening Activity

Click and follow the instructions.

1 ______ **2** ______ **3** ______

4 ______ **5** ______ **6** ______

7 ______ **8** ______ **9** ______

Activity — The euro

The euro is the currency of many countries in Europe. It uses **euros** and **euro cents**, where 1 euro = 100 euro cents.

The symbol for euro is €, and the symbol for euro cent is c.

What to do:

1 How much money is shown?

a c

b € ....

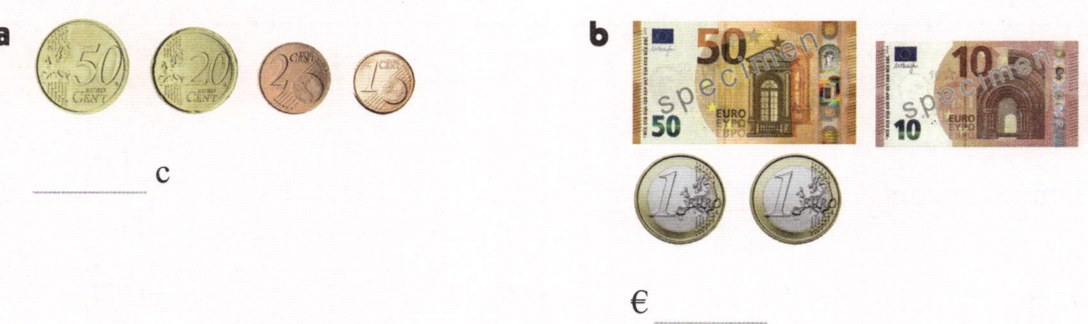

2 How much money is shown?

a € ....

b € ....

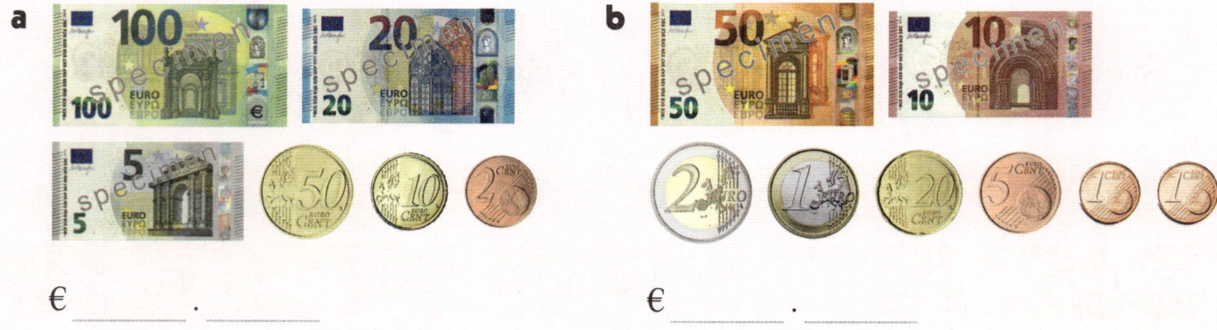

3 Divide these coins into two groups that have the *same value*. Circle the coins in one group in blue, and circle the coins in the other group in red.

Exercise 15

Use the correct operation to solve each problem.

a Tess bought a table for \$175 and a chair for \$68. How much did Tess spend in total?

Tess spent \$ .... in total.

b Courtney pays £43 each month to use her mobile phone. How much does Courtney pay over 6 months?

Courtney pays £ .... over 6 months.

c A computer game costs €87. The cost is shared equally between 3 friends. How much does each friend pay?

Each friend pays € ....

When we buy something in a shop, we do not always have the *exact* amount of money.

We hand over notes or coins worth *more* than we are being asked to pay.

The shop assistant gives back **change**, which is the difference between the two amounts.

To find the difference between two amounts, we *subtract* the *lesser* amount from the *greater* amount.

This plant cost \$7.

Ed paid with a \$20 note.

\$20 − \$7 = \$13

Ed paid \$13 *more* than the price of the plant.

Ed received \$13 back in change.

Exercise 16

a This cake cost $11.

Mel paid with a $20 note.

$20 – $11 = $ ______

Mel received $ ______ in change.

b This mirror cost €90.

Luna paid with a €100 note.

€ ______ – € ______ = € ______

Luna received € ______ in change.

Exercise 17

Use subtraction to find the change.

a Shane spent $23.

He paid with a $50 note.

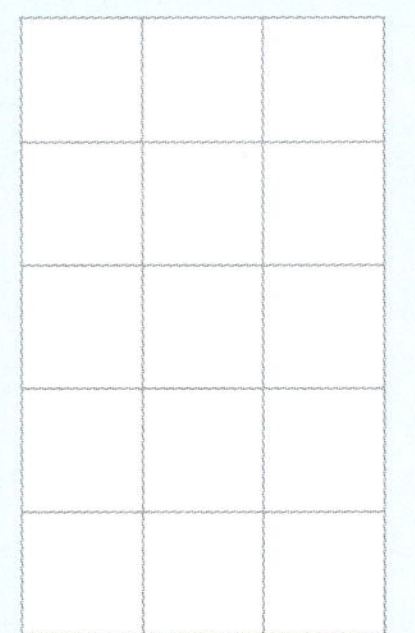

Shane received $ ______ in change.

b Cindy spent £108.

She paid £150 in total.

Cindy received £ ______ in change.

Discussion

Monica spent $8.05 at a shop.

- If she pays with a $10 note, how much change will she receive?
- Why might Monica choose to pay with a $10 note *and* a 5 c coin?

Activity — "Counting up" change

Instead of subtracting, we can "count up" to find the change.

Taili spent \$2.55 at the shop. She handed over a \$5 note.

The shop assistant said:

"\$2.55 and 45 cents makes \$3 .... and \$2 make \$5."

\$2.55 +  + 2 = \$5

So, Taili received \$2.45 in change.

In this Activity, you will use play money to practise counting up change with a partner.

Watch the video clip to see what you need to do.

Revision

You can click the icon to complete a set of Revision Exercises using United States currency.

1 How much money is shown?

a 50 10 5 5

______ c

b 50 20

\$ ______

2 How much money is shown?

a

\$ ______ . ______

b 5 1

\$ ______ . ______

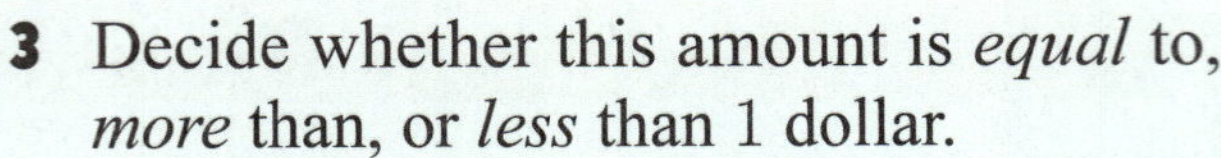

3 Decide whether this amount is *equal* to, *more* than, or *less* than 1 dollar.

____

4 Write in cents.

a $2.45 = ______ c

b $6.90 = ______ c

5 Write in $ as a decimal number.

a 320 c = $ ____ . ____

b 965 c = $ ____ . ____

6 70 c + 80 c = ______ c = $ ____ . ____

7 Complete:

a $1.20 + 50 = $ ____ . ____

b $3.55 + 5 20 = $ ____ . ____

8 Lucas bought 3 shirts. Each shirt cost $38. How much did Lucas spend in total?

Lucas spent $ ______ in total.

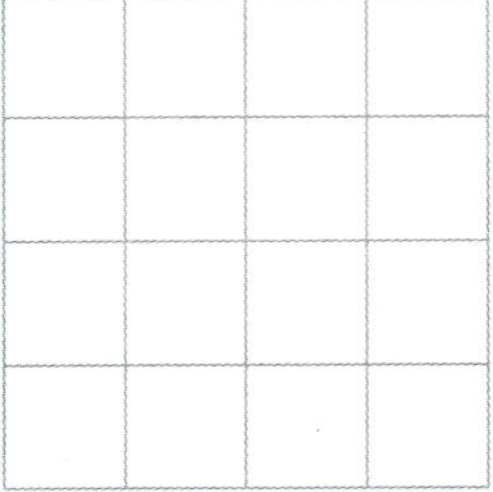

9 Shelley bought a television for £540. She will pay for the television in 4 equal payments. How much is each payment?

Each payment is £ ______ .

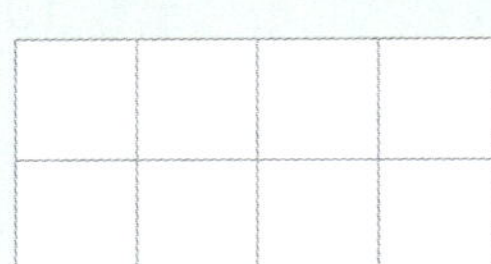

10 This belt cost $20.

Chelsea paid with a $50 note.

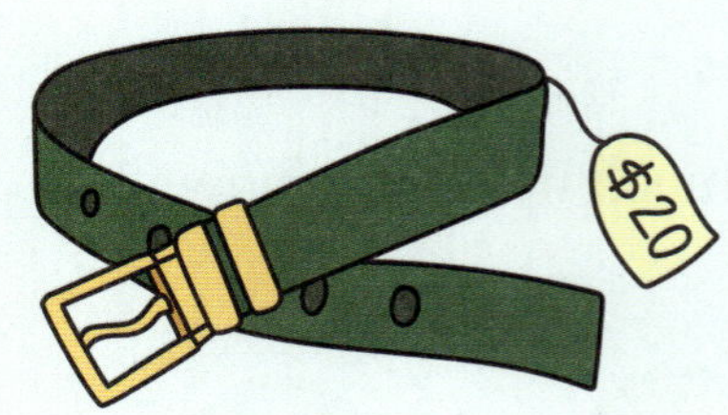

$ ______ − $ ______ = $ ______

Chelsea received $ ______ in change.

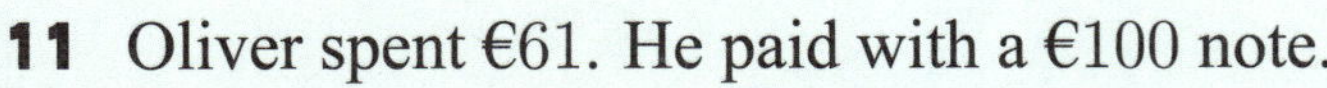

11 Oliver spent €61. He paid with a €100 note.

Oliver received € ______ in change.

CHAPTER 21: PROBABILITY

We often use words to describe the **chance** or **probability** of an event happening.

For example:

- If an event will definitely not happen, we say it is **impossible**.
- If an event will definitely happen, we say it is **certain**.
- An event has **50-50 chance** if it is as likely to happen as not happen.

Exercise 1

Circle the word or phrase which describes the *chance* of the event.

a "It is highly likely it will snow this afternoon."

b "It is unlikely that we will arrive at school on time."

c "There is a 50-50 chance the coin will land on heads."

d "It is highly unlikely that I will be at choir practice tomorrow."

Exercise 2

Amy will be blindfolded and then select a ball from this bag.

Use *certain*, *impossible*, or *50-50 chance* to describe the probability that Amy will select:

a a green ball ....

b a yellow ball ....

c a red ball ....

d a green ball *or* a red ball ....

Activity

What to do:

1 Working with a partner, order these words from impossible to certain.

likely	slightly less than 50-50 chance	highly unlikely
slightly more than 50-50 chance	highly likely	unlikely

certain

....

....

....

50-50 chance

....

....

....

impossible

2 Check your answers with the rest of your class.

Activity

Think ahead to tomorrow. Write down an activity which *for you*:

a is highly likely to happen

....

b is unlikely to happen

....

c has slightly more than 50-50 chance of happening

....

d is impossible.

....

Exercise 3

Use a word or phrase to describe the probability that the spinner will stop on red.

a

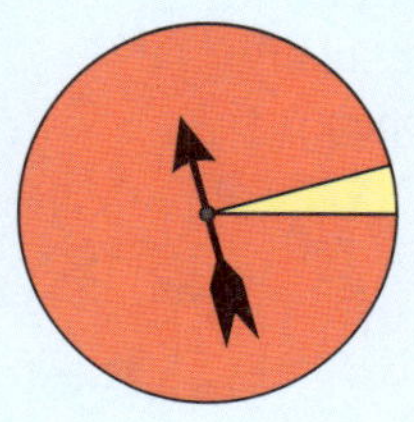

b

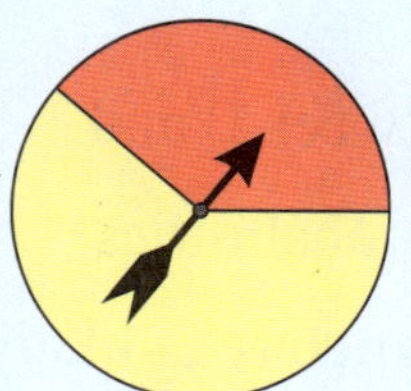

c

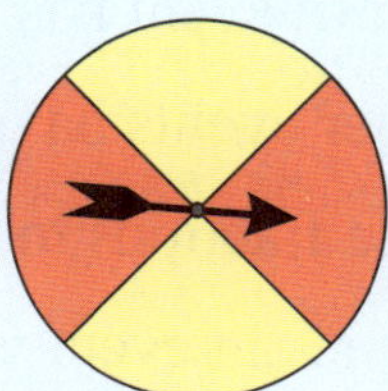

d

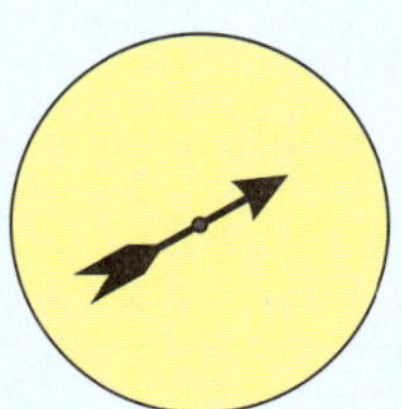

e

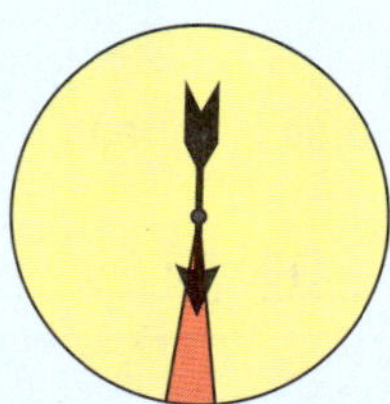

f

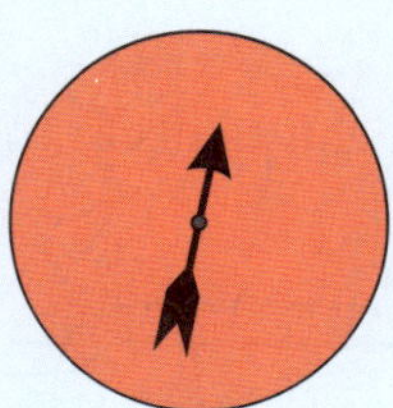

Discussion

A ball is to be *randomly selected* from the bag.

What do we mean by "randomly selected"?

We mean that every ball has ______ ______

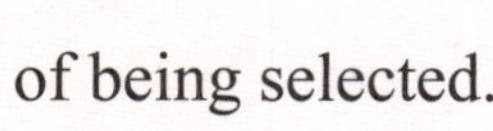

______ of being selected.

Exercise 4

Colour enough balls in each bag so the phrase underneath describes the chance of randomly selecting a coloured ball.

a

highly unlikely

b

50-50 chance

c

likely

d

slightly more than 50-50 chance

A more accurate way to describe probability is to use **numbers**.

- An **impossible** event has probability 0.
- A **certain** event has probability 1.
- All other events have probability between 0 and 1.

We can use fractions to describe these probabilities.

For example, an event with a **50-50 chance** is as likely to happen as not happen, so its probability is halfway between 0 and 1. Its probability is $\frac{1}{2}$.

We can display the range of probabilities on a number line:

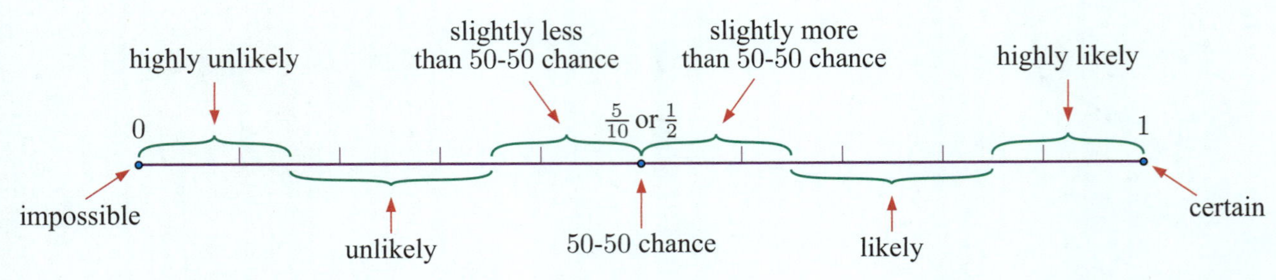

Discussion

Look at this cereal box.

How can we write this probability as a fraction?

Jarrod and Tony go to the same tennis practice.

Jarrod has probability $\frac{9}{10}$ of being on time. Tony has probability $\frac{3}{10}$ of being on time.

It is *highly likely* that Jarrod will be on time.

It is *unlikely* that Tony will be on time.

Exercise 5

Match each fraction with the most appropriate word or phrase:

0 $\frac{4}{10}$ $\frac{8}{10}$ $\frac{1}{2}$ $\frac{1}{10}$

likely | highly unlikely | impossible | slightly less than 50-50 chance | 50-50 chance

Exercise 6

The probability of rain tomorrow is $\frac{2}{10}$ in Paris, and $\frac{6}{10}$ in Rome.

Write a word or phrase to describe the probability that it will rain tomorrow in:

a Paris ______________________

b Rome ______________________

Before we can describe the probability of an event happening with a number, we first need to think about the possible **outcomes** of our experiment.

Exercise 7

When this spinner is spun, the 3 possible outcomes are

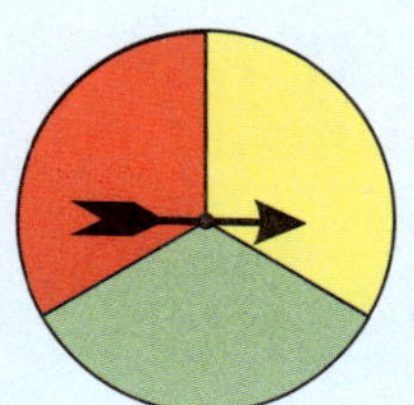

______________, ______________, and

______________.

Are the possible outcomes equally likely? ________

Exercise 8

When I roll a die, the 6 possible outcomes are

_____, _____, _____, _____, _____, and _____.

Are the possible outcomes equally likely? ________

Exercise 9

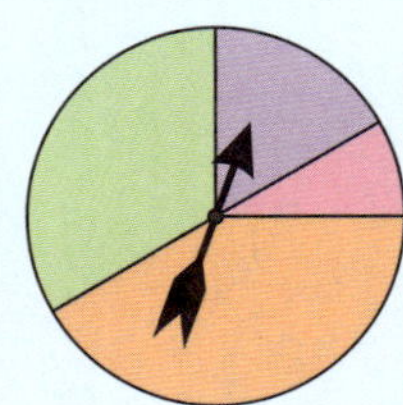

When this spinner is spun, the ______ possible outcomes are ______, ______, ______, and ______.

Are the possible outcomes equally likely? ______

This spinner is made up of 4 sectors of equal size.

Each of the sectors is equally likely to be spun.

Each sector has probability $\frac{1}{4}$ of being spun.

Notice that $\frac{1}{4}+\frac{1}{4}+\frac{1}{4}+\frac{1}{4}=1$.

$\frac{1}{4}$ of the spinner is red. So, the probability of spinning a red is $\frac{1}{4}$.

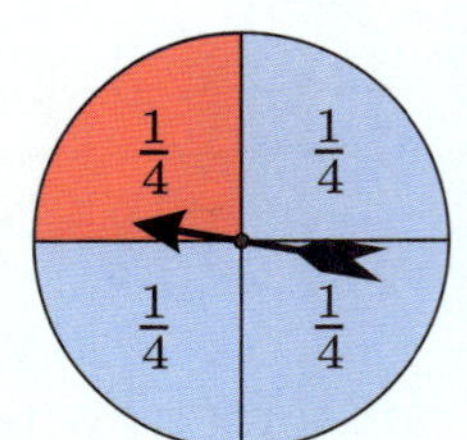

Exercise 10

When a coin is tossed, the possible outcomes are *heads* and *tails*.

There are ______ equally likely outcomes.

The probability of getting a head is ______.

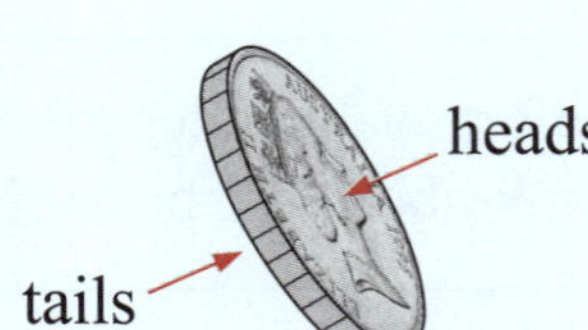

Exercise 11

When a die is rolled, there are ______ equally likely outcomes.

The probability of rolling a 5 is ______.

Exercise 12

A ball is to be randomly selected from this bag.

There are ______ balls in the bag.

The probability of selecting the yellow ball is ______.

Discussion

This spinner is made up of 3 sectors.

Why is the probability of spinning a red *not* $\frac{1}{3}$?

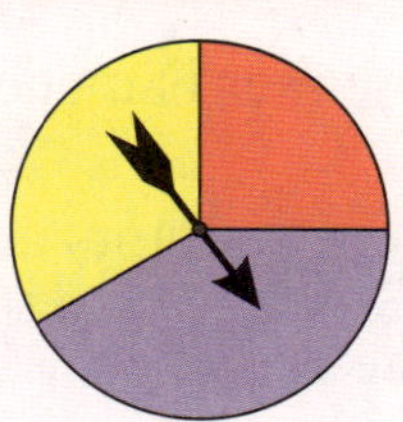

This spinner is made up of 6 sectors of equal size.

4 out of the 6 sectors are red, so the probability of spinning a red is $\frac{4}{6}$.

2 out of the 6 sectors are green, so the probability of spinning a green is $\frac{2}{6}$.

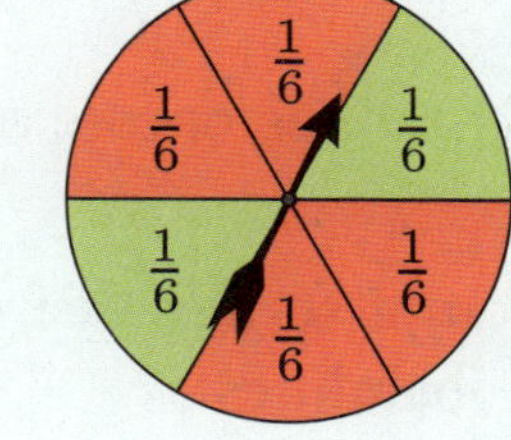

We can display these probabilities on a **tree diagram**.

The tree diagram shows all of the possible results.

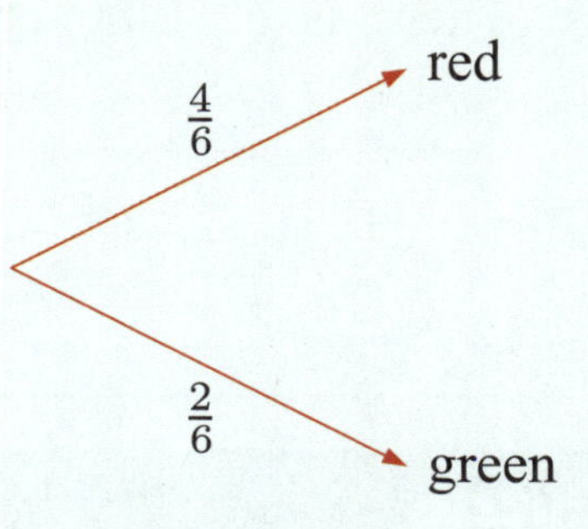

Exercise 13

There are 5 balls in this bag. A ball is to be randomly selected from the bag.

a ______ balls are blue, so the probability of selecting a blue ball is ______.

b ______ balls are yellow, so the probability of selecting a yellow ball is ______.

c Complete this tree diagram to display the probabilities.

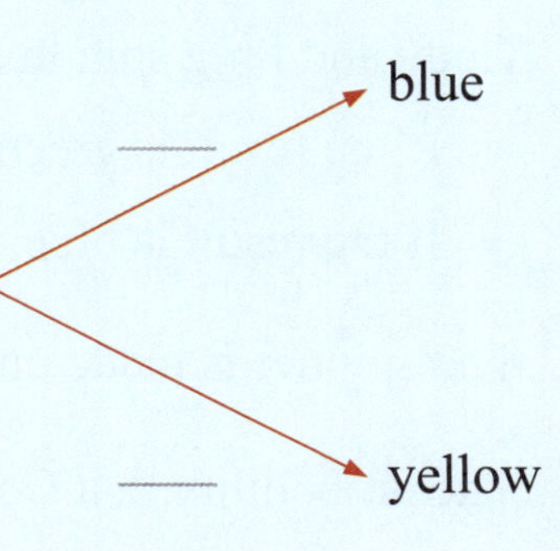

Exercise 14

This spinner is made up of ______ sectors of equal size.

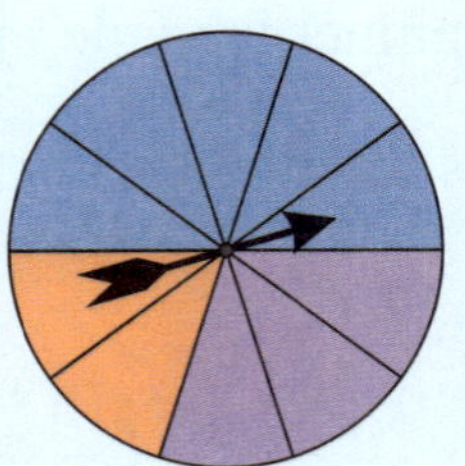

a ______ sectors are blue, so the probability of spinning a blue is ______.

b ______ sectors are purple, so the probability of spinning a purple is ______.

c ______ sectors are orange, so the probability of spinning an orange is ______.

d Complete this tree diagram to display the probabilities.

e There is a 50-50 chance of spinning a ______________.

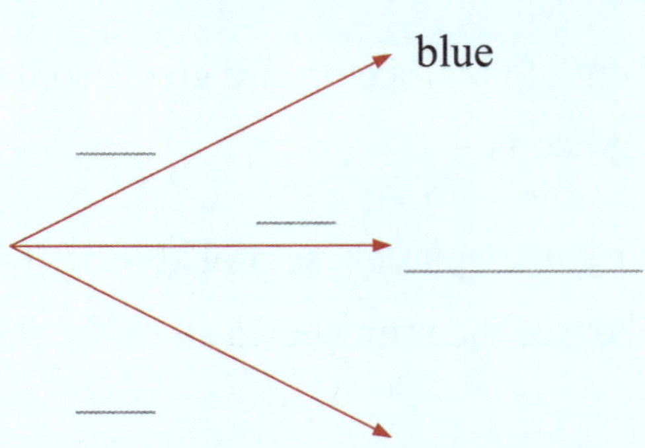

Puzzle

A ball is randomly selected from a bag containing red, blue, and green balls.

The tree diagram shows two of the probabilities.

What is the probability of selecting a green ball? ______

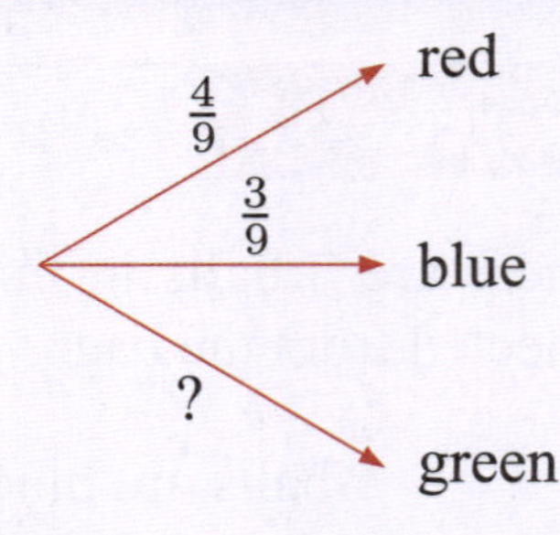

Activity — Fair games

A game is called **fair** if each player is *equally likely* to win.

1 Cody and Ping spin this spinner.

- If the result is green, Cody wins.
- If the result is blue, Ping wins.

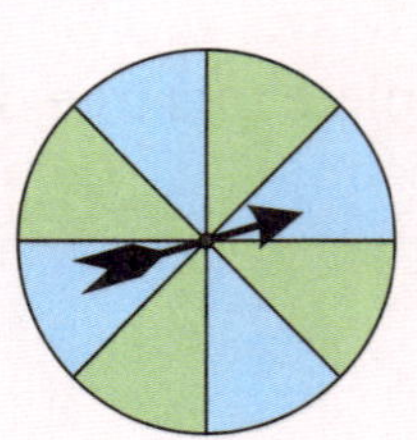

The spinner is made up of ______ equal sectors.

The probability that Cody wins is ______. The probability that Ping wins is ______.

Is the game fair? ________

2 A counter is randomly selected from this bag.

- If the counter is red, Ahmed wins.
- If the counter is yellow, James wins.

There are ______ counters in the bag.

The probability that Ahmed wins is ______. The probability that James wins is ______.

Explain why the game is not fair.

____

Describe how the game could be made fair by:

- adding a counter ______________________________
- removing a counter ______________________________

3 A ball is randomly selected from a bag containing 12 balls.

- If the ball is red, Amy wins.
- If the ball is blue, Boris wins.
- If the ball is green, Cindy wins.

Colour the balls so that the game is fair.

Discussion

- If you toss a fair coin, what is the probability of it landing "heads"? ______
- Suppose you toss a coin 10 times.
  - ► How many "heads" do you *expect* to get? ______________
  - ► Will you always get this number of "heads"?

  ____

Revision

1 Use a word or phrase to describe the probability of selecting:

a a red ball ______________________

b a white ball ______________________

c a black ball ______________________

2 Sasha and Wendy are practising archery. From past experience, Sasha has probability $\frac{7}{10}$ of hitting the bullseye, and Wendy has probability $\frac{4}{10}$ of hitting the bullseye.

Write a word or phrase to describe the probability of:

a Sasha hitting the bullseye ______________

b Wendy hitting the bullseye ______________

3 When this die is rolled, the possible outcomes are 1, 2, 3, 4, 5, 6, 7, and 8.

There are ______ equally likely outcomes.

The probability of rolling a 3 is ______.

4 This spinner is divided into ______ equal sectors.

a ______ sectors are yellow, so the probability of spinning a yellow is ______.

b ______ sectors are pink, so the probability of spinning a pink is ______.

c ______ sectors are green, so the probability of spinning a green is ______.

d Complete this tree diagram to display the probabilities.

yellow

CHAPTER 22: TRANSFORMATIONS

Discussion

Look at each pair of animal pictures.

How are the pictures related to one another?

In each pair of animal pictures, one picture has been **transformed** to make the other picture.

The bear has been copied and moved in a particular direction.

This is called a **translation**.

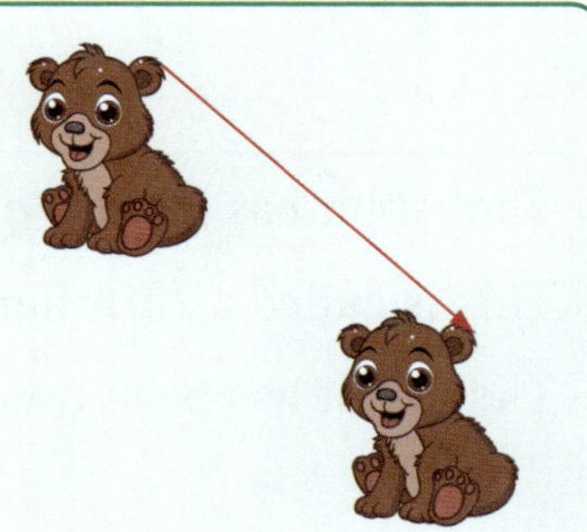

Exercise 1

Complete each translation:

a

b

c

d

Discussion

Explain why these figures are *not* transformations of each other.

....

....

....

Activity

Circle the figures that are a translation of the red figure.

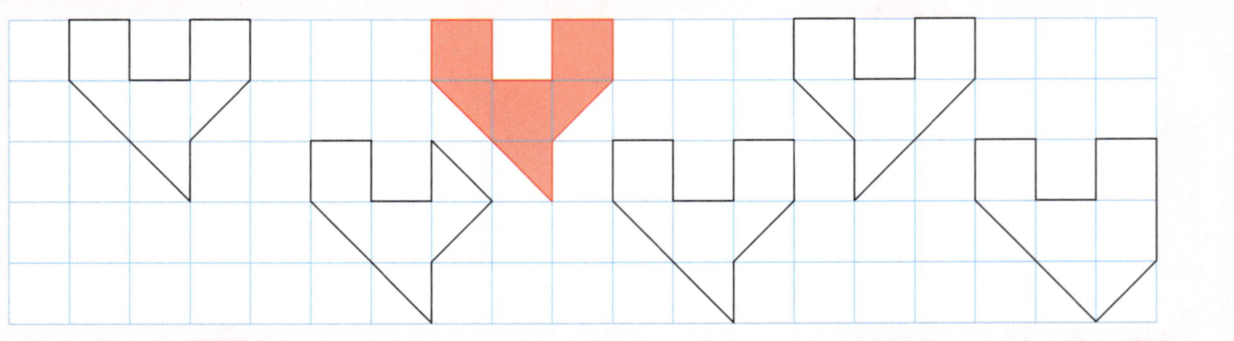

The rabbit has made a *quarter turn clockwise*.

This is called a **rotation**.

The rabbit has been rotated about the blue point.

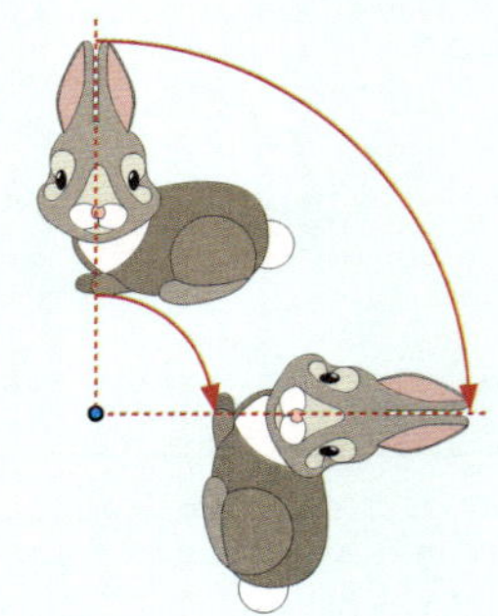

Exercise 2

Describe each rotation:

a

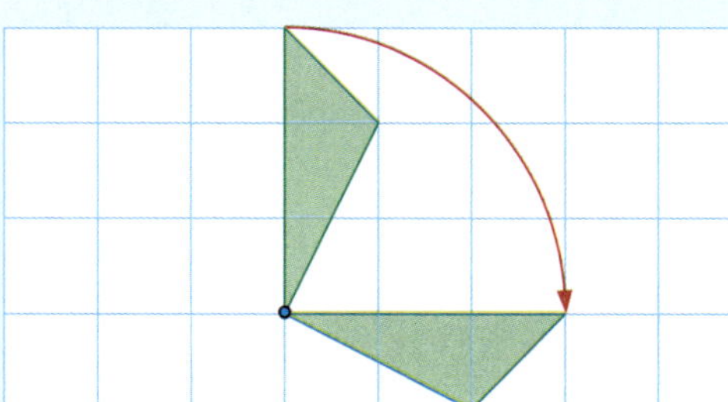

A quarter turn in a

.... direction.

b

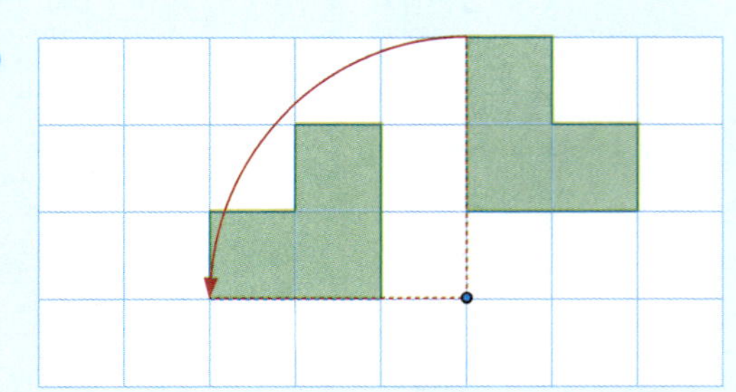

A .... turn in an

.... direction.

c

A half turn in a

________________ direction.

d

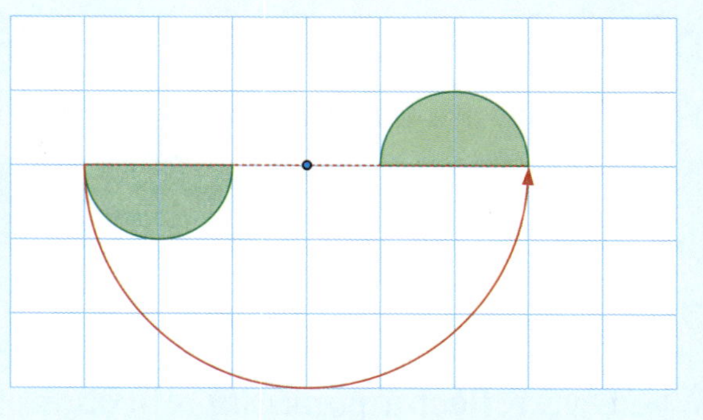

A ____________ turn in an

________________ direction.

Exercise 3

Complete each rotation:

a a quarter turn anticlockwise

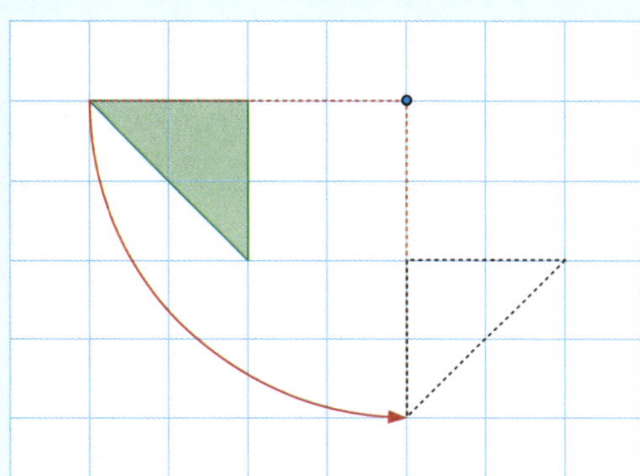

b a quarter turn clockwise

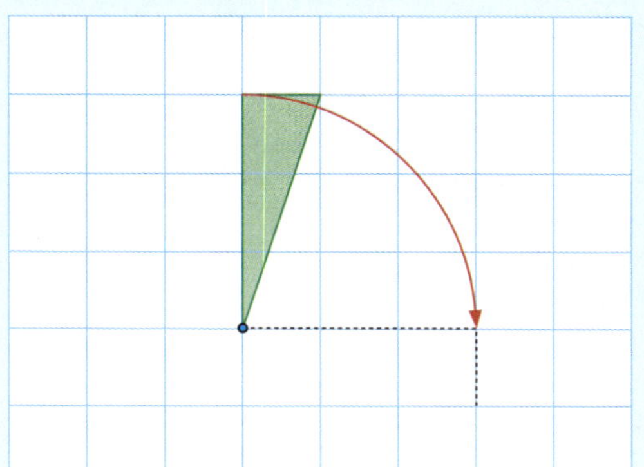

c a half turn clockwise

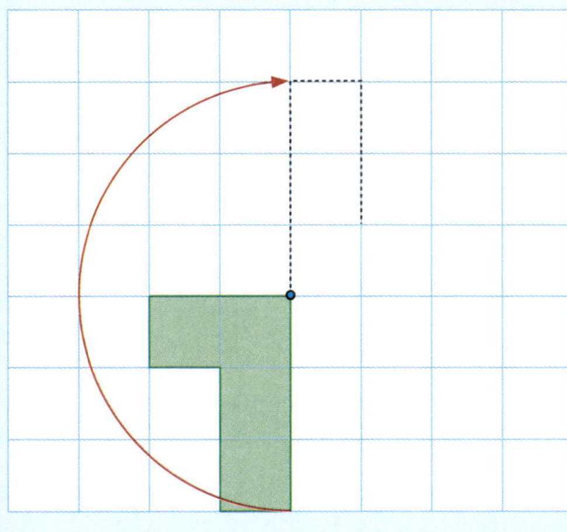

d a half turn anticlockwise

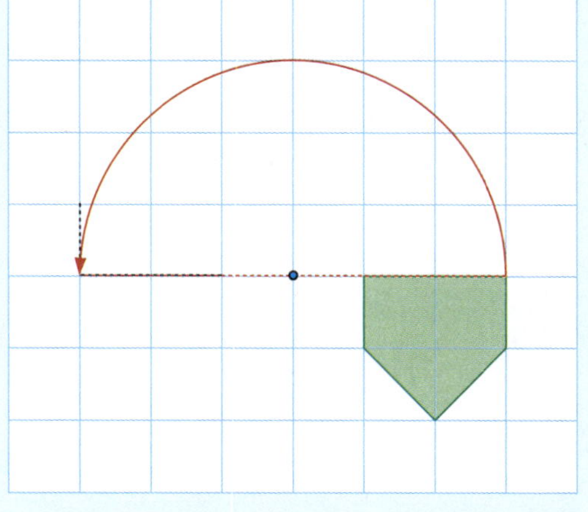

Discussion

When a figure is rotated by a half turn, does the *direction* of the rotation make a difference to the result?

____

The tortoise has been **reflected** in the **mirror line**.

The tortoises are **mirror images** of each other.

When we reflect a point, its reflection is the same distance from the mirror line, but is on the other side.

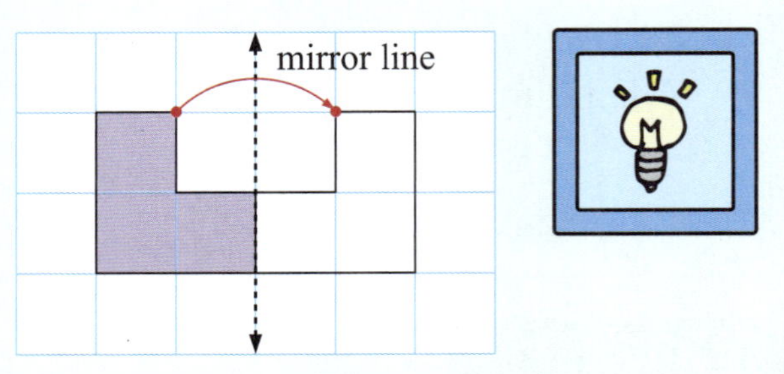

Exercise 4

Reflect each figure in the mirror line:

a

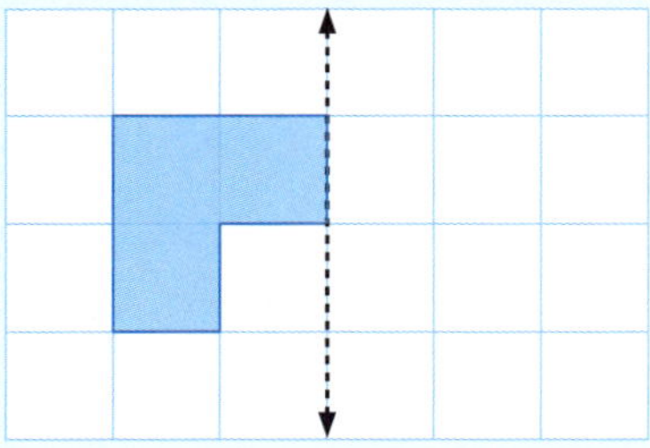

b

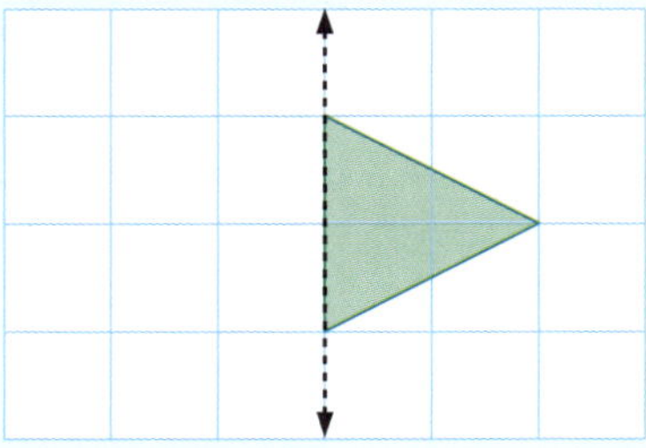

c

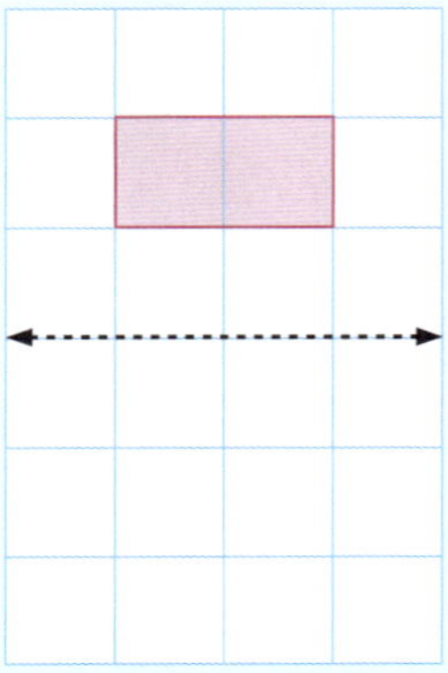

d

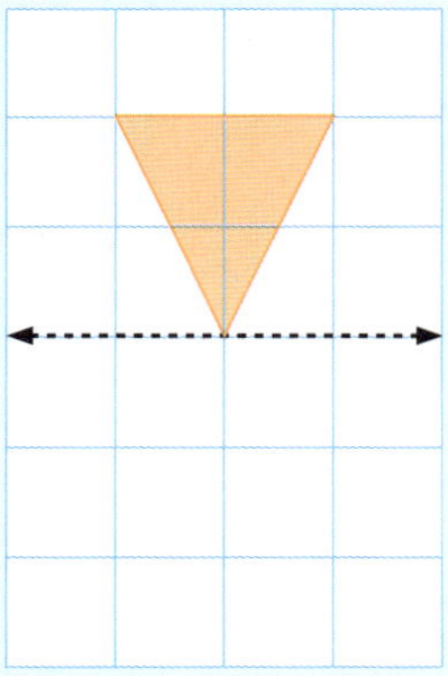

e

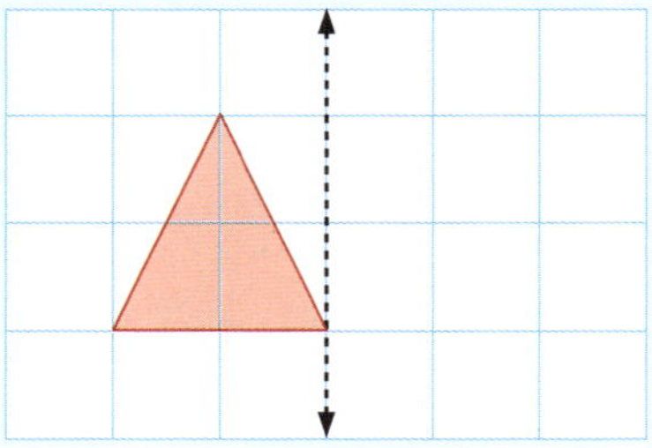

f

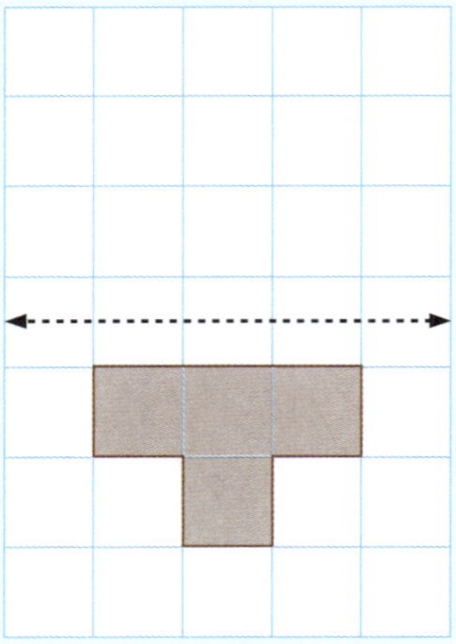

g

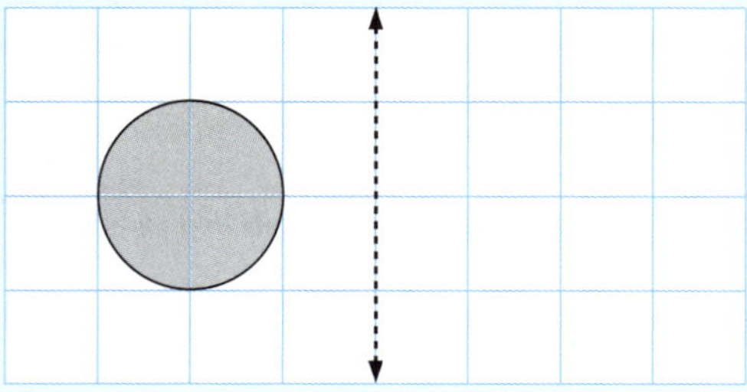

Activity

You will need: centicubes

What to do:

1. Click the icon.

   Print the grid paper.

2. Place the grid paper on your desk.

   Design a pattern of centicubes on one side of the mirror line.

3. Swap desks with your neighbour. Use centicubes to create the reflection of the pattern your neighbour has made.

mirror line

Make sure the position *and* colour of each cube is correct!

The parts of this tree either side of the red line are mirror images of each other.

We say that the tree is **symmetrical**.

The red line is called a **line of symmetry**.

We can use a mirror to show that one side is a **reflection** of the other side.

Exercise 5

This beetle is symmetrical.

Draw its line of symmetry.

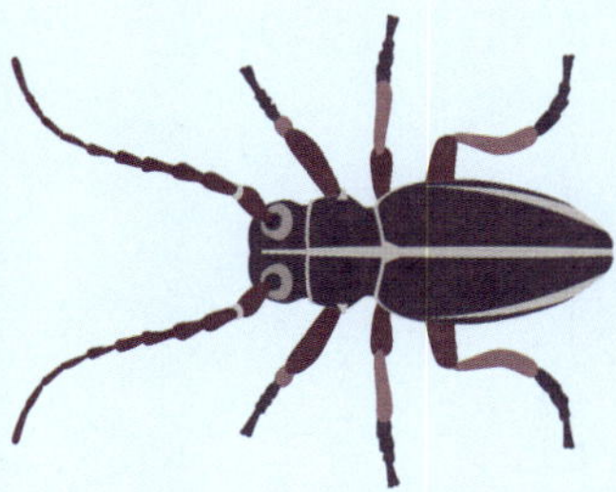

Exercise 6

Use a ruler to draw any lines of symmetry for these figures.

a

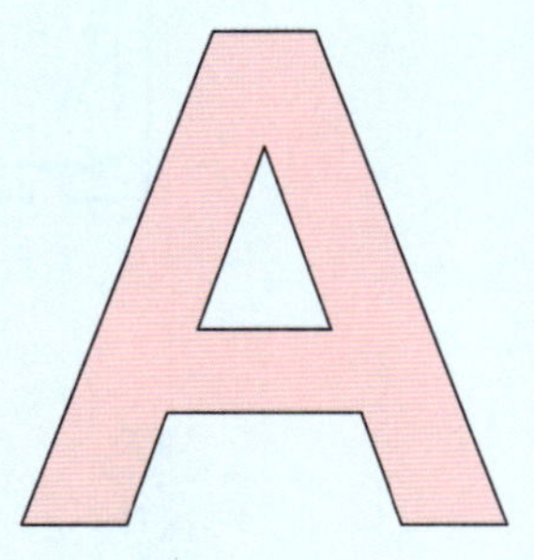

b

c

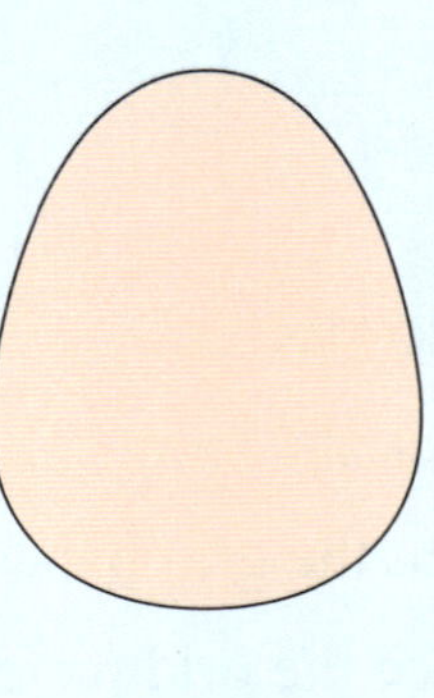

d

e

f

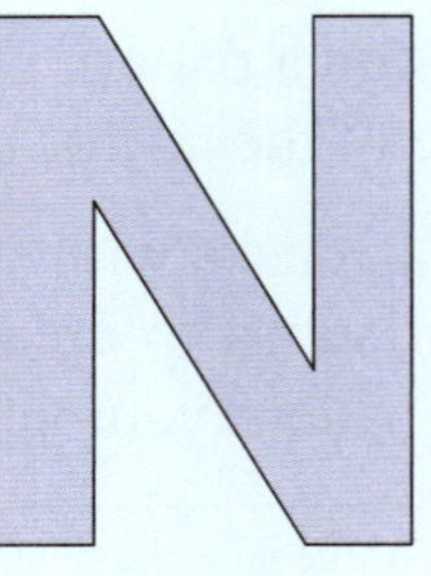

Discussion

Jake has drawn some lines of symmetry on this rectangle. He says:

"A rectangle has 4 lines of symmetry."

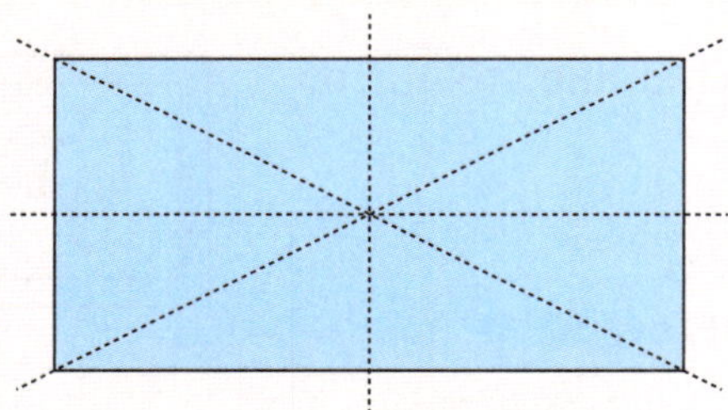

Is Jake correct? ....

Explain your answer.

....

....

Exercise 7

Name each shape, and draw any lines of symmetry.

a

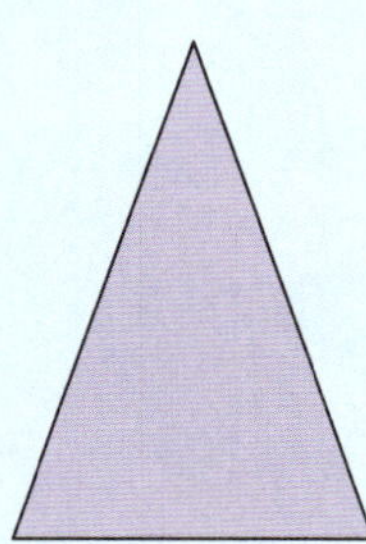

....

b

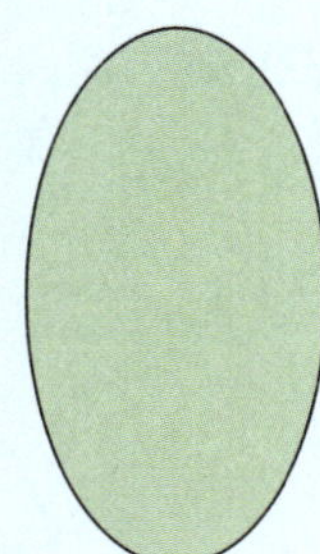

....

c

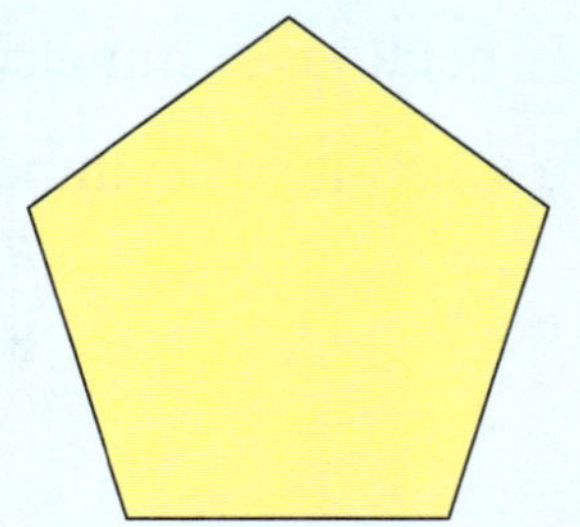

....

Activity

The photograph alongside shows the Taj Mahal, India.

Can you see where a line of symmetry could be drawn?

Click the photograph to reveal the line of symmetry.

Find other famous or interesting buildings which are symmetric.

Make a poster to show your work.

Activity

You will need: paper, ruler, pencil, scissors

What to do:

1. Fold your piece of paper in half 3 times. Use your ruler to flatten the edge of each fold.

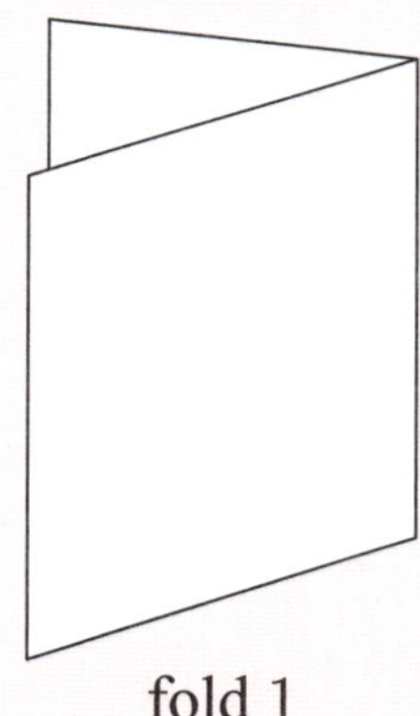

fold 1

fold 2

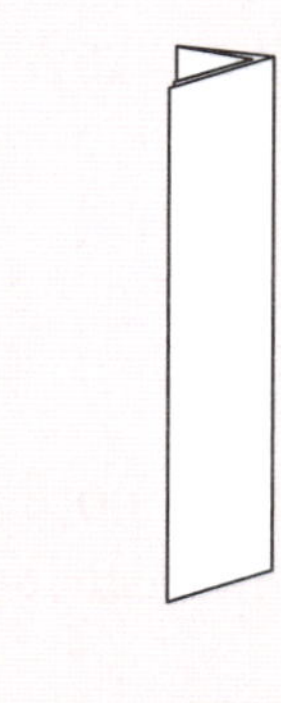

fold 3

2. 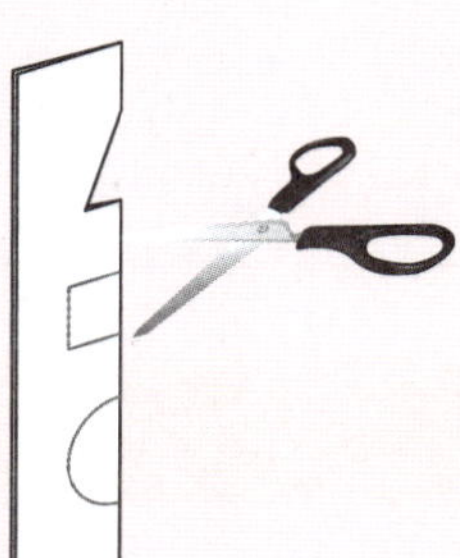

Neatly draw 3 shapes along the fold. Cut each shape out.

Be careful to not cut through to the opposite edge.

3. 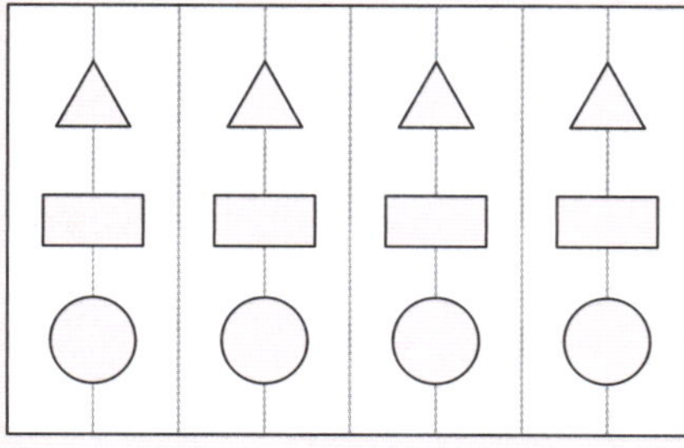

Slowly unfold your piece of paper to reveal your pattern. What lines of symmetry can you see?

We can sometimes use copies of a shape to make a pattern which completely covers a surface, with no gaps.

This pattern is called a **tessellation**.

Here is a tessellation made from identical triangles.

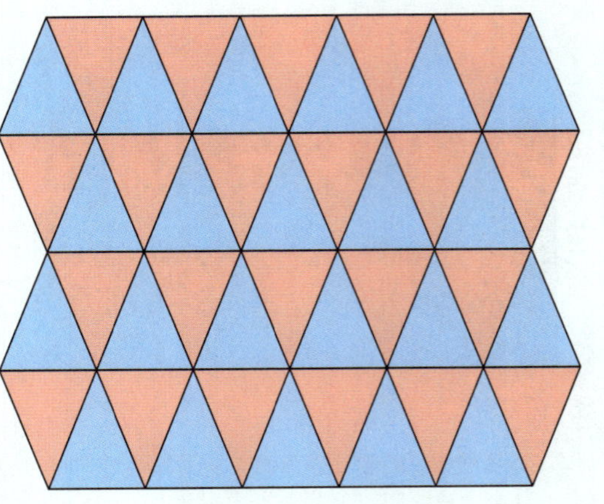

Exercise 8

a Draw identical copies of this square to cover the white surface, with no gaps.

b Draw identical copies of this trapezium to cover the white surface, with no gaps.

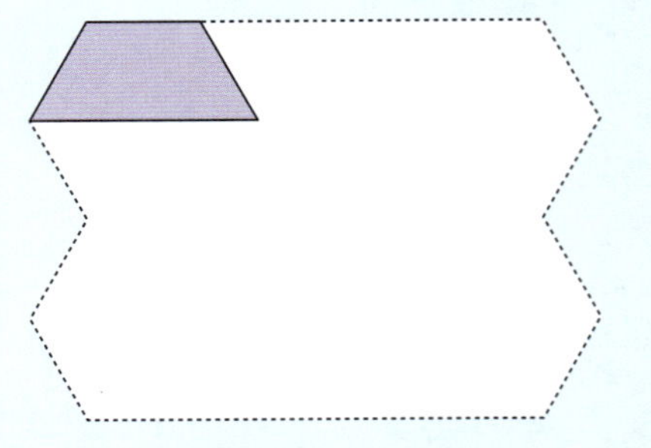

Activity

Draw copies of this shape to completely cover the grid.

Activity

Tessellations can also be made from two or more different shapes.

What to do:

1 Write down the shapes used in each tessellation:

a

b

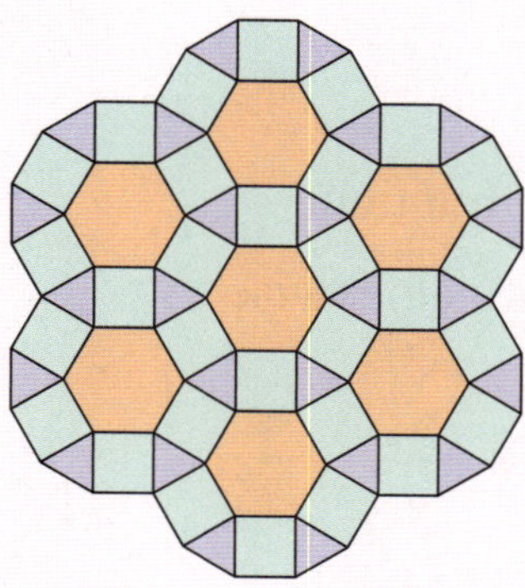

2 Draw a tessellation of your own, which is made from *hexagons* and *rhombuses*.

Activity — Fish tessellation

You will need: scissors, coloured pencils, glue

What to do:

1 Click the icon.

Print the 12 fish, then cut them out.

2 Arrange the fish in the space below. The fish should fit together exactly, with no gaps.

3 Glue the fish to the page. Use your coloured pencils to decorate them.

Revision

1 Complete each translation:

a

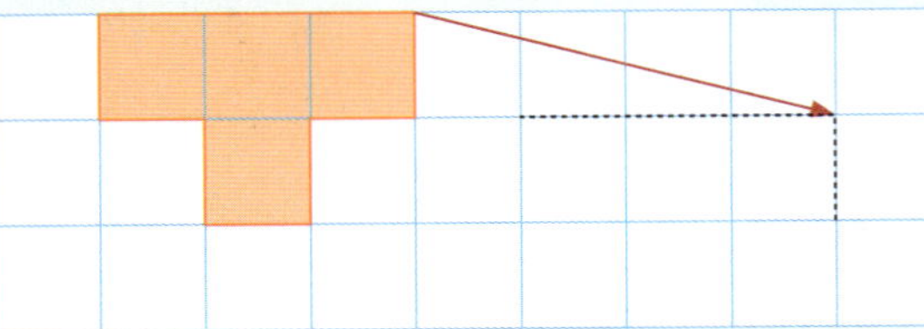

b

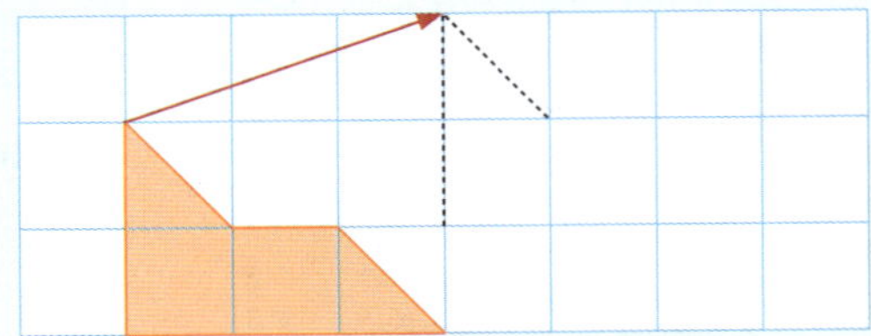

2 Complete each rotation:

a a quarter turn clockwise

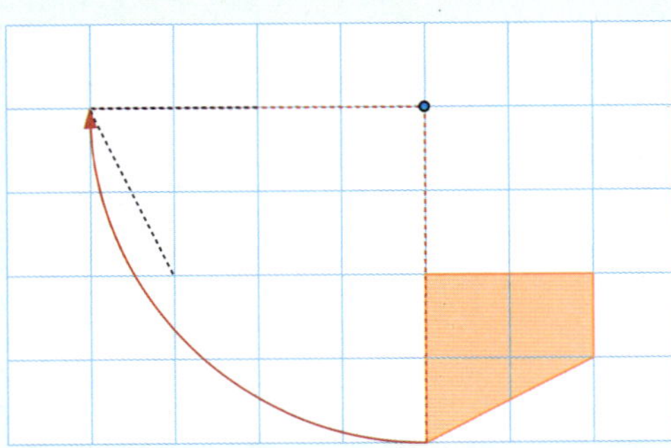

b a half turn anticlockwise

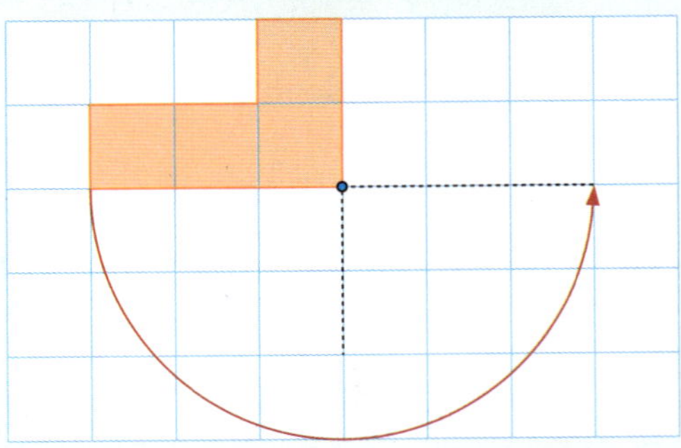

3 Reflect each figure in the mirror line:

a

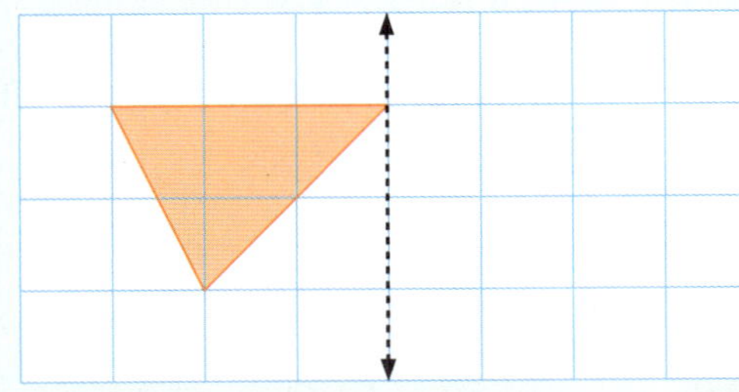

b

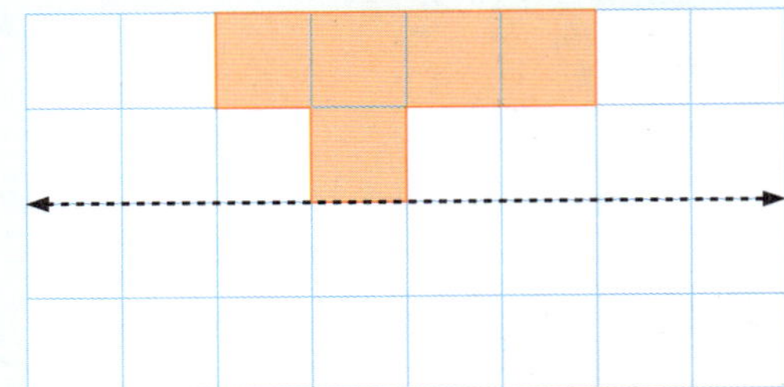

4 Name each shape, and draw any lines of symmetry.

a

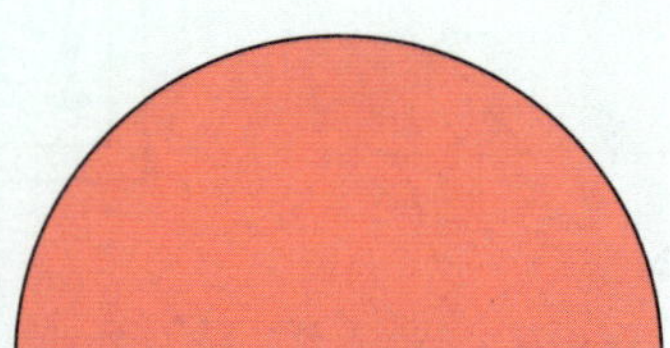

....

b

....

c

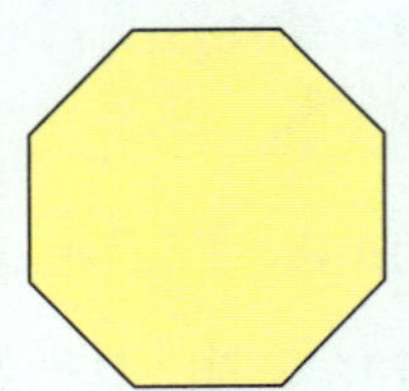

....

5 Draw identical copies of this quadrilateral to cover the white surface, with no gaps.

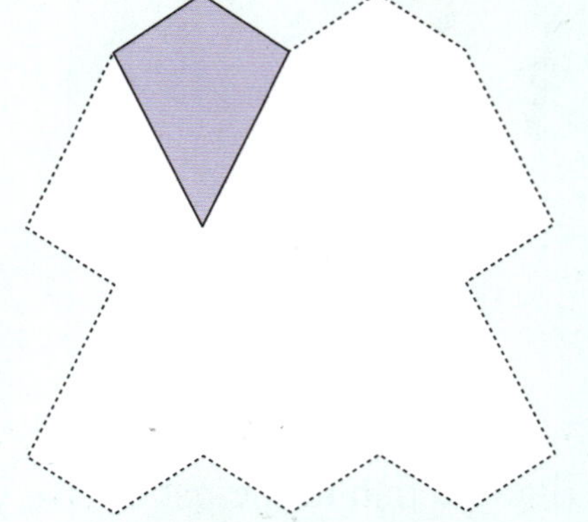